Proceedings

of the

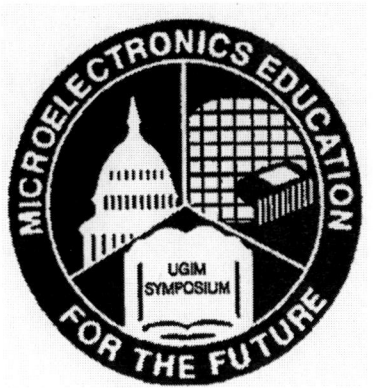

Sixteenth Biennial

University/Government/Industry

Microelectronics Symposium

June 25-28, 2006

San Jose State University

San Jose California

Sponsored by:

The SJSU College of Engineering

Optical Associates Incorporated

Intel

Synopsys

Cadence Design Systems

IEEE Catalog Number: 06CH37785

ISBN: 1-4244-0267-0

ISSN: 0749-6877

The papers appearing in this book comprise the proceedings of the meeting mentioned on the cover and title page. They reflect the author's opinions and are published as presented and without change in the interests of timely dissemination. Their inclusion in this publication does not necessarily constitute endorsement by the editors, The Institute of Electrical Electronic Engineers, Inc. or San Jose State University.

Copyright and Reprint Permission: Abstracting is permitted with credit to the source. Libraries are permitted to photocopy beyond the limit of U.S. copyright law for private use of patrons those articles in this volume that carry a code at the bottom of the first page, provided the per-copy fee indicated in the code is paid through Copyright Clearance Center, 222 Rosewood Drive, Danvers, MA 01923. For other copying, reprint or republication permission, write to IEEE Copyrights Manager, IEEE Operations Center, 445 Hoes Lane, P.O. Box 1331, Piscataway, NJ 08855-1331. All rights reserved. Copyright©2006 by the Institute of Electrical and Electronics Engineers.

IEEE Catalog Number: 06CH37785
ISBN: 1-4244-0267-0
ISSN: 0749-6877

Welcome to UGIM 2006

Sixteenth Biennial University/Government/Industry Microelectronics Symposium June 25-28, 2006 San Jose State University San Jose California

On behalf of the IEEE Electronic Devices Society, the San Jose State College of Engineering, and the Microelectronics Process Engineering Faculty it is my pleasure to welcome you to San Jose California, and to UGIM 2006. The UGIM Symposium provides a unique opportunity for educators and researchers to interact and discuss activities related to micro or nano fabrication, MEMS and Bio-MEMS. An outstanding technical program of over 50 presentations and posters is planned as well as two laboratory tours (Stanford's Nano Fabrication Facility and the MPE laboratory of SJSU).

The SJSU MPE faculty feels honored to host the UGIM Symposium, which has been a traditional meeting place for microelectronics professionals for over thirty years. After starting our program over ten year ago, we feel that it is our chance to "show off" to the microelectronics community.

The SJSU college of Engineering, Optical Associates Incorported, Synopsys, Intel, and Cadence design systems generously supported this symposium.

Special thanks goes to the SJSU organizing committee, Greg Young, John Lee, Stacy Gleixner, Emily Allen, and Koorosh Aflatooni. I give additional thanks to Greg Cibuzar, Karl Hirschman, Robert Pearson for always being available to help. I would like to thank Mary Tang and Paul Rissman for handling the Stanford tour. I would also like to thank Dipu Pramanik, Boris Polsky, Karen Bartleson, Troy Wood and the many other at Synopsys for helping to organize the meeting.

Dr. David Wahlgren Parent
General Chairman

Executive Committee

Event Logistics Chairs: John Lee and Stacy Gleixner, SJSU
Treasurer: Gregory Young, SJSU
Technical Program Chair: Koorosh Aflatooni, Foveon/SJSU

Technical Program Committee

Emily Allen, San Jose State
Greg Cibuzar, U. Minnesota
Karl Hirschman, RIT
Robert Pearson, RIT
Mary Tang, Stanford
John Shott, Stanford
Joel Kubby, UCSC
Jack Judy, UCLA
Stephen Parke, Boise State
Ali Shakouri, UCSC
Kevin Walsh, U. Louisville
Mark Crain, U. Louisville
James Zhou, WNLO-China
Paul Rissman, Stanford
Dipu Pramanik, Synopsys
Boris Polsky, Synopsys
Russel Martin, Foveon
Farnaz Parhami, Cypress S.
Bin Yu, NASA
Lili He, San Jose State

Session

2.1 Establishing Chip Scale Packaging (CSP) Capabilities at the University of Alaska Fairbanks: Lessons Learned in Tech Transfer, R&D, and Sustaining	Pramod C. Karulkar, David A. Bunzow, Lawrence Bowman, John Hunt, Philip Severs, David Thomas, Jami Warrick, John Dickinson	U. Alaska	1
2.2 Development of a Deep-Submicron CMOS Process for Fabrication of High Performance 0.25 µm Transistors	Michael Aquilino, Dr. Lynn F. Fuller	RIT	7
2.3 RIT MEMS Fabrication Course	Robert E. Pearson, Lynn F. Fuller, Ivan Puchades	RIT	13
2.4 Building the New Berkeley Microfab	A. William Flounders, Katalin Voros	Berkeley	19
2.5 25 Years of Microelectronic Engineering Education	Santosh K. Kurinec, Lynn F. Fuller, Bruce W. Smith, Rickard L. Lane, Karl D. Hirschman, Michael A. Jackson, Robert E. Pearson, Dale E. Ewbank, Sean L. Rommel, Sara Widlund, Joan Tierney, Maria Wiegand, Maureen Arquette, Charles Gruener, Scott P. Blondell	RIT	23
3.1 RF Physical Device Simulation For Wireless Applications	O.L. Hartin, Evan Yu	Freescale	33
3.2 Three-dimensional TCAD Process and Device Simulations	I. Avci, P. Balasingam, K. El Sayed, J. Gharib, M.D. Johnson, K. Kells, G. Kiralyfalvi, V. Koltyzhenkov, A. Kucherov, E. Lyumkis, O. Penzin, B. Polsky, V. Rao, S.D. Simeonov, N. Strecker, Z. Tan, L. Villablanca, W. Fichtner	Synopsys	41
3.3 Investigation of the Performance limits of III-V Double-Gate n-MOSFETs	Abhijit Pethe, Tejas Krishnamohan, Donghyun Kim, Saeroonter Oh, H.S. Philip Wong, Krishna Saraswat	Stanford	47
3.4 Simulation of Quasi-stationary and Transient Effects in GaN Based Heterostructure Field Effect Transistors	N. Braga, R. Mickevicius, V. Rao, W. Fichtner	Synopsys	51
3.5 A 2-Mask NMOS Process Design Fabricate and Test Module for Use in Microelectronics Instruction and Process Development	D.W. Parent	SJSU	57
4.1 Effect of Process and Layout on Strain Enhancement from Dual Stress Liners	Victor Moroz, Munkang Choi, Xi Wie Lin, Dipu Pramanik	Synopsys	63
4.2 Modeling Process Impact on Cu/Low k Interconnect Performance and Reliability	Xiaopeng Xu, Greg Rollins, Xiao Lin, Dipu Pramanik	Synopsys	65
4.3 Lithography Challenges toward Nano Scaled Device	HeeMok Lee, JinSoo Kim, YoungKeun Cho, SangCheol Jeon, KiNam Kim, KwangHee Kim, JaeSub Oh, HeeChurl Lee	NNFC	71
5.1 Academic Development in Test Engineering	Tamara A. Papalias, R. Bryan Gonzales, Frank Gurtovoy	SJSU	77
5.2 Low Budget Undergraduate Microelectronics Laboratory	David J. Hunt	Alfred State	81
5.3 Survey of University Micro/Nanotechnology Cleanroom Facilities as the First Phase in the Development of a U of L Business Model	Kevin Walsh, Mark Crain, Robert Keynton, Lisa Itamura, Scott Smith, Bruce Kemelgor	U. Louisville	89
5.4 National NanoFab Center (NNFC): Nanofabrication Facility	Jong Wan Park, Jeoung Woo Kim	NNFC	97
5.5 A University-Technical College Nanoscience Training Program	Greg Cibuzar, Steve Campbell, Greg Haugstad, Michael Flickinger, Deb Newberry, Karen Halverson, Micheal Opp	U. Minnesota / DCTC	101

Session

6.1 Low-Noise Amplifier Circuit for Embedded Electrophysiological Recording with Adjustable Gain and High-Pass Filtering	Shahin Farshchi, Jack W. Judy	UCLA	105
6.2 Si-based Resonant Interband Tunnel Diode/CMOS Integrated Memory Circuit	Stephen Sudirgo, David J. Pawlik, Karl D. Hirschman, Sean L. Rommel, Santosh K. Kurinec, Phillip E. Thompson, Paul R. Berger	RIT / NRL / Ohio State	109
6.3 Numerical and Analytical Results for the Polysilicon Gate Depletion Effect on the MOS Gate Capacitance	H. Abebe, E. Cumberbatch, H. Morris, V. Tyree	USC / MOSIS	113
6.4 Compact Models for I-V Characteristics of Double Gate and Surround Gate MOSFETs	H. Abebe, E. Cumberbatch, H. Morris, V. Tyree	USC / MOSIS	119
6.5 Evaluation of a Double Implanted Diffused MOSFET for Analog Operation	Eric J. Basham, David W. Parent	SJSU	125
7.1 Large Stroke Actuators for Adaptive Optics	Bautista Fernández, Joel A. Kubby	UCSC	131
7.2 Experimental Studies on the Effects of Geometric Parameters in a Planar Pneumatic Microvalve	K.J. Maung, J. Chan, S.J. Lee	SJSU	137
7.3 Advanced Studies into the Hermeticity of Micro-Electronic and Micro-Ordnance Devices	Karl K. Rink, George R. Neff, Jimmie K. Neff	U. Idaho / Iso Vac Eng.	143
7.4 Low Cost, Tailored Polymer-Metal Nanocomposites for Advanced Electronic Applications	Abhijit Biswas, Pramod C. Karulkar	U. Alaska	145
7.5 Wuhan National Laboratory for Optoelectronics and Its Collaboration with Georgia Tech	Zhiping Zhou	Georgia Tech	151
8.1 V_T Adjustment by L_{eff} Engineering for LSTP Single Gate Work-function CMOS FinFET Technology	Vidya Varadarajan, Tsu-Jae King Liu	Berkeley	155
8.2 Impact of Millisecond Anneals on CMOS Scaling - A Device Simulation Study	Sunderraj Thirupapuliyur	Applied Materials	159
8.3 Low Temperature Dopant Activation for Integrated Electronics Applications	Eric M. Woodard, Robert G. Manley, Germain Fenger, Robert L. Saxer, Karl D. Hirschman, David Dawson-Elli, J. Greg Couillard	RIT / Corning	161
9.1 Modeling and Analysis of the Charging Dynamics in Siquantum Dots Based Non Volatile Flash Memory Cells	Pavan Singaraju, Rama Venkat, Samar Saha	UNLV / Synopsys	169
9.2 Signal Enhancement of Time-resolved Magneto-optic Measurements on Individual Nanomagnets	Suqin Wang, Naser Qureshi, Mark A. Lowther, Aaron R. Hawkins, Sunghoon Kown, Alexander Liddle, Jeffrey Bokor, Holger Schmidt	UCSC	175
9.3 Integrated ARROW Waveguides for Molecule Specific Surface-enhanced Raman Sensing	Philip Measor, Leo Seballos, Dongliang Yin, John P. Barber, Aaron R. Hawkins, Jin Zhang, Holger Schmidt	UCSC	181
9.4 Phonon Confinement in Germanium Nanowires	Xi Wang, Ali Shakouri, Bin Yu, Xuhui Sun, Meyya Meyyappan	UCSC / NASA	183
9.5 The Design of MOS-BJT-NDR-Based Cellular Neural Network	Dong-Shong Liang, Yaw-Hwang Chen, Chun-Min Wen, Chun-Da Tu, Kwang-Jow Gan, Cher-Shiung Tsai	Kun Shan U.	189

Session

10.1 Leakage Current in DRAM Memory Cells	Jonathan Yu, Koorosh Aflatooni	SJSU	191
10.2 A Gain Control Low Power CMOS Power Amplifier for Ultra Wideband Applications	Jack C. Reed, Houshang Amir Aghahassan, Jane Chi, Albert Yen	UMC	195
10.3 A 2.4-GHz, Wide Tuning Range GmC VCO Using a Novel Load Biasing Technique	Vivek Verma, Tamara Papalias	SJSU	201
10.4 Lithography Solutions for a .35μm 25V PDMOS Technology	Brett Williams, Mike Thomason, Chuck Belisle, Bruce Greenwood	AMI Semiconductor	207
10.5 Advanced Process Simulation - Laser and Flash Annealing	Mark E. Law	U. Florida	213
11.1 6 Bit Decimation Filter in Sub-threshold Region	Ritu Jain, Pratibha Guttal, D.W. Parent	SJSU	215
11.2 Frequency Multiplier Design Based on Multiple-Peak R-BJT-NDR Devices Fabricated by SiGe Technology	Dong-Shong Liang, Kwang-Jow Gan, Chun-Da Tu, Cher-Shiung Tsai, and Yaw-Hwang Chen	Kun Shan U.	221
11.3 Studying the Etch Rates and Selectivity of SiO_2 and Al in BHF Solutions	Meow Yen Sim, Stacy Gleixner	SJSU	225
11.4 Implementation of On-line Error Delecting, Constant Delay, Carry Free Adder	Yugandhar Asmath	SJSU	229
11.5 Pixel Level Analog to Digital Converter	Nguyen Phong, Chung Joseph, Mariavanessa Pascua, Scott Tarkul, Eric Vasham, David Parent	SJSU	233
11.6 Improvement of a 4-mask Process Recipe	Kuang-Wai Tseng, Mariavanessa Pascua, Taslima Rahman, Siu Kuen Leung, Scott Echols	SJSU	237
11.7 Design of an Ultra-Low Power Receiver for 2.4GHz Applications in 90nm CMOS	Maryam Tabesh, Koorosh Aflatooni	SJSU	241

Establishing Chip Scale Packaging (CSP) Capabilities at the University of Alaska Fairbanks: Lessons Learned in Tech Transfer, R&D and Sustaining

Pramod C. Karulkar, David A. Bunzow, Lawrence Bowman, John Hunt, Philip Severs,
David Thomas, Jami Warrick, and John Dickinson
Office of Electronic Miniaturization, University of Alaska Fairbanks, AK 99775-8090

Introduction

The process of transferring a pilot fabrication technology in the area of microelectronics licensed from an industrial partner[1] to a small academic institution is a significant undertaking requiring numerous iterations coupled with clear and effective communications on both sides of the transfer. Many systems that are well established in the industrial sector need to be created in an academic environment that is usually unfamiliar with either the importance or the workings of industrial systems. Several limitations are encountered, including a dilute talent pool to undertake such work at a university and a limited local support for infrastructure systems needed to ensure technical viability of the prototype operations.

We describe the major components that played a key role in establishing a successful chip scale packaging (CSP) line at the University of Alaska Fairbanks (UAF) under the auspices of the Office of Electronic Miniaturization. (OEM). The critical needs in specific areas to establish a state-of-the-art center that achieved operational and structural success in a relatively short period of time are presented. UAF's capabilities now include a full range of licensed CSP technologies from our industrial partner[1], as well as being optimally positioned to develop new capabilities in support of our customers.

Capabilities[2-5]

OEM's Chip Scale Packaging (CSP) facility is designed for R&D and pilot production. It now successfully runs a wide range of packaging process variations as well as creating board-level systems that utilize CSP devices made within our facilities and those secured on the open market. Devices built in our facility are shown in Fig. 1. Our qualified processes include:

- Single layer CSP devices
- Stacked device structures
- Multiple device modules (single layer and stacked)
- Wire-bonded and lead-bonded versions
- Flip chip and face up CSP packages
- Eutectic and Lead-free solder variations
- SMT pick and place

Figure 1. Examples of Single Layer (Top) and Stacked (Bottom) CSP Devices Built at UAF's Office of Electronic Miniaturization

We are now running the line at a low throughput for R&D projects. However, depending on the size of the staff size and the number of shifts, our facility can run dozens of different designs at the same time and produce several thousand CSP's per week in a pilot production mode.

A design and engineering group supports the CSP facility. In addition, a materials research and microfabrication effort has been initiated in a separate facility to complement and enhance

1-4244-0267-0/06/$25.00 ©2006 IEEE

electronic miniaturization projects. A nanocomposite materials program is designed to develop and apply these materials to miniaturized electronics. (Please see a companion paper at this conference.) A microfabrication facility; that includes thin film deposition, photolithography, plasma etching, wet etching, high temperature anneals, and chemical vapor deposition; is being established to provide substrate, test structure, device fabrication support. Further, an array of metrology instruments (wafer probers, environmental chambers, x-ray irradiation system) has been added to strengthen our capabilities.

Clean Room Facilities

UAF/OEM's Class 100 CSP cleanroom was established inside an existing building. This 1500 sq. ft. exceeds minimum requirements for an assembly facility and allows flexibility for future technology expansion. Thus the facility will not become obsolete in a period of time prior to its useful life. The clean rooms cost about $1,000 per square foot to construct, and include access control, air shower protection, point-of-use environmental monitoring for temperature, humidity and air pressure, and secure wireless communications. Clean room class monitoring, verification and contamination controls were incorporated into the design and implementation of our clean rooms. Our labs are co-located in a three story facility with its own separate facilities support functions. Separate facilities systems for air supply and return, exhaust and vacuum were retrofitted into our clean rooms space.

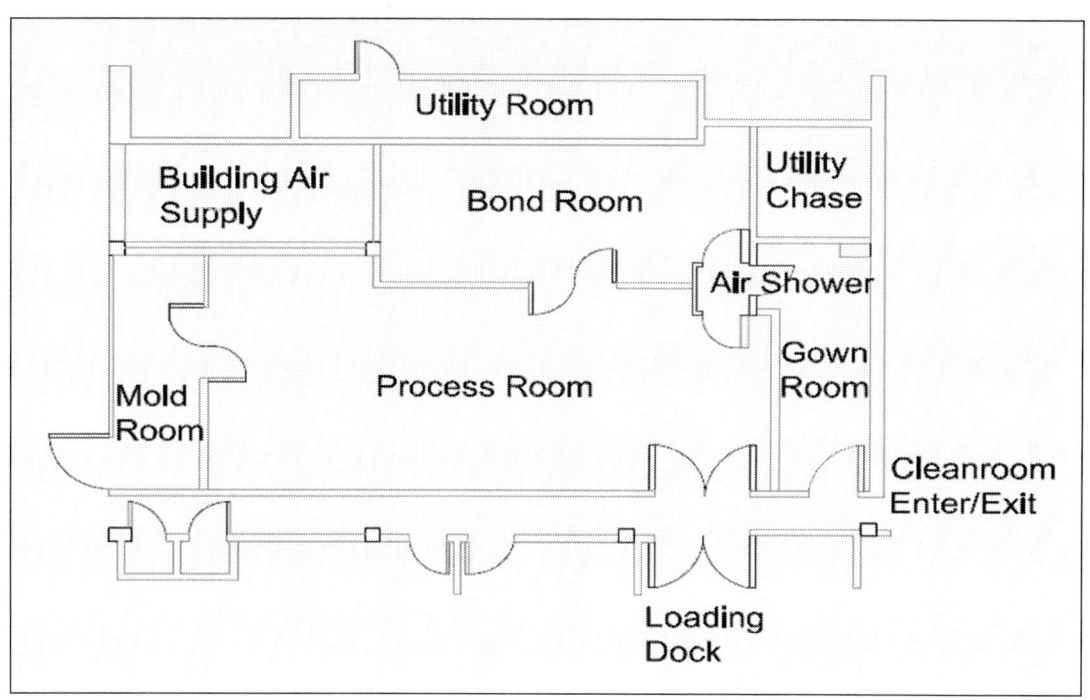

Figure 1. UAF's CSP Clean Room Facilities Layout

Guarding against ESD damage and ensuring reproducible and acceptable integrity and adhesion strengths at various package interfaces is very critical for sustaining high yields. To ensure this we have established temperature and humidity controls, and provided adequate grounding and bonding of equipment as well as of walking and working surfaces. We maintain a spec of 0.5 ohm to ground for our floors; and all tables, equipment and chairs are in intimate contact with this floor surface.

A thorough power and distribution study was performed during the facility design phase. We discovered electrical isolation/feedback problems from other parts of the existing infrastructure that supplied sharp kilovolt spikes back into sensitive controls circuitry of specific equipment. This was ultimately resolved by installing TVSS suppression devices at all points of use.

Support Facilities

The chip scale packaging (CSP) facility requires only limited quantities of de-ionized water, high purity nitrogen (or other inert gas) and clean dry air. We installed a small (10 gpm) RO system (18 megohm) with 500 gallon storage in the basement directly under our clean rooms. About 450 scfh of pure nitrogen during full operation at 100 psi are needed for the various equipment and operations in our clean rooms. After looking at more costly alternatives, our solution included two point-of-use nitrogen supplies that were capable of providing 99.9% pure nitrogen using a 50% duty cycle from redundant membrane systems and dual clean dry air supplies.

Operating the HVAC is challenging in a small facility such as ours located in the circumpolar north where energy costs are a significant portion of operating expenses (about $2.50 per square foot per month). Temperature and humidity controls, clean dry air supply, vacuum and electricity costs add another $1.20 per sq. ft./month.

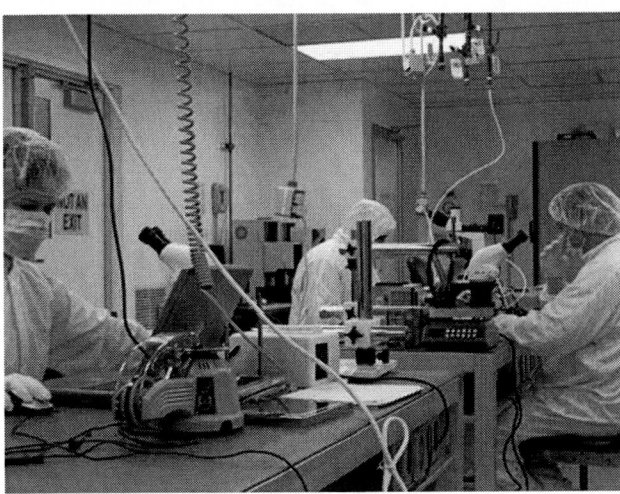

Figure 2. OEM's cleanroom

Technology Transfer & Implementation

Documentation supplied by a technology transfer partner is critical to the ultimate success of a prototyping facility. Technology transfer documents often are not written with a tech transfer process in mind. Further, the level of technical maturity of the receiving parties is often incapable of appreciating the nuances that may (or may not) be present in provided documentation. This reality can either significantly flatten the slope of an organization's learning curve. Transfer partners often supplement their own controlled documents with training documents that are to be used to educate the personnel receiving the transfer package.

We were fortunate in having a technology transfer partner who had gone through the technology transfer process before and had excellent documentation, training material, and skilled trainers/teachers. They were familiar with the intricate details and variables of the processes being transferred. Our team was very inexperienced when the training began. Hence it took significant amount of time for the training to take effect. Creating a local tech transfer guru would have helped but we could not do it because of the lack of qualified candidates. The tech transfer process was reinforced and verified when we fabricated a deliverable R&D part immediately after the training period.

Site Personnel

Employees have been a key to our success so far. A small staff (<6) is required to operate a small, successful prototyping CSP facility. However, Alaska's remoteness and the lack of any existing microelectronics capabilities created a challenge. No local talent in cleanroom microelectronics existed prior to the establishment of our facility. We are gradually addressing this issue by identifying trainable candidates and developing skills through training, experience, and self discovery. Formal hands-on training provided during the tech transfer period created the founding group. The skills are reinforced and new people are trained by exercising the process on a regular basis. Many employees enjoy the challenge of following work practices methodically and meticulously, learning new skills continuously, and understanding science and engineering behind every step. OEM has identified, attracted, employed and retained employees with backgrounds in electrical and mechanical systems as well as those in physical sciences and engineering. As with the other employers in this field, we have had to recruit continually to cope with turnover.

Process Equipment

Equipment selection for a CSP prototyping facility should be a process that ensures maximum flexibility (new vs. used) and minimized downtime. Our CSP facility is a mixture of new and refurbished equipment that operates on a one shift basis. When something breaks, it means the entire line is down. OEM thinks of itself as "centrally isolated" when it comes to vendor support. Our spare parts inventory strategy goes beyond what is typically found in an operation where services personnel and support is primarily local in nature. A simple vendor visit takes 2-3 days to complete.

Inventory Control

The control of valued inventory and maintaining its value deserves special attention. Once a process has been qualified, having strict policies on inventory age, disposal of out-of-date materials and inventory rotation are vital to ensuring a customer they will get the value they seek in their final products. OEM established an on-line entry, tracking and order/re-order system for line capabilities support. OEM uses an Access database to keep track of its materials inventory, location and reorder points. An incoming quality assurance program was established to help insure prototyping continuity and minimum delays in CSP processing.

Preventative Maintenance

All equipment will malfunction, break or cease to operate properly during the course of its life if it is not properly maintained on a regular basis by intelligent people. Preventative maintenance schedules that are typically based on operational hours and mean time between fails experience from industry data were used to set reasonable PM schedules for all equipment. These schedules often had to be modified to fit our prototyping activities versus similar needs for those in a manufacturing environment.

Document Control Systems

Early in its development, OEM made a decision to develop infrastructure systems that could be certified to ISO 9001 standards at a future time. A document system was developed to cover the broad range of activities that a prototyping facility encounters. It describes and shapes our activities in 11 key areas including product descriptions, tooling drawings, training programs and process controls. For example, there are 87 process documents currently under origination and change control available electronically within our clean rooms. We also maintain full tooling set drawings, produce training videos to help educate our students and employees, and define operating protocols for environmental monitoring and prototyping activities. We use an assigned documentation specialist for these critical activities; this individual also handles control of our intellectual property for customers and partners.

Device Testing Strategies

Parametric (open and shorts) testing that looks for assembly-related defects (contamination and ESD) occurs as a part of each device lot processing. These pass/fail tests provide the feedback needed to ensure a robust assembly process. A test socket that is custom designed for a particular CSP is needed to perform electrical.

Functional device testing is very complicated and requires specialized equipment and extensive knowledge of the integrated circuit. The original chip manufacturer and a few authorized foundries can provide the capabilities for full functional testing of a CSP. The costs of setting up an in-house capability for functional testing of integrated circuits are prohibitive. This fact can act as a major program inhibitor to new businesses wanting to migrate to a CSP solution. Usually the CSPs can be tested at the board level or at a test foundry that has access to functional testing procedures for a particular chip. OEM has been able to develop strategic alliances with both testing services houses and customers to develop creative ways to validate our final products.

Quality, Reliability and Failure Analysis

OEM leverages existing university capabilities as well as some newly established facilities for quality assurance and failure analysis through cross-sectioning, optical microscopy, scanning electron microscopy (SEM), transmission electron microscopy, x-ray fluorescence, laser particle counting, and x-ray microscopy, optical dimensioning (x, y and z directions), step height (profilometer and atomic force microscopy), film thickness and refractive index (ellipsometry), and micro-hardness testing. We also use vendor services for acoustical microscopy, reliability testing, and board-building. We may establish these capabilities in-house in the future if required by the work volume.

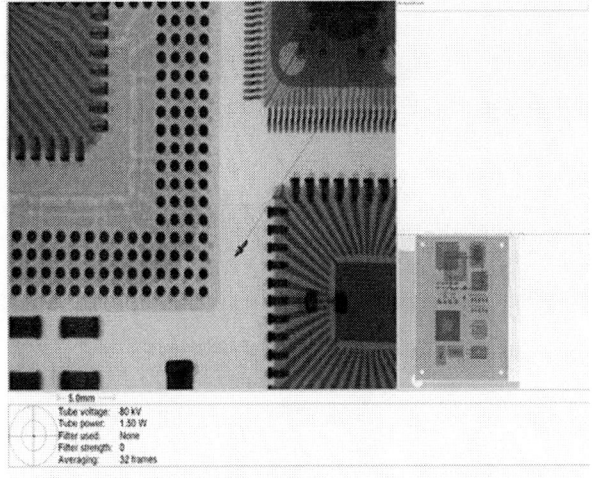

Figure 3 (a-b). Images obtained on OEM's X-Ray Inspection System. Top: Packaged parts. Bottom: stacked CSP device fabricated at OEM.

Summary

UAF's Office of Electronic Miniaturization (OEM) has successfully transferred technology and established a prototyping chip scale packaging line in Alaska. This was accomplished by carefully managing each operational interface in the academic environment consistent with our mission. Our program offers significant benefits to the University of Alaska system, its academic partners at other universities, governmental agencies that support our activities, the community and its many industrial partners.

Chip Scale Packaging is a very successful technology particularly in the consumer electronics sector. It has a proven track record. It has not penetrated as well in high value systems,

entrepreneurial technologies, and government sector electronics. Our facility offers an excellent opportunity to apply CSP technologies to many unexplored areas without requiring high volume commitments. OEM is keenly interested in collaborative R&D as well as pilot projects. We welcome inquiries (www.silicontundra.org).

Acknowledgements

This work is based on the research and development sponsored by the Defense Microelectronics Activity (DMEA). The United States Government is authorized to reproduce and distribute reprints for Government purposes, notwithstanding any copyright notations thereon. The views and conclusions contained herein are those of the authors and should not be interpreted as necessarily representing the official policies or endorsements, either expressed or implied, of the DMEA.

References

Tessera Technologies, Inc., www.tessera.com .
Pramod C. Karulkar, Lawrence Bowman, David A. Bunzow, John Hunt, Philip Severs, David Thomas, Jami Warrick, and John Dickinson, "Chip Scale Technology for 3-Dimensional Electronics," *Proc. IMAPS Device Packaging Conference, Scottsdale, AZ,* paper THA32, March 20, 06)
Pramod C Karulkar, "World Class Electronic Miniaturization Program at the Top of the World," *Electronic Packaging Quarterly (MEPTECH Quarterly) August 2005*
Pramod C. Karulkar, John Dickinson, and David Bunzow, "Electronic Miniaturization Program at the University of Alaska Fairbanks", Proceedings of the *IMAPS workshop on Electronic Packaging Issues for Military, Aerospace, Space, and Homeland Security, Paper 4, May 2005.*
A. Biswas, P. C. Karulkar, H. Eilers, M. Grant Norton and Daniel Skorski, C. Davitt, H. Greve, U. Schuermann, V. Zaporojtchenko and F. Faupel, "Vapor Phase Deposition of Nanostructured Polymer-Metal Composites for Advanced Technology Applications" *Vacuum and Coatings* Vol. 7, No. 4, p. 54, April 2006.

1-4244-0267-0/06/$25.00 ©2006 IEEE

Development of a Deep-Submicron CMOS Process for Fabrication of High Performance 0.25 μm Transistors

Michael Aquilino and Dr. Lynn F. Fuller, *IEEE Fellow*

Microelectronic Engineering, Rochester Institute of Technology, 82 Lomb Memorial Drive,
Rochester, New York 14623-5604, Email: mva6237@rit.edu or lffeee@rit.edu

Abstract –A process for fabrication of 0.25 μm CMOS transistors has been demonstrated. NMOS transistors with drain current of 177 μA/μm at VG=VD=2.5 V and a PMOS transistors with drain current of 131 μA/μm at VG=VD=-2.5 V are reported. The threshold voltages are 1.0 V for the NMOS and -0.735 V for the PMOS transistors. The mask defined gate lengths are 0.5 μm and 0.6 μm for the NMOS and PMOS, respectively. Through a photoresist trimming process, the poly gate lengths are 0.25 μm and 0.35 μm or smaller. Electrical extraction of the gate lengths should yield effective gate lengths of 0.25 μm or smaller. These are the smallest transistors ever fabricated in the SMFL at RIT. Large off-state leakage is reported for the NMOS due to drain leakage induced by implant damage or aggressive titanium silicide formation. A better understanding of this leakage is being investigated and process recommendations given.

1. INTRODUCTION

For over twenty years the Microelectronic Engineering Department at RIT has strived to continue the semiconductor industry trend of fabricating smaller and faster transistors. RIT is currently supporting 0.5 μm, 1.0 μm and 2.0 μm CMOS technologies that are fabricated daily in a student run factory. Students improve existing processes for current CMOS lines and develop new advanced processes that can be integrated into future technologies. The previous record for smallest transistor fabricated in the Semiconductor and Microsystems Fabrication Laboratory (SMFL) at RIT is an NMOS device with $L_{poly} = 0.5$ μm and $L_{eff} = 0.4$ μm which was developed by the author in May 2004.

The goal of this work is to develop a deep-submicron CMOS process to fabricate transistors with $L_{poly} = 0.25$ μm on 150 mm (6") silicon wafers. The device technology includes: 50 Å gate oxide with N_2O, shallow trench isolation by chemical mechanical planarization (CMP), dual doped polysilicon gates for surface channel devices, ultra-shallow low doped source/drain extensions using low energy As and BF_2 ions, Si_3N_4 sidewall spacers, $TiSi_2$ salicide source/drain contacts and gates, uniformly doped twin wells, 2 level aluminum metallization and is designed

to operate at a supply voltage of 2.0 V with a threshold voltage of ± 0.5 V. This technology will expose students to advanced CMOS processes and better prepare them for entry into the semiconductor industry.

Table I is a summary of industry scaling trends over the last 35 years and the recent advancements of the author's work at RIT over the same time.

TABLE I
TRANSISTOR SCALING HISTORY [1]

Date	Processor Family	Gate Length (μm)	Frequency	Transistors	Supply Voltage (V)
November 1971	4004	10	108 KHz	2,300	15
April 1974	8080	6	2 MHz	6,500	12
June 1978	8086	3	10 MHz	29,000	5
February 1982	Intel 80286	1.5	12 MHz	134,000	5
March 1995	Intel Pentium	0.6	120 MHz	3,300,000	3.5
**May 2004	RIT Submicron NMOS	0.5	-	-	3.5
August 1999	Intel Pentium 2 & 3	0.25	600 MHz	9,500,000	2.0
**May 2006	RIT Deep-Submicron CMOS	0.25	-	-	2.0
September 2001	Intel Pentium 3 & 4	0.18	2.0 GHz	45,000,000	1.7
February 2004	Intel Pentium 4	0.13	3.4 GHz	55,000,000	1.5
June 2004	Intel Pentium 4	0.09	3.8 GHz	125,000,000	1.4

2. DEVICE DESIGN

The NTRS Roadmap gives guidelines for scaling physical and electrical parameters to meet on and off state performance requirements at a particular technology node. [2] Physical parameters such as oxide thickness, source/drain junction depth and doping concentration, channel doping concentration and profile, and side-wall spacer length are scaled based upon the gate length, λ. Electrical parameters such as SS, DIBL, V_{DD}, V_T, and parasitic resistances must be controlled to achieve the required performance

1-4244-0267-0/06/$25.00 ©2006 IEEE

parameters such as I_{ON} and I_{OFF}. Some of the NTRS Roadmap guidelines for poly gate lengths of 0.25 µm are shown in Table II.

TABLE II
0.25 µM SCALING PARAMETERS [2]

I_{ON}	600 µA/µm	T_{ox}	40 - 50 Å
I_{OFF}	1 nA/µm	X_J (shallow LDD)	50 – 100 nm
Log(I_{ON}/ I_{OFF})	5.75 decades	N_D (LDD)	2 - 5 x10^{18} cm^{-3}
SS	85 mV/decade	R_S (LDD)	400 – 850 Ω/sq
DIBL	< 100 mV/V	X_J (contact)	135 – 265 nm
V_{DD}	1.8 – 2.5 V	N_D (contact)	1x10^{20} cm^{-3}
$|V_T|$	0.5 V	X_J (SSRW channel)	50 – 100 nm

A schematic of a transistor cross-section with physical parameters scaled according to gate length, λ, to achieve the performance parameters in Table II is shown in Figure 1.

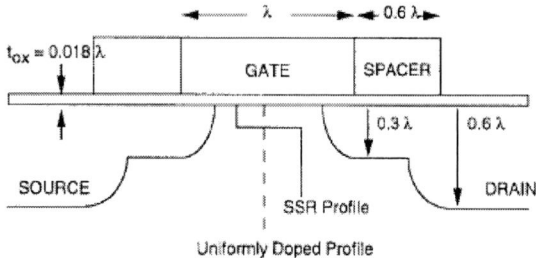

Figure 1: Scaling Guidelines as a function of Gate Length [3]

The junction depths of the shallow LDD region are made very small so the depletion region from the gate dominates the depletion region encroaching from the drain, thus giving the gate more control. Also, the gate oxide is made thinner so the gate electrode is closer to the channel and can have more influence in creating or stopping the inversion channel. Another way to look at this is the gate has a larger capacitance and causes a greater change in inversion charge per change in gate voltage. This leads to faster switching speeds and better off-current control by the gate. As the junction depths of the source/drain contact are made smaller their parasitic resistances will cause a decrease in drive current; therefore, they must be doped heavier to decrease the resistance. The channel region must be doped heavier to reduce the depletion regions from the source/drain from extending into the channel. This will decrease the control the drain has on the channel but will also decrease the mobility of the carriers at the surface since there is more dopant to cause scattering. A super steep retrograde well (SSRW) allows the channel to be lightly doped at the surface so the carrier mobility can be higher while having a higher doped sub-surface region to control the drain depletion region and short channel effects.

As the oxide thickness is decreased, the supply voltage must be decreased so the oxide doesn't break down. Also, the supply voltage must be decreased to reduce leakage between the reverse-biased drain and body terminal. Since the supply voltage is decreased, V_T must be decreased so that V_{GS} can be sufficiently higher then V_T. This is known as "gate over-drive" and is important since the drive current is proportional to $V_{GS} - V_T$. The problem that arises is there may not be enough gate voltage between 0 V and V_T to turn the device fully off. The sub-threshold swing, SS, is a measure of how much of a change in gate voltage, below V_T, is required to change the off-current in the device by a magnitude of 10. The equation for SS is shown in Equation 1 and has a theoretical limit of 60 mV/decade at room temperature. Where KT/q is the temperature dependent thermal voltage, C_D is the depletion capacitance which arises from the change in bulk charge for change in gate voltage, and C_{OX} which is the gate oxide capacitance.

$$ SS = \frac{KT}{q} \times \ln(10) \times \left[1 + \frac{C_D}{C_{OX}} \right] \qquad (1) $$

A 0.25 µm device needs a SS of 85 mV/decade to decrease the sub-threshold current 5.75 decades from on to off given the fact there is 500 mV between 0 V and V_T. It can be seen that as devices scale in length, the supply voltage and V_T must be scaled down. Also, the sub-threshold slope must be decreased by decreasing T_{OX}, and therefore increasing C_{OX}, so the SS is a value that can properly switch the device off.

3. PROCESS DESIGN

All fabrication, with the exception of Arsenic implants, is done at the Semiconductor and Micro-systems Fabrication Laboratory at the Rochester Institute of Technology. The bulk silicon process is done on 150 mm <100> p-type silicon wafers. The process includes 12 lithography levels which are exposed on a Canon FPA-2000i1 5x reduction stepper with i-line photoresist. A total of 76 steps are required to complete the devices through Metal 2, which takes approximately 6 months of processing. Fully scaled 0.5 µm design rules as well as non-fully scaled custom design rules are included in a new test chip.

The 0.25 µm CMOS process will yield transistors that have a cross section that is shown in Figure 3. Major process changes for this process are listed below and are advancements required to achieve the scaling guidelines discussed above

1-4244-0267-0/06/$25.00 ©2006 IEEE

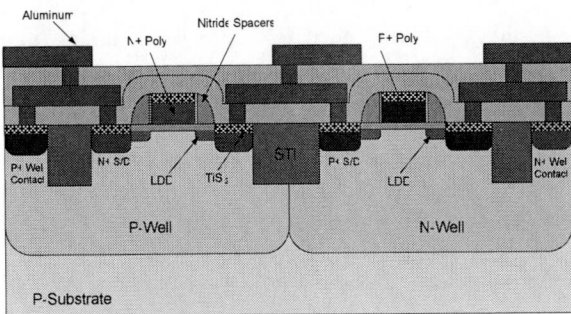

Figure 2: Final CMOS Crossection

- Shallow Trench Isolation
- 50 Å gate oxide w/ N_2O
- As/BF_2 Low Doped Source/Drain Extensions
- Nitride Sidewall Spacers
- Dual Doped Poly Gates
- Rapid Thermal Dopant Activation
- Titanium Silicide
- 2 Level Aluminum Metallization

Shallow trench isolation is performed on an IPEC/Westech 372 wafer polisher. Nitride over the active areas is used as a polish stop to planarize the TEOS oxide that is used to fill the shallow trenches. A SEM micrograph of a 1 μm wide shallow trench is shown in Figure 3.

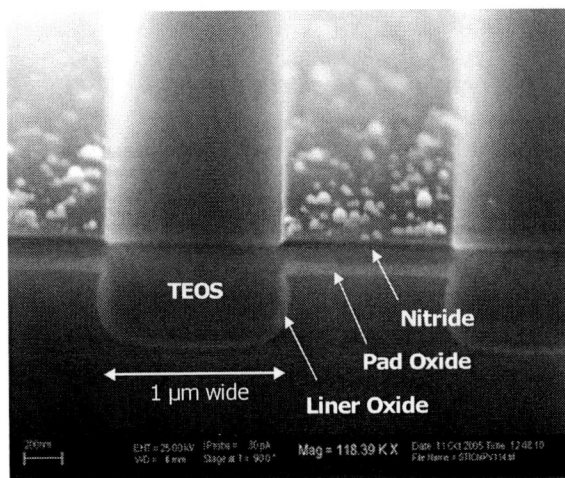

Figure 3: Shallow Trench Isolation after CMP

A 50 Å gate oxide is grown in a thermal furnace at 825°C with nitrogen incorporation from decomposition of N_2O gas. The nitrogen blocks boron from penetrating from the P+ poly gate into the PMOS channel during the source/drain activation anneal.

To form the 0.25 μm poly gates, a photoresist trimming process is used to reduce a 0.5 μm photoresist linewidth to 0.25 μm. The ASML 5500 DUV stepper has been down for the previous 3 years,

until recently, so the Canon i-line stepper is used. Photoresist trimming is performed in an oxygen plasma at a high pressure to create an isotropic etch. An SEM micrograph of 0.2 μm lines are shown in Figure 4.

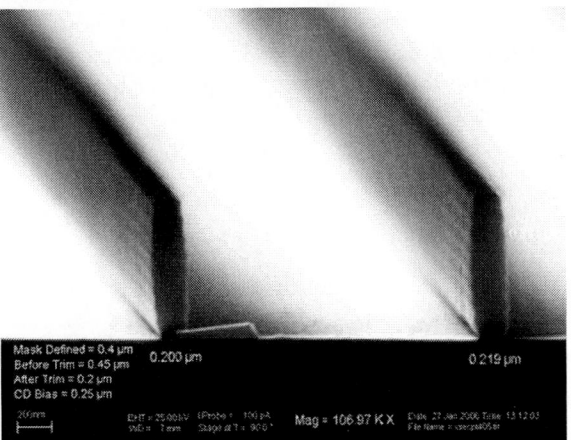

Figure 4: 0.2 μm Photoresist Lines After Trimming Process in O_2 Plasma

Nitride sidewall spacers are used to implant the source/drain contacts self aligned to the shallow LDDs. After etching of the poly gates, a 100 Å oxide is grown to repair damage to the edges of the gate oxide. This also provides a stopping layer for the spacer etch. An SEM micrograph of the spacers is shown in Figure 5.

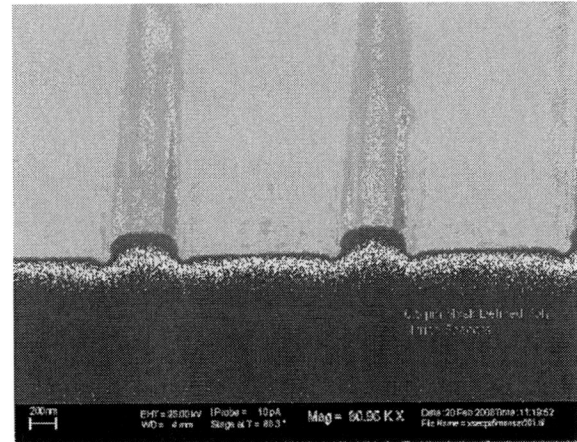

Figure 5: Silicon Nitride Sidewall Spacers

Low energy implants are required to form shallow junctions self aligned to the poly gate. The poly gates are etched in a Reactive Ion Etcher with chemistries that provide selectivity's high enough that the 50 Å gate oxide is sufficiently thick to stop on. 20 keV BF_2 and 25 keV As are used for the PMOS and NMOS, respectively. RIT currently does not have the capability to implant arsenic so a foundry service was used. Dual doped poly gates allow for surface channel NMOS and PMOS devices.

Dopant activation of the shallow source/drain extensions is performed with a rapid thermal processor. The wafers are heated to 1050°C for 10 seconds in an N_2 ambient. This will reduce the effect of transient enhanced diffusion and achieve the targeted shallow junction depths required for making small transistors. Titanium silicide is created over the poly gates and source/drain regions to reduce the parasitic resistance inherit to shallow junctions and increase the on-state current drive. It is a 2-step silicidation process which is also performed in a rapid thermal processor.

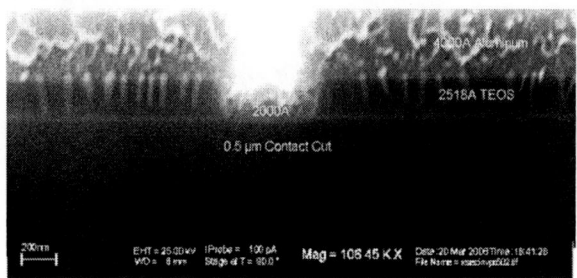

Figure 6: Aluminum Fill of 0.5 μm Contact Cuts

The contact cuts are etched in an Applied Materials P5000 RIE chamber. The process is designed for 2 levels of aluminum metallization. Figure 6 is an SEM micrograph of a 0.5 μm contact cut after Aluminum metallization.

4. RESULTS AND DISCUSSION

Electrical results reported in this section are preliminary data recently obtained from a finished CMOS lot. Figure 7 shows an ID-VD "family of curves" plot for gate voltages up to 2.5 V. The drain current at VG=VD=2.5 V is 177 μA/μm.

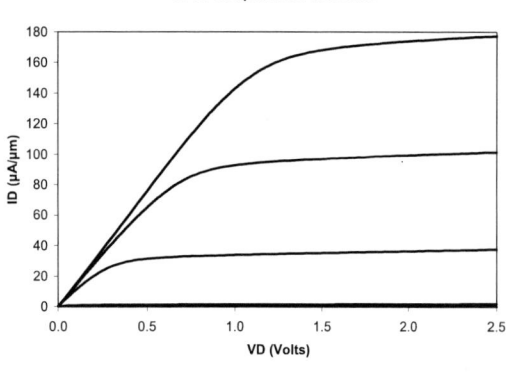

Figure 7: 0.25 μm NMOS ID-VD

The exact effective gate length has not yet been extracted but the mask defined gate length was 0.5 μm

with 0.25 μm of resist trimming. In turn, this device is most likely smaller then 0.25 μm when taking into account poly etch bias, and lateral diffusion of the source/drain extensions.

The threshold voltage for this device is 1.0 V which is higher then the 0.5 V target. As a result, the on-state current is lower then it could be and the off-state current is lower than it should be.

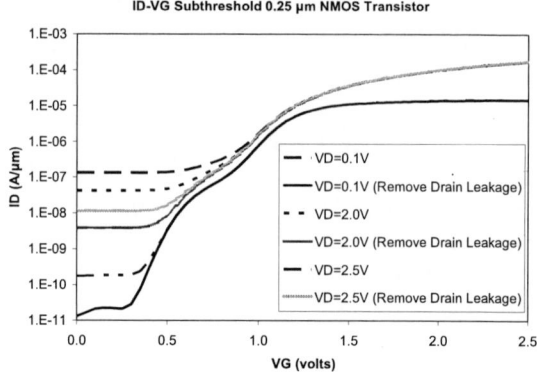

Figure 8: 0.25 μm NMOS ID-VG

A large amount of sub-threshold leakage current is observed as the drain is increased from 0.1 V. At first glance this looks like drain induced barrier lowering, DIBL, where the drain is influencing the amount of carriers allowed to leak from source to drain before the device is at threshold. However, Figure 9, which is a plot of the current through the drain diode, shows large reverse bias leakage. This leakage accounts for a significant portion of the off-state leakage and is subtracted from the ID-VG plot in Figure 8 to give a better estimation of the off-state performance of the device if this damage is reduced. The 2nd slope shown in Figure 8 before threshold, is most likely caused by plasma damage from the numerous reactive ion etch steps.

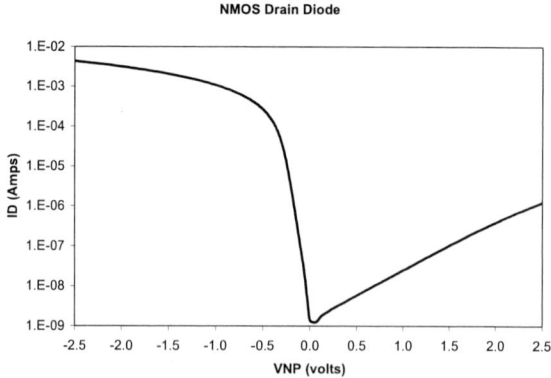

Figure 9: NMOS Drain Diode Leakage

This leakage could be from implant damage caused by the arsenic ions for the source/drain contact implant or the TiSi$_2$ consuming too much of the source/drain silicon. The exact mechanism for this leakage is being investigated.

A PMOS transistor ID-VD "family of curves" is shown in Figure 10. The gate length is reported as 0.35 µm because the mask defined gate length is 0.6 µm and 0.25 µm of resist trimming result in a 0.35 µm poly length. However, etch bias from the poly RIE and other processing factors may allow an electrical extraction of 0.25 µm to be found. This extraction is underway and will be reported.

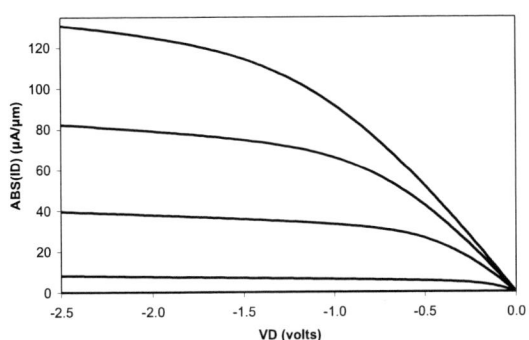

Figure 10: 0.35 µm PMOS ID-VD

The drain current at VG=VD=-2.5 V is 131 µA/µm. The threshold voltage is -0.735 V which is slightly larger then target of -0.5 V. Figure 11 shows the sub-threshold ID-VG plot for the PMOS transistor. The sub-threshold slope is 75 mV/decade at VD=-0.1 V and increases to 85 mV/decade at VD=-2.5 V. A DIBL value of 8 mV/V is calculated.

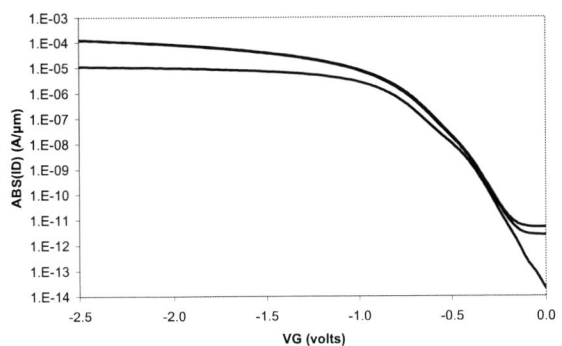

Figure 11: 0.35 µm PMOS ID-VG

The leakage in the sub-threshold region at drain biases of -2.0 V and -2.5 V can be attributed, in part, to leakage from the PMOS drain diode which is under reverse bias. Note the absolute value of the current is displayed for logarithmic plotting purposes.

5. CONCLUSION

A process for fabrication of 0.25 µm CMOS transistors has been demonstrated. Additional electrical testing is underway to fully characterize the transistors and recommend improvements for higher on-state current and lower off-state current. A number of advanced processes have been integrated to produce the final CMOS devices. Each process can be improved upon by students in the RIT factory for enhanced performance of the devices.

Successful fabrication of these deep-submicron devices will provide the SMFL with essential process qualifications and improve the teaching capabilities of the Microelectronic Engineering Department at RIT. Students will have devices which can be tested in labs as well as SEM crossection and process data that can be studied so improvements can be implemented. This work is funded by the Microelectronic Engineering Department at RIT for the author's Master of Science Thesis in Microelectronic Engineering.

REFERENCES

[1] "Microprocessor Quick Reference Guide." http://www.intel.com/pressroom/kits/quickreffam.htm

[2] L. Wilson, ed., "The National Technology Roadmap for Semiconductors: 1997 Edition", Semiconductor Industry Association, San Jose, California

[3] I. De and C. M. Osburn, "Impact of super-steep-retrograde channel doping profiles on the performance of scaled devices," *IEEE Trans.Electron Devices*, vol. 46, no. 8, pp. 1711–1717, Aug. 1999.

Michael Aquilino, originally from Liverpool, NY, received a B.S. degree, *Summa Cum Laude*, in Microelectronic Engineering from the Rochester Institute of Technology in 2004. He obtained co-op work experience at Atmel Corporation in Columbia, MD and Integrated Nano-Technologies in Henrietta, NY. He is currently pursuing an M.S. degree in Microelectronic Engineering at RIT. Michael will be joining IBM in East Fishkill, NY in June 2006 as a 300 mm Advanced Controls & Routing Engineer.

Dr. Lynn Fuller, completed his BS and MSEE degrees at Rochester Institute of Technology in 1970 and 1973, and received his Ph.D. in Electrical Engineering at the State University of New York at Buffalo in 1979. Dr. Fuller is the founder and Motorola Professor of the Microelectronic Engineering Department at RIT. He is also core faculty in the Microsystems Engineering Department at RIT. Dr. Fuller is a Fellow of the IEEE and has received the RIT Distinguished Alumnus Award and the IEEE Region 1 Award for "Prolific Contributions in Education and Research in Microelectronic Engineering".

1-4244-0267-0/06/$25.00 ©2006 IEEE

1-4244-0267-0/06/$25.00 ©2006 IEEE

RIT MEMS Fabrication Course

UNIVERSITY GOVERNMENT INDUSTRY MICROELECTRONICS SYMPOSIUM

June 2006

Robert E. Pearson, *Senior Member, IEEE*, Lynn F. Fuller, *IEEE Fellow,* and Ivan Puchades

Abstract— This paper describes a simple, reliable; bulk micro-machined, micro-electro-mechanical system (MEMS) process flow for the fabrication of a wide variety of devices which has been implemented within the Microelectronic Engineering Department at the Rochester Institute of Technology. The fabrication and testing results for thermopiles, pressure sensors, micro-speakers, micro-pumps, accelerometers and inductors in a ten week fabrication course are reported.

Index Terms—**Bulk MEMS, fabrication, thermopile, pressure sensor, inductor, flow sensor, micro-pump.**

I. INTRODUCTION

THE Microelectronic Engineering Department at Rochester Institute of Technology offered its first MEMS fabrication course in 1995 [1], [2]. Since 1995 over 150 graduate and undergraduate students have taken the course [3].

Device designs for the winter quarter (2005-2006), MEMS fabrication class (EMCR-870) included, three different thermopile designs, a micro-speaker, a piezoelectric pressure sensor [4], a piezoelectric accelerometer, a gas flow thermal sensor, silicon bio-probes, an electro-magnetic pump [5], a RF inductor and two different L-C intra-ocular pressure (IOP) sensors. The MEMS fabrication course can be part taken as part of a sequence of courses that starts with a MEMS theory course and ends with a MEMS evaluation and testing course.

II. DEVICE AND PROCESS DESIGN

Unlike conventional integrated digital electronic circuit technology which has come to be dominated by a small number of process flow variations (logic and memory), Micro-Electro-Mechanical-Systems has a vast array of possible process flows depending on the target device. Since the course described in this paper is a foundation course on which students can expand on in research pursuits and other courses, the single process flow must provide the most flexibility and ease of manufacturing. Bulk micro-machining, where the back side of the wafer is etched away to form thin membrane or diaphragm regions has proven to provide a wide variety of possible devices. Transistors, resistors and electronic circuits were included in the mask layout for test purposes and to demonstrate the possible integration path for MEMS and electronics. Front and backside alignment marks, resolution targets and alignment verniers were provided.

Each student is given a 5mm by 5 mm area for their design, with a 3mm by 3 mm maximum diaphragm size. Fourteen different design cells made up one die, and two cells were reserved for alignment marks and standard characterization structures. There are five mask levels used in the process. The first level is the P+ diffusion mask; the second is the back-side diaphragm mask. The third level is the poly-silicon mask, the fourth level is the contact cut mask and the fifth level is the metal (aluminum) mask. Five 1X masks were written using RIT's MEBES III electron beam mask making system. Each mask contained twelve copies of the die layout to be exposed on four inch wafers. Eight of the die extended beyond the edge of the wafer resulting in partial die. Special mask alignment marks were placed at the extreme corners of the masks for use in the backside alignment process.

The design phase of the course lasted three weeks. During this time the students had to pick a device to make. They had to research the device in the notes and reference materials and develop a model of predicted device behavior from which layout design parameters could be explored and extracted. The given process can not be used to do "everything" but as will be demonstrated, a surprising number of devices can be created using this process. Many students thought that the difficult part of the course had been overcome when their design were submitted. This illusion was shattered when the students began final testing of their finished devices after processing.

R. E. Pearson is an Associate Professor in the Microelectronic Engineering Department at the Rochester Institute of Technology, 82 Lomb Memorial Dive, Rochester, NY, 14623 USA (phone: 585-475-2923; fax: 585-475-5041; e-mail: repemc@ rit.edu).

L. F. Fuller, is Motorola Professor in the Microelectronic Engineering Department at the Rochester Institute of Technology, 82 Lomb Memorial Dive, Rochester, NY, 14623 USA (phone: 585-475-2035; fax: 585-475-5041; e-mail: lffemc@ rit.edu).

I. Puchades is a Ph.D. student in Microsystems at the Rochester Institute of Technology, 82 Lomb Memorial Dive, Rochester, NY, 14623 USA (phone: 585-475-2923; fax: 585-475-5041; e-mail: ixp6782@ rit.edu

1-4244-0267-0/06/$25.00 ©2006 IEEE

III. PROCESSING

A. Backside Polish and initial Processing

Previous course laboratory experience has shown that nitride layers used to protect against the effects of the long duration (85C) KOH etch step did not perform that well over surfaces with topography. The back surface of the starting wafers had only a chemical polish (as done by the wafer manufacturer) and had an estimated RMS surface roughness of approximately 20 microns. For this reason the first processing step was to chemically and mechanically planarize (CMP) the back surface of the wafer.

A 5,000 Angstrom thick oxide was grown on the wafers, which was patterned on the top side to act as a diffusion mask for the boron spin on dopant diffusion. After the diffusion the spin on glass and the masking oxide were etched off the wafers.

A thin (500 Angstrom) pad oxide was then grown on the wafer followed by a low pressure, chemical vapor deposited silicon nitride layer. The nitride layer on both sides of the wafer is used to protect the silicon surface from being attacked by the KOH etch used to create the diaphragm. The smooth wafer surfaces allow a reasonably thin (1500A) nitride to be used.

B. KOH Membrane Etch

The thickness of the starting wafers was measured using a focus technique on a conventional optical microscope. Membranes between 25 and 45 microns thick were fabricated. Test grade wafers were used for this course. It was observed that on a small percentage of membranes (2-5%) that there were localized regions within a single membrane that etched abnormally fast. The etch rate was faster than could be explained by variations in wafer thickness or etch bath temperature. The regions that etched through the membrane did so in rectangular or square shapes that indicated a crystallographic dependence that could be due to a volume defect in the wafer such as stacking faults. The next time the course is offered prime wafers will be used. Fig. 1 shows the back side of a wafer after the KOH etch step. The wafer shown in Fig. 1 has few membrane defects and has good edge integrity.

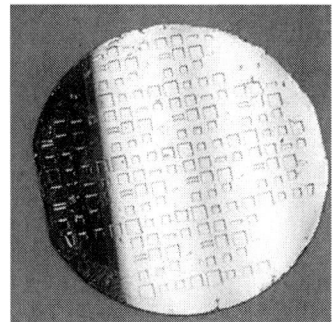

Fig. 1. Backside of the wafer after KOH etch.

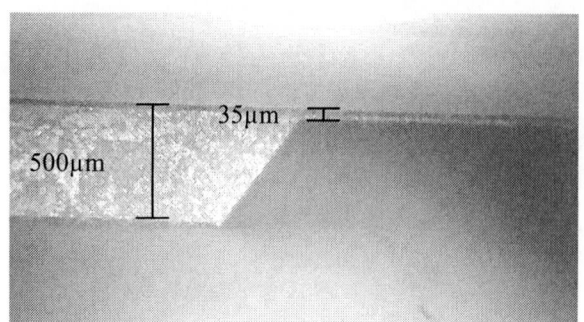

Fig. 2. SEM cross-section of the 35µm Si diaphragm.

Fig. 2 show a scanning electro-micrograph of a cross-section of a diaphragm that has been fabricated. New optical measurement capability has been obtained that will allow for much more accurate determination of the diaphragm thicknesses.

C. Photolithographic Processing

Once the back-side diaphragm etch step has been done, the wafers will not be held properly by either the automated resist coating track or the manual resist spinner. To solve this problem, the four inch diameter device wafers were taped at the edges to six inch diameter wafers for priming and resist coating. The tape was removed after coating before the resist pre-bake step to avoid melting the tape.

The transfer arm of the Karl Suss contact aligner will not properly pick up wafers that have had the back-side diaphragm etch. The work-around that was devised to solve this problem was the adhering of the four-inch device wafer to another four inch wafer by use of a drop or water and the capillary action of the two flat surfaces. The wafer flats of the device and carrier wafers were carefully aligned. The aligner is able to handle the pair of wafers without any problems. After exposure the wafers were carefully separated using tweezers and the device wafer manually developed (immersion process). The top "device" wafer was not observed to move relative to the carrier wafer during the alignment process and good alignment was observed for eight of the nine wafers.

D. Finished Cross-Section of one Possible Device

Fig. 3 shows a cross-section of a device which used the backside silicon etch, a from-side diffusion and metal patterning. This particular device did not make use of the poly-silicon layer.

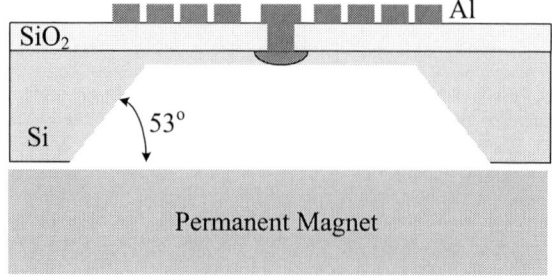

Fig. 3. Cross-sectional view of the micropump actuator

IV. DEVICES

Figure 4 contains a micrograph of an accelerometer. A piezo-resistive bridge-type pressure sensor was constructed. A lead proof mass was attached to the membrane using an epoxy. The differential voltage output of the bridge circuit is monitored to determine the plate deflection due to acceleration of the proof mass. The second version of the device has a coil on the membrane and an external circuit is used to provide a current through the coil. The magnetic field induced by the coil interacts with an external permanent magnet below the diaphragm to restore the diaphragm to its original "null" position, thus providing a coil current signal that is proportional to the acceleration of the proof mass.

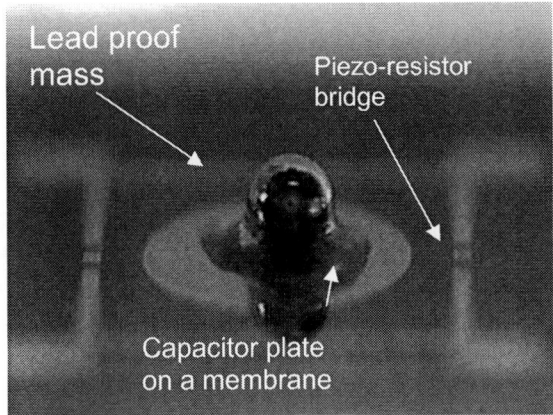

Fig.4. Photo-micrograph of a piezo-resistive pressure sensor combined with a proof mass and a electro-magnetic coil for a force-balance acceleration feedback circuit.

Figure 5 contains two different views of a micro-speaker. 5a shows the speaker coil with no current flowing, while 5b shows the speaker coil energized and deflected.

(a)   (b)

Fig.5. Photo-micrograph of a micro-seaker and micro-pump. 5a shows the un-energized coil (no deflection) and 5b shows the deflection observed for the energized coil.

The micro-speakers were audible within the testing room over the typical audio frequency range. The coil resistance was higher than the original design assumptions predicted due to the type of metal cross-over used. A process improvement that is being implemented during the spring of 2006 would use double-level metal. A laser pointer was used to determine the membrane deflection distance

during operation at a particular power input level. The laser was bounce off of the coil onto the fall wall of the lab and the spots on the wall observed. Geometry was used to solve for the vertical membrane deflection distance. A value of two microns was calculated and seems reasonable. A new real-time optical thickness measurement tool may be able to verify this calculation.

A micro-pump can be constructed as a variation on the micro-speaker. Figure 6 shows a electro-coil placed on top of a prepared cavity which is in turn on top of a permanent magnet. The asymmetrical design of the cavity provides for non-equal outflow and suction which in turn leads to overall flow in one direction only as shown in Fig 7.

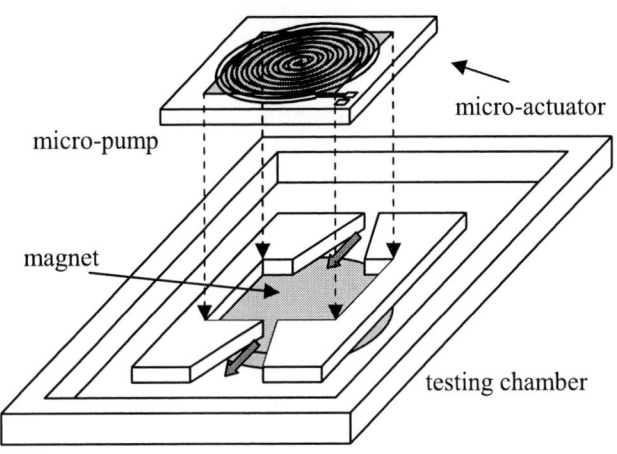

Fig. 6. Design of a micro-pump system

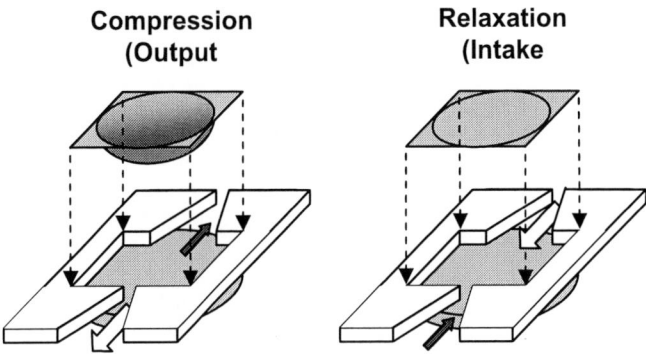

Figure 7 Directional Flow

Another category of devices made using this process flow is the thermopile. Figure 8 contain a micrograph of a thermopile optical power transducer temperature sensing device. The device consists of thin wires of aluminum and poly-crystalline silicon which contact each other on the ends forming a single resistive path. When one end of a rod is heated to a temperature above the other end a voltage is developed across the contact junction between the two metals (Seebeck effect). By increasing the number of junctions that exist in both the hot and cold regions, the output voltage that is developed can be increased. The design pictured in figure 2 consists of 24, aluminum to n+ poly-silicon junctions in the region heated by optical power. The optical power was supplied by the illumination microscope on the wafer probing station. The center junctions were placed over the thin silicon membrane so that the absorbed heat would have a higher thermal resistance

The optical pyrometer was difficult to characterize since its temperature response to ordinary optical (microscope) lamp intensity was rather low.

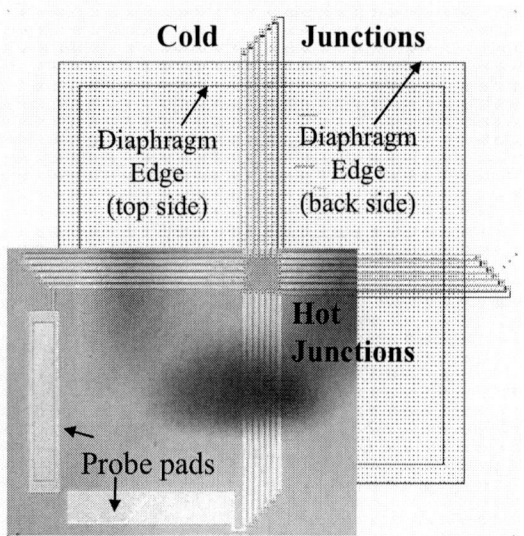

Fig.8. Photo-micrograph of one of the three thermopiles sensors fabricated. Total sensor dimensions are 5000 microns by 5000 microns. Shadow is a result of wafer vacuum chuck deflecting the membrane (not part o the device function)

Another student created a different type of thermopile design. In his design a separate heater was buried below the linear thermopile array as shown in Fig 9.

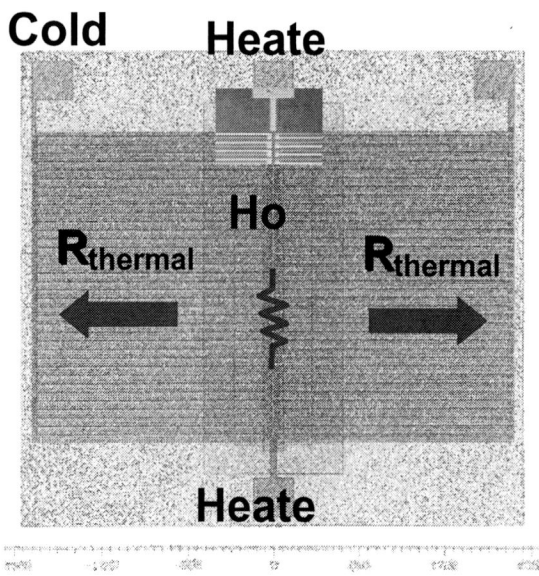

Fig. 9. Linear thermopile array with a heater.

A project that was of interest to the students was a wireless intra-ocular pressure sensor for detecting/treating glaucoma as shown schematically in Figure 10. An LRC impedance is created on a MEMS membrane with a variable capacitance

when coupled to a separate capacitor located inside the etched membrane cavity.

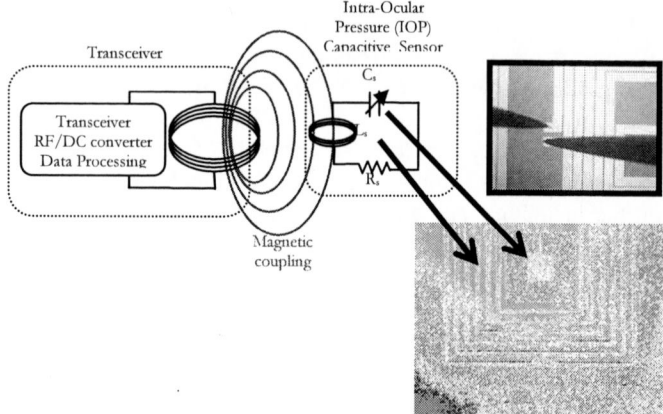

Figure 10 Intra-ocular pressure sensing system

V. TESTING

Wafers were first characterized for p+ diffused layer sheet resistance, contact resistance, n+ poly-silicon sheet resistance and aluminum sheet resistance.

The performance of a 40 junction (aluminum to n+ poly-silicon) thermopile with a built-in test heater was characterized. The hot junctions of the linear thermopile array was built on the center a silicon membrane that was approximately 30 microns thick, 780 microns wide and 3000 microns long. The cold junctions were located over the thicker bulk silicon portions of the wafer. The thermopile output voltage was measured as the heater power was swept from zero to one-quarter Watt as shown in figure 11.

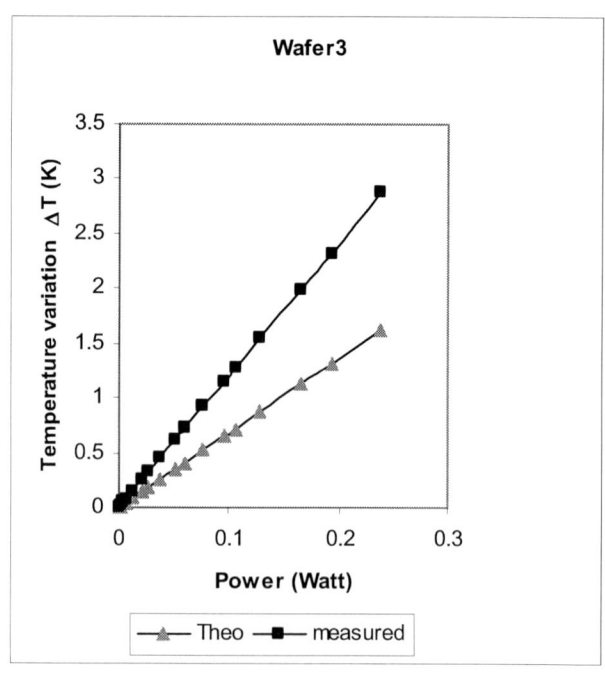

Fig. 11. Thermopile heater response testing

The output was found to be linear over the input power range with a value of 1.2 milli-Volt per Watt. Using the Seeback Coefficients for the junction materials the expected output voltage could be calculated as a function of the junction temperature. A junction temperature (over the heater) of two degrees centigrade above ambient at 24 Volts on the heater was determined. Theoretical calculations of the heater temperature rise using the input power and calculated thermal resistance paths predicted a 1.6 degree centigrade temperature rise. The accuracy of the prediction was found to be very sensitive (as expected) to the accuracy of the membrane thickness measurement.

The device was also tested as a gas flow sensor. The heater for the thermopile was biased at a fixed power level and a calibrated gas flow was directed across the surface as the thermopile voltage output was measured. The deviation of the output voltage with gas flowing compared to the output voltage with zero flow was recorded for various flow rates.

The output voltage characteristics of two other optical input pyrometers (thermopiles) were obtained and calibrated by comparison to the design with a known heater values as shown in Figure 12.

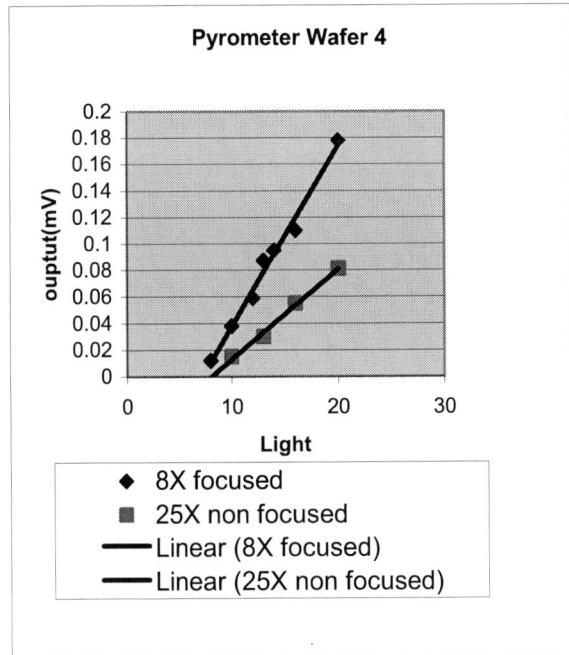

Fig. 12. Thermopile response for various lamp intensities

The increased response observed for the application of a heat absorbing "blackening" layer.

The micro-speaker was tested over various power and frequency ranges. A video recording showing the visual deflection of the membrane at low frequencies has been made. The audio portion of the recording illustrates the volume level obtained from the small area speaker. The micro-pump testing is currently awaiting the construction of the necessary micro-fluidic channels.

VI. CONCLUSIONS

A meaningful MEMS fabrication experience can be successfully undertaken in a brief ten week academic setting. A closed loop learning cycle, including some basic theory, design, fabrication through testing and comparison to predicted results is invaluable to students in such a diverse multidisciplinary experience such as MEMS fabrication. Students experience first-hand, the lack of fundamental models for some devices, processes and structures. They encounter process variability effects, testing and packaging issues and yield loss mechanisms associated with MEMS fabrication.

ACKNOWLEDGMENT

The authors would like to thank all of the students who have taken the course over the last eleven years and in particular those enrolled during the winter and spring quarters of the 2005-2006 academic year.

REFERENCES

[1] L. F. Fuller, "A Multi-project MEMS Lecture and Laboratory Course" Proceedings of the 12th Biennial University/Government/Industry Microelectronics Symposium, Rochester, NY, July 20-23, 1997.

[2] L. F. Fuller, A. Pham, and P. Merwah, "MEMS Activity at Rochester Institute of Technology" Proceedings of the 13th Biennial University/Government/Industry Microelectronics Symposium, Minneapolis, MN, June 20-23, 1999, pp 146-149.

[3] K. Munger and L. Fuller, "Fabrication of Polysilicon Surface Micromachined MEMS Structures" Proceedings of the 14th Biennial University/Government/Industry Microelectronics Symposium, Richmond, VA, June 17-20, 2001, pp 163-166.

[4] L. F. Fuller and S. Surdigo, "Bulk Micromachined Pressure Sensor" Proceedings of the 15th Biennial University/Government/Industry Microelectronics Symposium, Boise, ID, June 30- July 2, 2003.

[5] J. Getpreecharsawas, I. Puchades, R. Hournbuckle L. Fuller, R. Pearson and S. Lyshevski, "An Electromagnetic MEMS Actuator for MicroPumps". MEMSTECH'2006, May 24-27, 2006, Lviv-Polyana, UKRAINE

Robert E. Pearson (M'78–SM'02) Obtained his B.S.E.E. in 1981 and his M.S.E.E. in 1986 from Rochester Institute of Technology in Rochester, New York. He received his Ph.D. in Electrical Engineering from SUNY at Buffalo in 1995. He has worked at RIT from 1982-1997 and at Virginia Commonwealth University (1998-2003) and is now again at RIT where he is an Associate Professor in the Microelectronics.

Lynn F. Fuller (M, SM, F) Dr. Lynn Fuller completed his BS and MSEE degrees at Rochester Institute of Technology (RIT) in Rochester, New York in 1970 and 1973. He obtained his Ph.D. in Electrical Engineering at SUNY Buffalo in 1979. Dr. Fuller returned to RIT and created the Microelectronic Engineering Program. Dr. Fuller is the Motorola Professor of Microelectronic Engineering at RIT.

Ivan Puchades was born in Barcelona, Spain in 1975. He obtained the B.S. in Microelectronic Engineering in 1999 and the M.S. in Electrical Engineering in 2000, both from Rochester Institute of Technology in Rochester, NewYork. He is currently pursuing a Ph.D. in Microsystems Engineering from this same institution.

From 2000 to 2005 he worked as an RF device engineer and BiCMOS technology development engineer for Freescale Semiconductor in Phoenix, Arizona.

1-4244-0267-0/06/$25.00 ©2006 IEEE

1-4244-0267-0/06/$25.00 ©2006 IEEE

Building the New Berkeley Microlab

A. William Flounders, Katalin Voros
Microfabrication Laboratory
University of California
Berkeley, CA 94720-1770

Abstract—The University of California at Berkeley is proceeding with construction of a new nanofabrication laboratory – the CITRIS Nanolab. This new facility is a key component of the College of Engineering's CITRIS research center - the Center for Information Technology Research in the Interest of Society. The new lab will enable world class faculty research and stimulate creative partnerships with industry to address nanoscale CMOS electronics, nanoelectromechanical systems (NEMS), integration of opto and bioelectronics, and nano/micro/macro interface technologies. This laboratory will be the successor to The Berkeley Microlab and will continue and expand the Microlab tradition of a professionally managed, shared laboratory resource open to all academic researchers and supported on a recharge basis to insure the lowest possible barrier to entry.

This presentation will provide an overview of the design and planning process with specific commentary on some of the challenges unique to the academic laboratory. Sample questions that will be addressed are as follows. How do you define facility and utility needs when future research requests are unknown? How do you efficiently translate the extensive industry experience of design consultants to a university situation? Is it possible to communicate directly with the architect and design team when working within the procedural confines of a large state university? Construction budget versus fit-up budget – are there strategies to take advantage of the Capital Projects process? Value engineering – does it provide either? And finally, is it possible to build a research laboratory this decade without including the word nano in the facility name?

I. INTRODUCTION

THE University of California, Berkeley takes great pride in the fact that it opened the first university integrated circuit lab in 1962. This modest 1200 ft2 facility utilized ¾ - 2 inch silicon wafers (cut and polished in-house), spin on dopants, rubylith masks, manual alignment and thermal evaporators to construct a wide range of IC devices. As David Hodges, former Dean of the Berkeley College of Engineering describes, "It was pretty primitive in those days... but we got some working circuits and we learned an awful lot." This facility was upgraded and expanded in 1982 to a 10,000 ft2 Class100 cleanroom – The Berkeley Microlab. Since 1983, the Microlab has supported over 2200 graduate and post-graduate researchers and more than 70 local companies. For many years, the Berkeley Microlab, along with MIT and Stanford,

has set the standard for shared university integrated circuit and general microfabrication facilities. The Microlab's success is thanks to dedicated management and staff; cross disciplinary faculty support of the shared facility model; and, department and college level recruiting to maintain IC and microfab related research at critical mass. However, maintaining critical mass does have repercussions and the Microlab is exploding at the seams. Several utility systems are running at or in excess of capacity. Recent tool deliveries to the lab have required verifying elevator load limit, cutting and re-welding equipment frames, removal of multiple wall systems, expanding support chases, redesigning room air flow and addition of electrical and chilled water capacity. Quite simply, Berkeley needs a new and better equipped facility to enable next generation nanofabrication research.

In 2005, UC Berkeley broke ground for the headquarters of the Center for Information Technology Research in the Interest of Society - CITRIS. For a compelling introduction to the societal scale issues CITRIS research has already been addressing, see: http://www.citris-uc.org/. A key component of the CITRIS headquarters building will be the new CITRIS NanoLab – the successor facility to the Berkeley Microlab. Construction completion is anticipated in 2008 followed by equipment transition from the Microlab to the new CITRIS NanoLab. An artisitic rendering of the CITRIS Headquarters and the CITRIS Nanolab is shown in Figure 1.

Fig. 1: CITRIS main building (right); CITRIS Nanolab (left)

1-4244-0267-0/06/$25.00 ©2006 IEEE

II. PLANNING

Most universities build the facility and then hire the manager. The manager spends the first few months or even years walking through the lab wishing they had been consulted during design. One of the simplest and best recommendations is to bring the new lab manager in as early as possible during the design phase. No one has a greater vested interest in the lab design than the person that will be responsible for future day to day operations. Usually, the campus planning department or capital projects office will not fund this individual. They will hire a project manager whose focus is construction cost and schedule; this project manager has limited interest in future operational efficiency or costs. If the construction project can achieve a cost savings by transferring expense to future operations, it usually will. The future owner – academic department, college, or research unit – must hire their own individual to counter this trend.

New building design is performed by the building architect in concert with a large array of design consultants. Consultants with specialties ranging from landscape design to elevators even to door hardware appear from nowhere at the request of the architect. The most important consultants for the lab manager are mechanical/electrical/plumbing (MEP) and code/life safety. At UCB, we had both a building wide MEP consultant and a laboratory design consultant. This created some coordination conflicts. A later strategy was to have a main building MEP consultant and a laboratory systems MEP consultant. The university project manager prefers a generic MEP consultant for the main building to keep costs down. The lab manager wants an MEP consultant that can design specialty systems for the lab – but this consultant will cost more for design of 'standard' building systems. With hindsight, the best solution is to have a single MEP consultant and have the lab manager verify they have detailed specialty laboratory design experience. Such specialists do exist but are less common.

III. FACILTIES

As seen in Figure 1, the laboratory is an independent wing of the main, seven story classroom, auditorium and office tower. The separate laboratory wing contains a mechanical support space in the basement with two levels of cleanroom above, each with a separate air handling interstitial space. Though the lab has two levels, it is a single lab with a single gowning area and an elevator and stairwell connecting the chase areas of the two lab levels.

Hazard classification, vibration control, EMI control, clearance/clear span issues and clean class were the primary facility issues in order of prioritization. The CITRIS Nanolab must be able to welcome all semiconductor processing and therefore requires an H6 occupancy classification. Though some university nanotechnology labs have attempted to accept less stringent classifications (e.g., H8) this strategy is not recommended. Even if your lab is not focused upon semiconductor processing, the hazardous material volume limits of less stringent classifications will preclude support of many nanotechnology research efforts requiring more than lecture bottle quantities of pyrophoric and Class1 toxic materials; e.g., nanowire growth, MOCVD and ALD. Vibration and EMI control are critical to enabling satisfactory performance of electron beam tools. The lower lab level meets Vibration Criteria Level E (VC-E, 125 □inch/sec) and EMI limits of <3mG AC and <1mG DC. The upper level is rated VC-C with no defined EMI criteria. Air handling requirements are lower level Class 100, upper level Class 1000. The payoff of the two lab levels now becomes clear. Most stringent (and expensive) criteria are imposed on the lower level and cost savings are achieved on the upper level.

IV. UTILITIES

The most important components of a laboratory are its specialized utilities. Total utility capacities are typically accomplished by starting with a detailed tool list, capturing utility needs of every tool from that tool's specification sheet and totaling all tool needs. This total is then 'diversified' (scaled down) to estimate actual lab wide demand. In an academic setting, this strategy is completely artificial. For a new university lab, dependent upon tool donations and future faculty recruiting, the lab manager often has no idea what the actual tool list will be. In this case, the existing Berkeley Microlab tool set could be used for baseline sizing but expansion and growth estimates had to be made. In this case, it was verified that most tools could be upgraded from 150mm through 200mm and up to 300mm and the utility systems would still be sufficient.

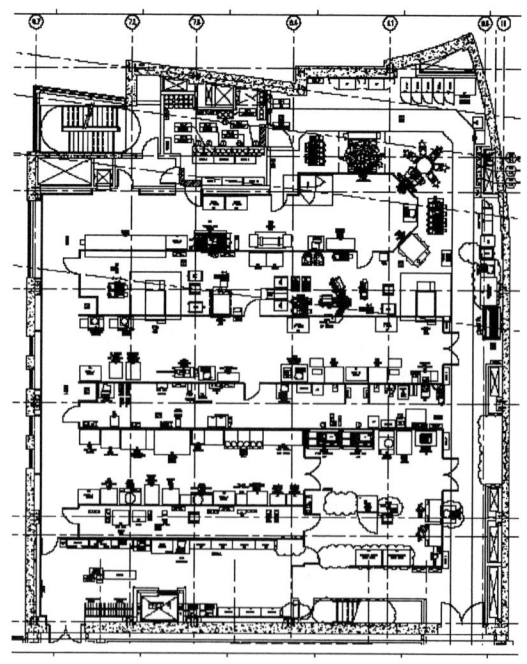

Fig 2: Upper lab level

V. LAYOUT

The lab will be a bay and chase design with sidewall air return. This is a proven flexible concept for academic lab settings. Layout is a straightforward interleaved set of bay and chase fingers. Unique aspects are the chase can be entered separately without entering the lab and full access from the mechanical space to the chase to the chemical storage rooms is possible without entering the clean areas. Layout of the upper lab level is shown in Figure 2.

VI. CONTINGENCIES

PCW (process cooling water) – design emergency bypass to city water in the event of PCW service interruption. Consider, just because one or two plasma tools have a cooling water resistivity requirement (e.g. in the range 50k-ohm to 1 M-ohm) is no reason to impose this requirement on the entire PCW loop. Imposing resistivity requirements on the entire PCW system will lead to specification of 304 ss as the piping material and large DI columns (with regular maintenance needs). Copper piping and small point of use DI cartridges for the one or two tools that need deionized cooling water are recommended as a more cost effective solution.

CDA (compressed (clean) dry air) – design emergency bypass to house N2 in the event of CDA service interruption.

DI (deionized water) – design RO storage tanks to enable limited operation in the event of DI service interruption. DI make up rate does not have to be directly tied to maximum or even diversified DI consumption rate. RO storage can also enable the lab to accept a slower DI water make up rate. During high consumption times tank depletion will exceed makeup; during low consumption; tanks will be replenished. Install DI columns in two sets that are plumbed for both serial and parallel flow.

AWN (acid waste neutralization) – even if using a continuous flow through system, a pretreatment holding tank is recommended. The holding tank will enable some limited operation in the event of AWN service interruption.

Windows – include the largest possible freight elevator the project can accommodate, then use windows as backup. In the rare event of delivery of exceptionally large equipment - make certain lab access windows are as large as possible and can be removed to enable equipment delivery. Review the load limits of the plaza outside the potential delivery window to make certain crane and associated rigging can be supported.

Roof – put a hatch in your roof; size it to the largest serviceable component. It is much less costly and less disruptive to operations to deliver an exhaust fan motor by chain hoist to the roof then by crane or helicopter. It is admittedly preferable to simply have the freight elevator service the roof level - but this will add cost.

Floor Penetrations – a vibration resistant waffle slab is a significant slab of concrete – 24 – 48" thick depending upon vibration class and column layout. At any location where pipe penetrations were already planned through the floor, additional penetrations of a wide range of diameters were designed. These unused penetrations are essentially fire rated and approved empty and capped conduit paths. These penetrations will make future piping installs much easier.

VII. CONCLUSION

Laboratory planning in an academic environment is challenging. The only assurance is that the demands on the facility will constantly change and evolve. The first Berkeley Microlab served the EECS department and beyond for twenty years; the present Berkeley Microlab has served the entire College of Engineering and far beyond for almost 25 years. The new CITRIS Nanolab is poised to serve the entire Berkeley campus and partner campuses at Davis, Santa Cruz and Merced for the next twenty years following the same successful model of its predecessors – a laboratory is meant to be a dynamic environment that constantly modifies the facility to respond to the needs of the researchers. Though construction may be complete in 2008, the CITRIS Nanolab will not be finished until it is retired by its successor facility.

1-4244-0267-0/06/$25.00 ©2006 IEEE

25 Years of Microelectronic Engineering Education

Santosh K. Kurinec, Lynn F. Fuller, Bruce W. Smith, Richard L. Lane, Karl D. Hirschman,
Michael A. Jackson, Robert E. Pearson, Dale E. Ewbank, Sean L. Rommel, Sara Widlund,
Joan Tierney, Maria Wiegand, Maureen Arquette, Charles Gruener and Scott P. Blondell
Department of Microelectronic Engineering
Rochester Institute of Technology
82 Lomb Memorial Drive, Rochester, NY 14623-5604
www.microe.rit.edu

Abstract— Rochester Institute of Technology started the nation's first Bachelor of Science program in Microelectronic Engineering in 1982. The program has kept pace with the rapid advancements in semiconductor technology, sharing 25 of the 40 years characterized by Moore's Law. The program has constantly advanced its integrated circuit fabrication laboratory in order to graduate students with state-of-the-art knowledge, who become immediate and efficient contributors to their company or graduate program. Today, this facility serves as a key resource for research in semiconductor devices, processes, MEMS, nanotechnology, and microsystems. This has led to the creation of the first PhD program in engineering at RIT, a doctorate in Microsystems Engineering. The department enjoys strong support from the semiconductor industry through its industrial affiliate program. Recently the department received a $1 million department level reform grant to address the imminent need for a highly educated workforce for the US high tech industry that is on the verge of nanotechnology revolution.

I. INTRODUCTION

THE semiconductor industry, with the invention of the transistor in 1947 at ATT Bell Labs, and the debut of the integrated circuit (IC) at the beginning of the 1960s, was born as a promising and soon to be a formidable industry. From this modest beginning in which ICs were used in only a limited number of specialized applications, has grown a technology that is pervasive in today's world. The introduction of the personal computer (PC) by IBM in 1980 made semiconductor microchips a household term. This large-scale integration has continued over the decades due to innovations, process advancements in manufacturing, and rapid implementation into new applications.

The semiconductor industry consists of many groups of companies and institutions, all of which contribute to its vitality. At the center are the chip-manufacturers; but they are supported by a large number of outside organizations including manufacturers of chip-processing and metrology-tools, suppliers of materials and chemicals, analytical-laboratories, industry-associations that provide manufacturing standards and organize co-operative research efforts, and colleges and universities that provide technically trained

workers.

Semiconductors are dominated by silicon electronics, and about eighty percent of that is complementary metal oxide semiconductor (CMOS) technology.

Microelectronics fabrication today probably employs the most highly trained engineering workforce of any manufacturing industry. As the density of integrated circuits rises (and therefore device feature size decreases) and as industry shifts to large wafer sizes, the complexity of microelectronic fabrication processes creates a demand for an ever more highly educated and trained workforce.

According to the Semiconductor Industry Association (SIA), the US semiconductor industry employs ~ 225,000 trained workers worldwide at present times. A $226B world wide semiconductor market forecasted to grow to $309B by 2008, an 11% Compound Annual Growth Rate (CAGR) [1]. With US to keep the dominant market share and maintain the innovation edge, it is imperative to invest in education of highly skilled workforce. The educational programs in Microelectronic Engineering at Rochester Institute of Technology have been designed to meet this critical need. This paper describes a quarter century of commitment, results, and challenges in sustaining this program at RIT.

II. EDUCATIONAL PROGRAMS

A. The BS Program

The Bachelor of Science program in Microelectronic Engineering at RIT started in 1982 after a study revealed a critical national need for engineers suitably qualified to drive the PC revolution that had just begun [2]. The lithography lab at that time consisted of a house hold blender for photoresist spin coating on 2 inch wafers and Rubylith for masks. Presently, it has automated resist coaters, ASML deep UV and Canon I-line wafer steppers, and Perkin Elmer MEBES III electron beam mask writer. Today the program supports a complete 4 and 6 inch CMOS line equipped with diffusion, ion implantation, plasma and CVD processes, chemical mechanical planarization (CMP), electron microscopes and device design, modeling and test laboratories. With the successive advancement of the semiconductor industry, the program has evolved to meet the changes and challenges of the industry. The Microelectronic Engineering program at RIT

1-4244-0267-0/06/$25.00 ©2006 IEEE

remains the first ABET accredited Bachelor of Science level program granting a degree in Microelectronic Engineering. The laboratories at RIT include the largest university clean room for integrated circuit fabrication in the United States (world). The program has gone through several curriculum changes in response to the technological developments of the industry.

The current five-year BS program consists of 196 quarter credit hour coursework and 15 months of mandatory co-op experience integrated throughout the final three years. The program combines an essential electrical engineering curriculum with optics, lithography, semiconductor processing, and manufacturing. Students are given 'cleanroom' experience right from the first quarter they join the program. The program received full accreditation in 2004 under the new ABET Engineering Criterion EC 2000.

B. Affiliate Program

In 1986, a 56,000 sq. ft square feet building was constructed with the support from US government, RIT and donations from industrials affiliates. The industrial affiliates allowed us to create a $1m endowment for building/facility maintenance and $1m endowment from Perkin Elmer for MEBES III mask making facility operation. Additionally, the affiliates continue to provide vital input to the curriculum and hire co-op and full time graduates [3]. Figure 1 shows the increase in affiliate membership of the program over the last 24 years.

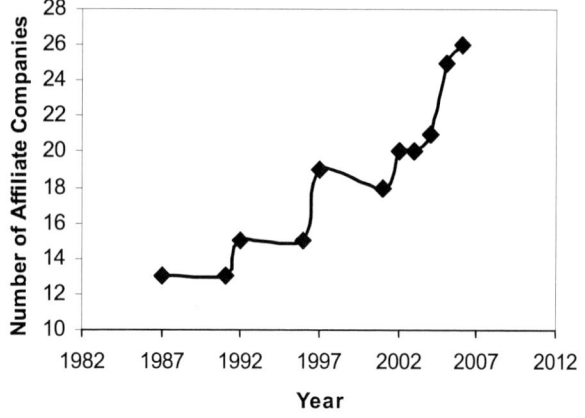

Fig. 1. Affiliate Membership of the Microelectronic Engineering Program.

Three professorships have been funded that provide support for the program – Motorola Professorship, Intel Professorship and most recently, Micron Professorship.

C. Graduate Programs

Success of the BS program and establishment of a well developed cleanroom facility led to the graduate programs – Master of Engineering in Microelectronics Manufacturing Engineering in 1987 and Master of Science in Microelectronic Engineering in 1995. A unique educational and research program that leads to a Ph.D. in Microsystems Engineering was instituted in 2002.

1) ME Program

The Master of Engineering degree is awarded upon successful completion of an approved graduate program consisting of a minimum of 45 credit hours [4]. The program consists of one transition course, seven core courses, two elective courses and a minimum of 5 credits of internship (professional work experience in the semiconductor industry). Under certain circumstances, a student may be required to complete more than the minimum number of credits. The transition course is in an area other than that in which the BS degree was earned. The core courses are microelectronics (processing) I, II, and III, microelectronics (manufacturing) I, II, and microlithography materials and processes and microlithography systems. The two elective courses are graduate-level courses in microelectronics or a related field. Elective courses may be selected from a list that includes courses such as metrology and failure analysis, semiconductor process and device modeling, and nanoscale CMOS. The courses delivered on campus have strong laboratory components. The laboratories teach basic principles involved in each of the core courses and most of the elective courses.

2) Online ME Program

The ME program also became available in 1998 entirely via distance delivery [5]. As our targeted online audience comes from the semiconductor industry, almost all students have access to the cleanroom fabrication/manufacturing environment in their respective companies. They primarily need fundamental understanding of the principles involved in engineering and do not need the base level laboratory instruction. The laboratory component is substituted with a self study paper or a research report under the faculty guidance. Distance learning courses at RIT have continuously evolved as new techniques and tools became available. For example, our new learning management system, Desire2learn allows our faculty to explore variations in online exams as alternatives to proctored exams, streamlines how assignments are managed through drop boxes, provides timely feedback through grade books, and facilitates good online discussion in small groups and team projects. We have also seen continuous improvement in our approach to course media. In the early 1990's we recorded lectures in front of a live classroom and delivered them in subsequent quarters via VHS videotapes to distance learning students. Our distance learning students would typically receive 20 hours or more of VHS lectures along with the textbooks ordered from our bookstore. Since then, our course lecture material has moved onto CD's and now most of that is also available online through streaming media, offering students an even greater degree of flexibility. And now Breeze Presenter enables faculty to produce their own high-quality voice annotated lecture material from PowerPoint directly on their own personal computer and publish it to the Breeze server where students can access it immediately. Instructors are no longer tied to a videotaped classroom or studio where materials have to be created weeks and months in advance of a course offering. The improvements in our course management tools have increased faculty ability to interact with students, and now with Breeze Presenter our faculty can create up to date content appropriate for the current course offering. In the year 2001, RIT produced its first graduate of the online Master of Engineering in Microelectronics Manufacturing Engineering. The total number of graduates is 12 by the year 2006.

3) MS Program

The Master of Science program started in 1995. The objective of the Master of Science program is to provide an opportunity for students to perform a master's level research as they prepare for entry into the semiconductor industry or a Ph.D. program. The program requires strong preparation in the area of microelectronics takes two years to complete and requires a thesis. Table I lists the placement of our MS graduates.

TABLE I
PLACEMENT OF MS STUDENTS

Year	Graduates	Placement
1995	1	Motorola
1997	1	Motorola
1998	2	Intel
1999	1	Motorola, CIDTEC
2002	3	IBM, Kodak, PhD(RIT)
2003	3	IBM, Kodak, PhD
2004	2	IBM, PhD(RIT)
2005	5	Cypress, National Semiconductors. Freescale, RF Micro Devices, IBM
2006	4	IBM, Micron, PhD (UC Berkeley), PhD (Delft)

4) BS-MS Program

A modern solid-state device is an ensemble of a variety of materials that include semiconductors, dielectrics, conductors, and polymers manufactured using advanced processes and miniaturized with extreme precision. The technology roadmap projects a critical need for the development of new materials. A combined Bachelor of Science in Microelectronic Engineering / Master of Science in Materials Science and Engineering program was approved and implemented in the year 2003-4. It is interdisciplinary between two colleges (College of Engineering and College of Science). This five-year program consists of completion of 225 credits that include a minimum of 36 graduate credits. It substitutes a co-op quarter by graduate thesis work. Students with interest in materials science aspect of microelectronics find this program very attractive. Table II lists thesis topics and the placement of our first and recent graduates of this program.

TABLE II
THESIS TOPICS AND PLACEMENT OF BS-MS STUDENTS

Year	Thesis Topic	Placement
2005	Al alloy films for microreflective applications	Texas Instruments
	Development of NiSi process	IBM
2006	Self aligned metal gate structures	AMD
	Low temperature dopant activation	IBM
	Hafnium oxide gate dielectric	IBM

5) Minor in Semiconductor Processing

We have developed a five course minor in Semiconductor processing for students of other science and engineering disciplines who desire exposure and experience to the exciting world of nanotechnology. We believe that this minor may do more to increase the number of women students with engineering experience at RIT utilizing the large number already enrolled in the College of Science programs, as opposed to separate recruitment strategies geared solely toward engineering. This program is designed to provide basic knowledge to students from other engineering and science disciplines interested in a career in the semiconductor industry that include design, manufacture, equipment, chemicals, and software sectors. The minor consists of five courses: three core and two electives as given in Table III.

TABLE III
SEMICONDUCTOR PROCESSING MINOR CURRICULUM

Level	Courses
Freshmen Level	Intro to Microlithography
Sophomore Level	IC Technology
Senior Level	Thin Film Processes
Two Electives	Process Integration
	CMOS Processing Lab
	Microlithography Materials & Processes
	Microlithographic Systems
	Process and Device Modeling
	Nanoscale CMOS
	Microelectronics Manufacturing
	Microelectromechanical Systems

The prerequisites for each of these courses are basic university level math, physics and one course in chemistry. The courses are multidisciplinary in content so there is an enormous knowledge value for students of every science/engineering program.

These five courses will equip students from other disciplines to work in the semiconductor industry or go to graduate programs in emerging fields of MEMS, nanotechnology. For instance- electrical engineering students with fabrication and processing knowledge will be better circuit designers, understand the tools and relationships between electrical data and process conditions. Computer engineers will be better chip designers. Mechanical engineers, largely employed by the equipment industry and packaging industry will be at an advantage by knowing the processes involved. Similarly, chemistry students will find better job opportunities with chemical industries that support semiconductor fabs. The industrial engineers will be exposed to the fab layout, wafer flow, lot tracking and other manufacturing issues. Physics/materials science majors will be the top choice for operating and interpreting electron microscopy, surface analysis, Raman and other spectroscopic techniques.

Fig. 2 and Table IV summarize curriculum development and technological achievements in microelectronic engineering at RIT over the last 25 years.

1-4244-0267-0/06/$25.00 ©2006 IEEE

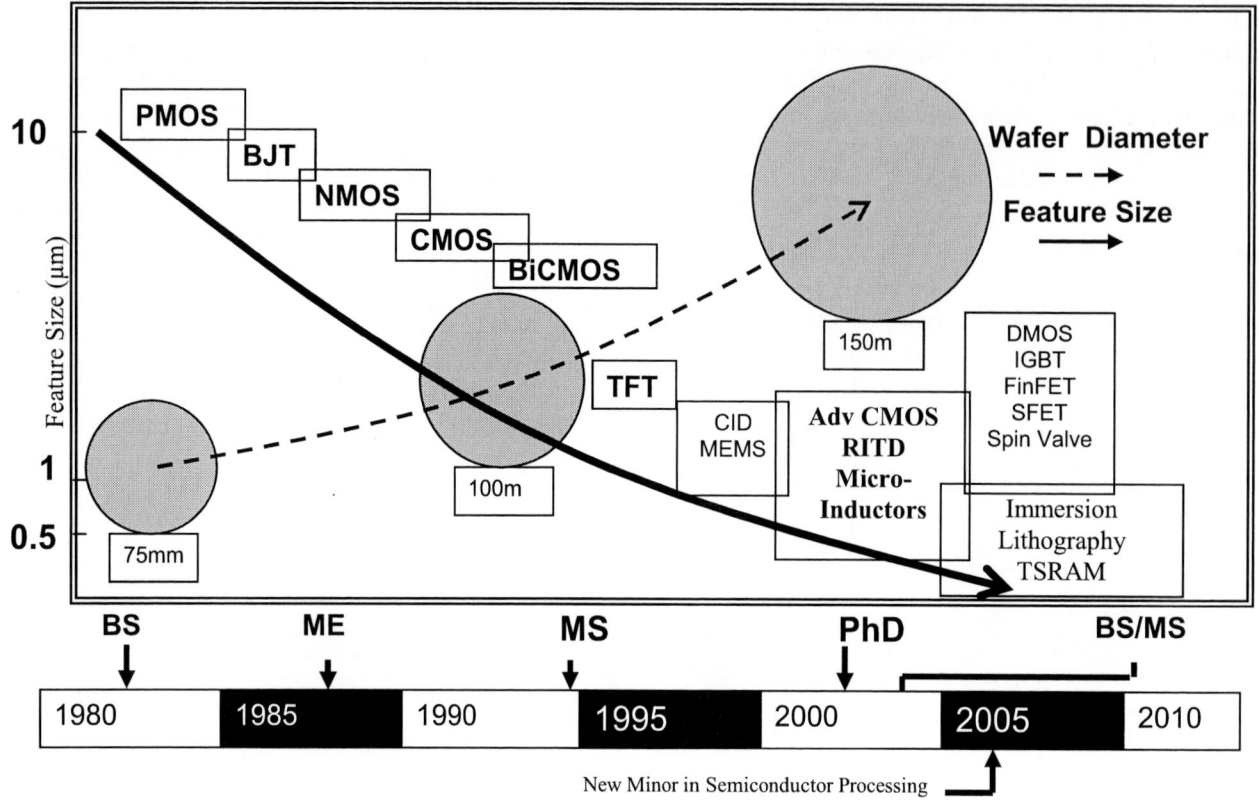

Fig. 2. Curriculum and technology evolution over the last 25 years in Microelectronic Engineering program at RIT

6) PhD Program in Microsystems Engineering

This multi-disciplinary program builds on the strengths in microelectronic fabrications, photonic, imaging and micro-power research programs at the institute. Students are involved in cutting edge research and have access to modern facility, the largest of its kind in any academic institution. The program has graduated six students in the last four years. Five of these students are hired by semiconductor industry – Kawasaki Microelectronics (1), Intel (2), and IBM (1).

III. PLACEMENT

Table V lists top employers of co-op and full time graduates of Microelectronic Engineering (MicroE) in descending order of numbers employed. The figures for full time graduates are subject to change as graduates may make transitions between their employments that are not tracked by RIT. Nevertheless, the figures reflect the nature of employment of our graduates.

TABLE IV
CURRICULUM DEVELOPMENTS LED BY THE MICROELECTRONIC
ENGINEERING DEPARTMENT AT RIT

Year of Introduction	Programs	Requirements (quarter credits)
1082	BS (Microelectronic Eng.)	196 credits + 15 months of Co-op
1987	ME (Microelectronics Manufact. Eng. (also offered Online)	45 credits include 5 credits for Internship
1995	MS (Microelectronic Eng.)	45 credits include 9 credits of thesis
2003	BS–MS (Microelectronic and Materials Science and Eng.)	225 credits with 9 credits of thesis
2005	Minor Semiconductor Processing	20 credits of courses
2002	PhD Microsystems Engineering (Institute wide)	92 credits of graduate course work, 24 credits in Dissertation research.

TABLE V
TOP EMPLOYERS OF MICROE CO-OP AND GRADUATES

Co-Op	Graduates
Infineon (Qimonda)	Motorola (Freescale)
Micron	Intel
Photronics	IBM
National Semiconductor	National Semiconductors
Fairchild Semiconductor	AMD
Eastman Kodak Company	Micron
Motorola / Freescale	Infineon (Qimonda)
Intel	Texas Instruments
IBM	Cypress Semiconductor
AMD	Analog Devices
Integrated Nanotechnologies	Eastman Kodak
Univ of Rochester	Xerox
	HP
	Graduate Schools

IV. SEMICONDUCTOR AND MICROSYSTEMS FABRICATION LABORATORY (SMFL)

A. Facilities

In the year 2001, the laboratories developed by the department of microelectronic engineering became central institutional facility, named as SMFL for research in addition to serving for the microelectronic engineering curriculum. The RIT SMFL offers a complete capability in microelectronics and MEMS fabrication. The SMFL has an extensive toolset for IC fabrication, including e-beam maskmaking, i-line and deep-UV microlithography, ion implant, plasma etching systems, diffusion furnaces, LPCVD and PECVD systems, sputtering and evaporation systems, and chemical-mechanical planarization. The SMFL has several classes of users, including undergraduate and graduate level laboratory sections, graduate thesis student researchers, faculty researchers, and corporate users. The SMFL facilities are opened to all RIT researchers who have undergone appropriate training, as well as users from other universities who wish to travel to our facility. Masks are created on-site with the MEBES-III electron beam exposure system from standard GDS format files. A complete tool set of the SMFL is listed on web site- http://smfl.microe.rit.edu.

Laboratory support is provided by eight full-time staff members, including four equipment technicians, a facilities manager, one process engineer, an operations manager, and a technical director. A number of student hourly workers provide janitorial support and perform various equipment / process support functions.

B. Processes

RIT is supporting four different CMOS process technologies (Table VI). The older p-well CMOS has been phased out. The SMFL-CMOS process is used for standard 5 Volt Digital and Analog integrated circuits. This is the technology of choice for teaching circuit design and fabricating CMOS circuits at RIT. The Sub-CMOS and Advanced-CMOS processes are intended to introduce our students with process technology that is close to industry state-of-the-art. These processes are used to build test structures and develop new technologies at RIT.

TABLE VI
IC PROCESSES AT RIT

Process	Design Rule (λ, μm)	$L_{min}(\mu m)$
Metal gate PMOS	10	20
RIT p-well CMOS	4	8
RIT SMFL-CMOS	1	2
RIT Subμ-CMOS	0.5	1
RIT Advanced-CMOS	0.25	0.5

In addition to standard CMOS processes, triple implanted bipolar process and application specific MOS processes can be made available. The faculty has developed unique processes such as low temperature CMOS process for silicon on glass and on-chip inductors integrated with MOS circuitry by developing electroplating of ferrite and copper with metal gate MOS process. MEMS processes involve surface and bulk micromachining.

V. RESEARCH

The primary objective of the microelectronic engineering has been to support education of workforce needed for the semiconductor industry. With the advances made in graduate curricula and laboratory facilities, research activities have significantly grown in various fields such as lithography, devices, advanced processes, MEMS and Microsystems. In addition, research activities attract quality faculty and inspire undergraduates towards emerging areas. It is particularly critical at present times as the technology has entered nanoscales. In this section, key research programs are briefly described for which RIT has made outstanding contributions.

A. Nanolithography

The microelectronic engineering program has facilitated the growth of expertise in the area of micro- and nano-lithography. Today, RIT is well known in nanolithography through the leadership provided by Dr. Bruce Smith and his students with support from the Semiconductor Research Corporation (SRC), DARPA/AFRL, International SEMATECH, ASML, Finle KLA/Tencor, Exitech, Photronics, Intel, IBM and others. The research performed by the group has enhanced lithography capabilities in the SMFL as well as provided vital know how to the industry. The group is actively engaged in leading edge technologies, including liquid and solid immersion lithography and a new approach to sub-32nm imaging known as evanescent wave lithography

As the Microelectronic Engineering program started, an emphasis was placed on microlithography which, at the time, was supporting one micron technologies and larger. As device technology continued to shrink, the curriculum and capabilities in the program kept pace. With the initial emphasis on UV optical electron beam lithography, this expanded to include excimer laser lithography, X-ray technology, extreme UV (EUV), projection electron beam imaging, and vacuum UV technology. As optical techniques continued to be pushed toward smaller dimensions, resolution enhancement approaches including phase shift masking, off-axis illumination, optical proximity correction, polarization, and immersion lithography have gone from experimental ideas to full commercial application. Some of the fundamental research and development in these fields has been carried out at RIT. The most recent is the evanescent wave lithography technology that has been utilized to achieve a world record in optical lithography. Using a 193nm ArF excimer laser, 26nm imaging in photoresist has been achieved, leading to the capability required for IC devices over the next 10-13 years.

1-4244-0267-0/06/$25.00 ©2006 IEEE

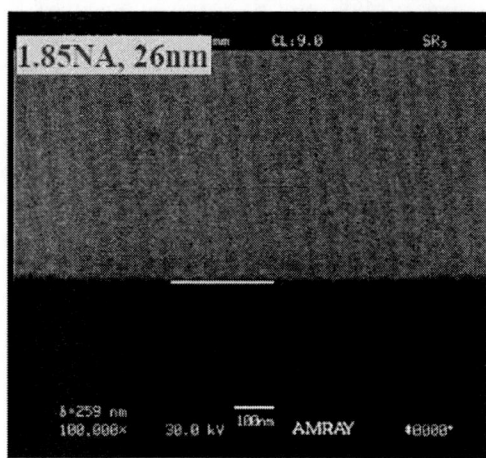

Fig. 3. 26nm half-pitch lines imaged at 1.85NA with a 193nm source using interferometric solid immersion lithography. The technique employs evanescent wave coupling to exceed the conventional material limits due to the refractive index of photoresist. [6]

B. Device Research

A new type of memory devices based on Si/SiGe resonant interband tunnel diode (RITD) has been fabricated. The first Si/SiGe tunnel diodes fabricated through openings in the field oxide and on top of p^+ implanted regions were realized in early spring 2002 [7]. This result was a strong indication that integration with CMOS could be possible. Recently, a fully-integrated tunneling-based SRAM (TSRAM) has been demonstrated [8]. Fig. 4 shows the SEM micrograph of the memory array with an inset of a single cell TSRAM.

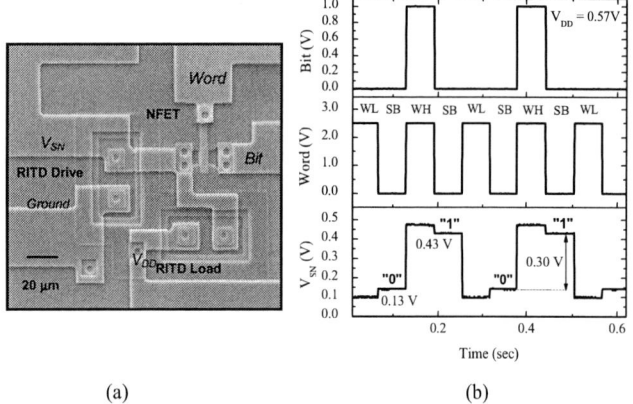

(a)   (b)

Fig. 4. (a) A micrograph of a single TSRAM; (b) Time diagram of T-SRAM cell during standby (SB), write high (WH) and low (WL) operations at power supply voltage of 0.57V [8].

Other devices include Charge Injection Devices, Charge Couple Devices, FinFets, strain silicon MOS, Schottky CMOS, and magnetic tunnel junctions.

C. On Chip Inductors

A fabrication process has been developed to physically realize inductors and transformers with microscale dimensions using copper embedded in a thick PECVD SiO_2 film. Circuits have been made that integrate the inductor with capacitors and PMOS transistors. PMOS transistors

have been implemented to provide a variable resistance leading to two varieties of LC tank circuits (parallel and series). RF measurement techniques have been developed for wafer level testing of micro-inductors on silicon. Data have been captured using an Agilent 8363B network analyzer with a frequency range from 10 MHz to 40 GHz, in conjunction with the Cascade Microtech GSG (ground-signal-ground) probes and the 9100 probe station. A calibration procedure has been developed for full two port measurements and a methodology has been optimized for measuring the impedance [Z] matrix and the scattering [S] matrix. The input impedance is extracted from the [Z] matrix and Q has been calculated. There is agreement between experimental results, numerical results from HFSS, and analytical results from the desegmentation and segmentation techniques.

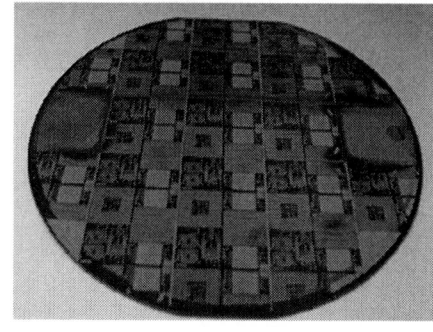

Fig. 5. Photograph of a 4" wafer with ferrite inductors integrated with MOS circuitry [9].

D. Thin Film Transistors (TFTs)

In addition to conventional CMOS, various types of devices/processes have been developed through senior projects or graduate research. Work on thin film transistors began in 1991 – with successful fabrication of TFTs on polysilicon using SSIC (seed selection with ion implantation) process followed by TFTs on bonded and etch back SOI wafers.

In a recent project a low temperature process has been developed to fabricate high-performance TFTs on a special glass, Corning's new material under development. Results are projected to have a major impact on the flat panel display industry.

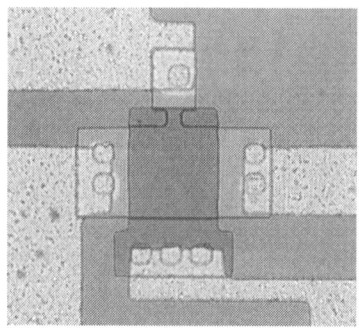

Fig. 6. A micrograph showing a fabricated thin-film transistor (TFT), made on Corning's new substrate material [10].

E. MEMS and Microsystems

With the development of IC and MEMs processes, RIT is well equipped to integrate on-chip electronics with MEM and sensor devices. An intense laboratory based graduate level course on Microelectromechanical Systems is offered twice in a year that allows students to design, build and test a wide range of devices as depicted in Table VII and Fig. 7. This course serves the PhD students to learn about design and fabrication issues in Microsystems.

TABLE VII
MEMS DEVICES FABRICATED AT RIT

Type	Devices
Pressure Sensors	Pressure sensors
	Microphone
	Speakers
	Chemical sensors
Flow sensors	Gas flow sensors with resistor anemometer and two resistors
Accelerometer	Diaphragm with mass in center
	Diaphragm actuator with coil
	Cantilever accelerometers
Gyroscope	Piezoresistor sensors or coil and magnet sensors
Optical Pyrometer	Thermocouples, Thermopile
	Heater plus temperature sensors

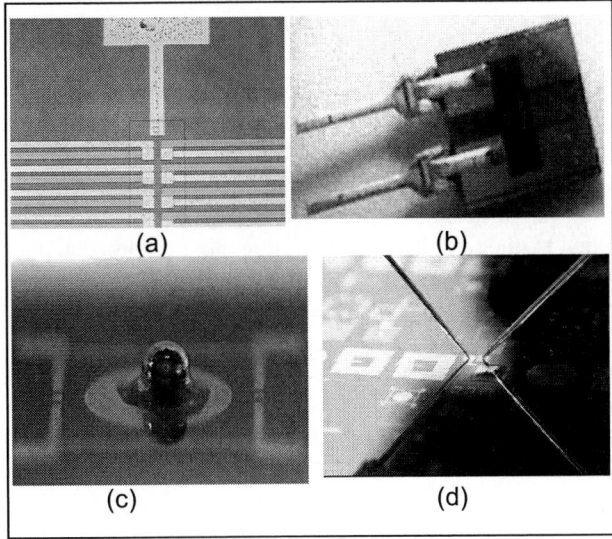

Fig. 7. Examples of MEMs devices fabricated (a) Thermopile; (b) packaged chemical sensor; (c) accelerometer; and (d) a microphone under test.

VI. ANNUAL MICROELECTRONIC ENGINEERING CONFERENCE

The department organizes a two day Annual Microelectronic Engineering Conference each year. The conference consists of Affiliate Meeting, curriculum Advisory Board Meeting and technical presentations by industry, faculty and graduating students. This annual event provides an excellent opportunity to assess performance and plan for the coming years. Senior students present their capstone projects and graduate students present their research activities. The entire conference is video taped and conference proceedings called *Journal of Microelectronic Research* are published that are made available online. Twenty four proceedings have been published to date.

VII. ALUMNI

Over 600 graduates from our undergraduate and graduate programs are working in the semiconductor industry. It is very likely that any semiconductor product we use today (or will use in the future) has involved an RIT Microelectronic Engineering student during its inception, development, or manufacturing. There are several examples of graduates providing leadership roles for the semiconductor industry. The RIT Kate Gleason College of Engineering honored two of the Microelectronic engineering graduates as outstanding alumni; Steve Carlson in 1999, Senior Vice President of Technology at Photronics in Dallas, in Texas and Louis Anastos in 2001, Program Manager for the photolithography and metrology areas for IBM's new facility in East Fishkill, New York.

Many alumni of the program have gone on to post graduate programs in fields such as PhD, MD, Education, Law and Business.

VIII. INDUSTRY SHORT COURSES AND OUTREACH

The department has developed several short courses for industry – IC Processing, Microlithography and Chemical Mechanical Planarization. These are typically one week long laboratory oriented courses with hands on instruction in the cleanroom. This program has over 1000 attendees from a wide range of semiconductor companies including overseas companies.

The department has provided mentorship to several other universities in promoting microelectronics laboratory development. Recently, Alfred State College has developed an IC processing laboratory for their Electrical Engineering Technology programs in collaboration with RIT.

The department of microelectronic engineering organizes numerous K-12 activities where students actually perform basic lithography experiments to learn about microfabrication and involve high school interns.

As an example of such an activity, a chemical sensor was designed, fabricated, and tested as part of a Science Research Program between Naples High School, NY and RIT Microelectronic Engineering. The chemical sensor consisted of a polymer carbon film in contact with interdigitated gold electrodes. The electrical resistance is measured using an ohmmeter. Sensed chemical vapors cause the polymer to swell and results in an increase in electrical resistance. The sensor response as a function of time was measured while presenting various amounts and types of chemical vapors [11].

The faculty participates in the local Science Educators Conference and Science fairs each year.

1-4244-0267-0/06/$25.00 ©2006 IEEE

IX. TURKMAN SCHOLARSHIP

Professor Ibrahim Renan Turkman has been a Professor at RIT since 1984. Professor Turkman had a tragic accident in March 2001 and has been on long term disability since then. Professor Turkman has been responsible for innovating semiconductor device physics and technology curricula both at undergraduate and graduate level. He developed advanced semiconductor processes and test facilities at RIT and advised graduate students in research. He was often invited for consultation and for teaching courses by leading semiconductor companies including Motorola and National Semiconductor where he also spent time on sabbaticals. To recognize his service in inspiring his students with his semiconductor device knowledge, RIT has instituted a scholarship in his name, the "Prof. I. Renan. Turkman Scholarship" in the year 2002. The scholarship is awarded to one student each year demonstrating top performance in the understanding of the physics of semiconductor devices through academic performance and a seminar presentation. All the recipients of this award since its inception have gone on to higher studies in semiconductor field – one of them has received the PhD degree in 2006 from RIT's new Microsystems Engineering program and is joining Intel in October 2006.

X. FUTURE DIRECTION

The students entering into our BS program in 2006 will graduate in 2011, - projected to be sub 45nm CMOS node. The end of conventional CMOS is in sight. We have to prepare students for understanding manufacturing issues at sub 45 nm nodes and 450mm wafer diameters, learning new concepts such as quantum confinement, subthreshold logic, 3D integration, spintronics, self assembly and quantum computing.

We have recently received a support from the National Science Foundation to lead our program to the next level by introducing state-of-the-art educational material into the curriculum and new methodologies for learning through experiential co-op employment that incorporate service learning. We have initiated steps to enhance faculty expertise and laboratories by reformulating our programs to incorporate nanooelectronics, MEMS and nanotechnology content in a reduced number of required courses. In order to accomplish this, we have reformulated the BS curriculum to accommodate more elective courses and restructured some courses by eliminating legacy material [12].

XI. CHALLENGES

There are tremendous challenges in sustaining a program that is continually advancing not only to keep pace but to lead the industry it serves. Operation of IC fabrication facility that is used to educate students starting from the freshmen year requires utmost dedication. The cost of facilities operation and maintenance increases rapidly as the laboratory advances. This directly translates to a need for more funding, industry support and tuition revenue. Since most of the tools are acquired through donations, obsolete tools become problematic because of lack of accessories and spare parts. That was the reason for moving towards a six inch wafer line.

The biggest challenge is increasing student enrollment. Fig. 8 shows the enrollment trends over the last 24 years.

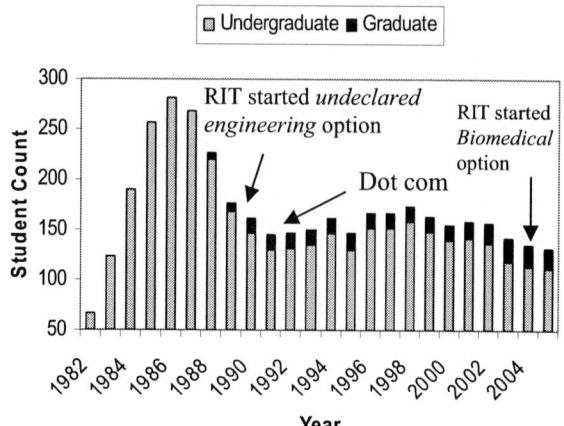

Fig. 8. Enrollment trends in microelectronic engineering program at RIT

The program started with very impressive enrollment numbers that peaked when the building was inaugurated. The PC revolution had just begun. In the year 1989, RIT started an 'undeclared engineering option' that severely affected enrollment in Microelectronic Engineering. Another factor that influences enrollment is the cyclic nature of high tech industry. The advent of the Internet and dot com in nineties resulted in attracting students towards information technology and web design related disciplines. Then by the turn of the century, dot com bust and cell phone growth brought some confidence back in 'hardware'. At present times, biomedical programs are appealing more to high school seniors even though there is no clear evidence of high manufacturing base creating biomedical engineering jobs.

The term *nanotechnology* is resonating with young college bound students. This phrase is over played and has been used indiscriminately that often misleads students. Is nanoelectronics nanotechnology? Or can nanotechnology prevail without semiconductor electronics? In a list of major nanotechnology and MEMs companies, given in a recent nano business journal, traditional semiconductor companies are among the top [13]. These are the questions that engineering/science educators have to convey to community at large.

XII. CONCLUSION

Pioneered by Dr. Lynn Fuller, IEEE Fellow and Motorola Professor, and by other faculty who have dedicated their professional careers, the program has served the semiconductor industry extremely well. The program is well positioned to educate engineers for the 21st century, truly believing in its motto "*Mindpower for Tomorrow's Technology*"

1-4244-0267-0/06/$25.00 ©2006 IEEE

ACKNOWLEDGMENT

The authors express sincere appreciation for the support provided by the affiliate companies. These include: Air Products, Advanced Micro Device, Analog Devices, ASM Lithography, Canon USA, Inc., Eastman Kodak, ETEC, IBM, Intel, Micron, Motorola, National Semiconductor, NEC, Nikon, Photronics, Rohm & Haas, Silvaco, Synopsys, Texas Instruments, and Xerox.

Without their support this program could not have sustained for quarter of a century and staying strong to take on the future challenges.

REFERENCES

[1] www.semico.com/mediacov/nov05/ SemicoMediaCoverage_11-16-05_EETimes.pdf -

[2] Lynn Fuller, "Microelectronic Engineering: A New Program at Rochester Institute of Technology,"Proceedings of the 5th IEEE/ISHM University/Industry/Government Microelectronics Symposium, Texas A&M University, College Station, Texas, May 1983.

[3] L.F. Fuller, R.E.Pearson, S.K.Kurinec, I.R.Turkman, M.A.Jackson, B.W.Smith, R.L.Lane, "Microelectronic Engineering at RIT - The First 10 Years", Proceedings of the 10th Biennial University Industry Government Microelectronics Symposium, May 1993, Durham NC.

[4] L.F.Fuller, R.L.Lane, R.E.Pearson, B.W.Smith, I.R.Turkman, K.H.Hesler, S.K.Kurinec, M.A.Jackson, "A New Program at RIT: Master of Engineering in Microelectronics Manufacturing Engineering", Proceedings of the 8th IEEE/ISHM University/Industry/Government Microelectronics Symposium, Westborough, MA, June 12-14, 1989.

[5] Santosh Kurinec, Dale Ewbank, Daniel Fullerton, Karl Hirschman, Michael Jackson, Robert Pearson, Sean Rommel, Bruce Smith and Lynn Fuller , Joeann Humbert, Leah Perlman , Ian Webber, "Online Master of Engineering Program in Microelectronics Manufacturing Engineering: A Valuable Resource for Engineers in Semiconductor Industry", 9th International Conference on Engineering Education, San Juan, Puerto Rico, July 2006, TIA1-5.

[6] B. W. Smith, Y. Fan, M. Slocum, L. Zavyalova, "25nm Immersion Lithography at a 193nm Wavelength," Proc. SPIE 5754, 2005.

[7] S. Sudirgo, R.P. Nandgaonkar, B. Curanovic, R. Saxer, J. Hebding, K.D. Hirschman, S.S. Islam, S.L. Rommel, S.K. Kurinec, P.E. Thompson, N. Jin, and P.R. Berger, "Monolithically Integrated Si/SiGe Resonant Interband Tunnel Diode/CMOS Demonstrating Low Voltage MOBILE Operation," *Solid-State Electronics*, vol. 48, pp. 1907-1910, Oct.-Nov., 2004.

[8] S. Sudirgo, *et al.*, "NMOS/SiGe Resonant Interband Tunneling Diode Static Random Access Memory," *Submitted to 2006 Device Research Conference*.

[9] Cody Washburn, Daniel Brown, Jay Cabacungan, Jayanti Venkataraman and Santosh K. Kurinec, "Application of Magnetic Ferrite Electrodeposition and Copper Chemical Mechanical Planarization for On-Chip Analog Circuitry", Proc Mat. Res. Soc. Symp. : Materials. Integration and Technology for Monolithic Instruments, pp 157-162, 2005

[10] Eric M. Woodard, Robert G. Manley, Germain Fenger and Karl D. Hirschman, David Dawson-Elli and J. Greg Couillard , "Low Temperature Dopant Activation for Integrated Electronics Applications', UGIM-2006.

[11] Elizabeth Greff and Lynn Fuller, "Fabrication and testing of a resistive chemical sensor", 14th Undergraduate Research Symposium, RIT, August 2005.

[12] Santosh Kurinec, Dale Ewbank, Lynn Fuller, Karl Hirschman, Michael Jackson, Robert Pearson, Sean Rommel Bruce Smith and Surendra Gupta Maureen Arquette and Maria Wiegand, "Microelectronic Engineering Education for Emerging Frontiers, , 9th International Conference on Engineering Education, San Juan, Puerto Rico, July 2006, TIA1-5.

[13] http://www.nanotechwire.com/news.asp?nid=1209&ntid=119&pg=9

1-4244-0267-0/06/$25.00 ©2006 IEEE

1-4244-0267-0/06/$25.00 ©2006 IEEE

RF Physical Device Simulation for Wireless Applications
O.L. Hartin, Evan Yu

Invited Paper
Freescale Semiconductor
Tempe AZ

Abstract

In the RF TCAD community routine large signal RF TCAD simulation has been a goal for some time. Several methods of doing large signal simulation from TCAD are discussed and compared. An example is shown from one approach based on extraction of compact models from TCAD data. This method has several advantages not the least of which is allowing the full capability of the circuit simulator to be used.

Introduction

TCAD is used in the engineering community to satisfy several objectives. Probably the most typical objective is prediction. Simulation almost always provides understanding. In some cases simulations can be use to get a handle on something that can't be easily measured in the laboratory. Increasingly, simulation is used to improve manufacturing yields.[1]

Prediction provides a significant advantage in technology development, when done reliably. The technologist never wants to abandon an approach based on TCAD results only to find out later that the TCAD gave an inaccurate prediction. While trends and quantities that are predicted occasionally conflict with measured data, TCAD very often provides the best prediction possible based on available data.

Problems in the RF arena can be split into two main areas, small and large signal. TCAD based prediction at small signal may at times be difficult because of complexities in duplicating the device characteristics in the simulator. Under large signal conditions distortion occurs as a result of operating limits and non-idealities of the device. In instances in which we are able to accurately simulate small signal solutions it may be possible to simulate large signal solutions as well.

Theory

Let us consider a practical problem shown in Figure 1. This is applicable to both small and large signal amplifiers. Under small signal conditions the class of the amplifier is typically class A, conjugate matches are assumed, and the design is formulaic. Under large signal conditions the class of operation of the amplifier may be AB, or even B and matches chosen for best performance under those conditions are typically not conjugate matches. The design of such a large signal amplifier is done by optimization, often in the test lab, of input and output matching networks, input bias, and harmonic terminations. If these parameters are considered along with intrinsic device design parameters a large search space results. This may be exacerbated by multiple stages often using different technologies.

1-4244-0267-0/06/$25.00 ©2006 IEEE

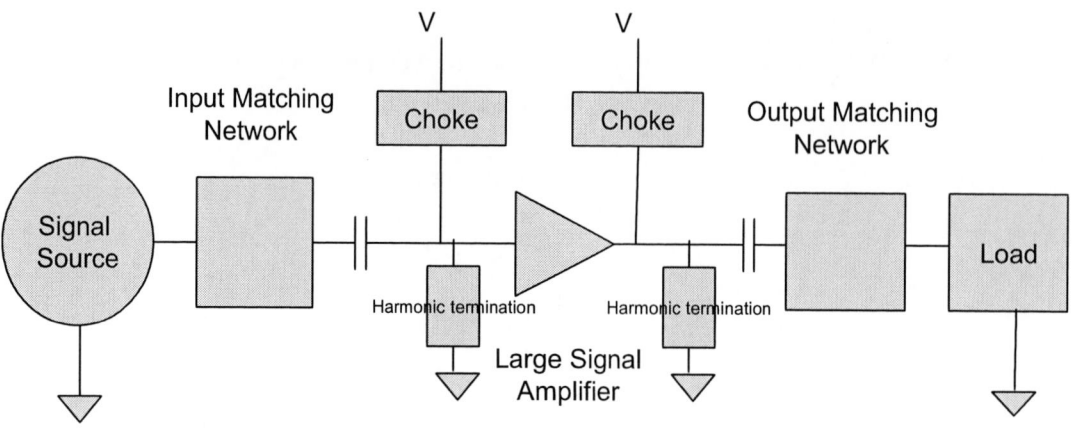

Figure 1: RF amplifier circuit.

In order to comprehend the performance with different device designs they must be compared not at the same match, termination, and bias as all the device designs under consideration, but at those values that give the best performance for that device design. By comparing the peak performance condition for each device design within the amplifier the choice of the best performing device design can be made.

This makes a clear case for a full large signal simulation of this circuit and the device under consideration. If the technologist wishes to evaluate new devices for which measured data doesn't exist then TCAD solutions must be used. Large signal simulations from TCAD have been done in the literature by four methods.

- Transient simulation in TCAD using mixed mode[2],
- Large signal simulations directly in the TCAD tool using integrated harmonic balance [3][4],
- Special tool to integrate composite simulation results with circuit design[5], and
- Extract large signal compact models from TCAD data and

use those models to understand large signal characteristics.[6]

It is important to discuss the possibility of doing the simulation using the most basic approach, mixed mode. In mixed mode a transient circuit simulation is done directly in the TCAD software. Typically these large signal simulation problems are not solved by circuit designers using transient simulation but rather by using harmonic balance. The challenge is to describe the RF circuit accurately in the workbench and then perform a TCAD simulation using a simple single (or dual) frequency carrier wave (CW) input swept over power. The transient simulation must settle to steady state condition, which may take several cycles. These simulations become more numerically challenging as the device goes into compression. The time sequence simulated has to be long enough to describe the lowest frequency needed. For two, or multi tone problems this may require quite a long transient. As a result, running a power sweep may be quite time consuming. In addition mixed mode work benches tend to be quite poorly designed for this missing key items like complex impedances. The main issues with this approach are speed and convergence for realistic circuits.[2][5] In addition the entire time penalty occurs on every circuit simulation.

The second method implements harmonic balance within the TCAD simulation tool. The harmonic balance method is more correctly called KCL-HB, or Kirchhoff's current law harmonic

$$F(V) = [I_s + Y \cdot V] + j \cdot \Omega \cdot Q + I_G = 0, \qquad (1)$$

which describes the relationship between linear and nonlinear circuit currents, where the term in the bracket is the linear part and the rest is the non-linear part. I_s is the current from the source, Y is the linear circuit admittance matrix, V is a vector of internal node voltages, Ω is a matrix with angular frequencies on the diagonal, Q is the charge vector in the frequency domain, and I_G is currents of the non-linear circuit in the frequency domain. This solution has converged when the currents between linear and non-linear circuit balance. This may be solved, for instance, by Newton's method.[7]

Implementation with a TCAD requires significant development. Although there has been significant research and university codes, a robust tool has not been offered on the market.[3][4] Harmonic balance is a method used for large signal RF problems typically within circuit simulation tools. Harmonic balance is a non-linear frequency domain steady state simulation.

Linear circuit components are solely modeled in frequency domain. Non-linear components are modeled in time domain and converted to the frequency

balance. It is used in Agilent's Advance Design System (ADS), Cadence's Spectre-RF and other circuit simulators commonly used in analog and RF. Harmonic balance is a nonlinear frequency domain technique. It is used to determine quasi-periodic steady state solutions for systems with widely varying frequency content.[3] This method solves the equation

domain at each step.[7] Algorithms limit the number of harmonics retained in this process typically to 7-11. The memory requirement to solve 11 harmonics is 4-8 Giga bytes, not including the memory required for simulation of the device. Iterative solution techniques requiring less memory may be used.[3] These memory requirements result in restrictions due to these resource limitations and analysis of multiple stages is probably not currently possible using this technique. Sweeps might be expected to take a several hours, much more for realistic devices.

A third method was explored by Loechelt in 2000. It is Computational Load Pull (CLP). In this method simulations (or measurements) of large signal transients are used to describe the intrinsic device and a tool was used to pull all this together and construct circuit evaluations. This method has several strengths. Once the dataset has been constructed describing the intrinsic device it can be used in multiple circuit simulations. There are disadvantages as well. An RF workbench is to be constructed within the CLP tool limiting the design to what is implemented there.[5]

The problems that we have sited so far with these methods are speed, functionality of the RF workbench, performance, and setup time, are described in Table 1. The fourth method is to extract a compact model from TCAD simulation data. The main advantage is that the simulation based model uses the same procedures, extraction, and can be dropped into the same designs as measurement based models. This allows use of the very powerful RF circuit simulation capability and previous RF designs that have already been developed. The

disadvantages are the time required to run the TCAD, time required to extract the model, and limitations of the compact model used. This is an important restriction because the TCAD simulation may contain physics that cannot be reflected in the compact model. There are two remedies, one is to create a user defined version of the model with better physics, and the other is to use a table based model.[8] In order to make this method practicable automated extractions must be created which allow for a number of device models to be extracted in a very quickly.

TABLE 1

	Time		Suitablity for load plane evaluation	Flexability	Convergence	Key limitation
	Setup Time	Load Pull (20 powers)				
Mixed Mode Simulation	no special setup	24-48 hours	no	single stage/device, limited circuit elements	convergence problems at high power	time and difficulty
TCAD Harmonic Balance	no special setup	10-12 hours	no	single stage/device, limited circuit elements	convergence both in device, and HB algorithms	time and limitations of simulator
Computation Load Pull	24-48 run	minutes	yes	multiple stage/devices, limited circuit elements	convergence problems likely	limitations in circuit step, incompatibility with designers
Model Extracted from TCAD	24-48 TCAD simulations, and model extraction (~hour)	minutes	yes	no issue	moderate in Initial TCAD, simulation and typical HB in circuit simulator	limitations of model available

Since from Figure 1 we know that the best performance occurs at an undetermined source and load match, simulations must be done across the source and load plane searching for the peak performance point. Assuming 60 source states and 60 load states that must be alternately searched, perhaps 300 or so power sweeps must be done to determine the peak performance point.[9]

Example of Large Signal Simulations from TCAD

TCAD simulations were run for a device using Synopsys tools.[10] A model was extracted using automated techniques from that simulated data and comparisons of Forward, Reverse Gummels, I/Vs, and CV characteristics shown here in Figure 2. TCAD data is shown in blue and model data is shown in red. The match shows that the model accurately reflects the original TCAD data.

1-4244-0267-0/06/$25.00 ©2006 IEEE

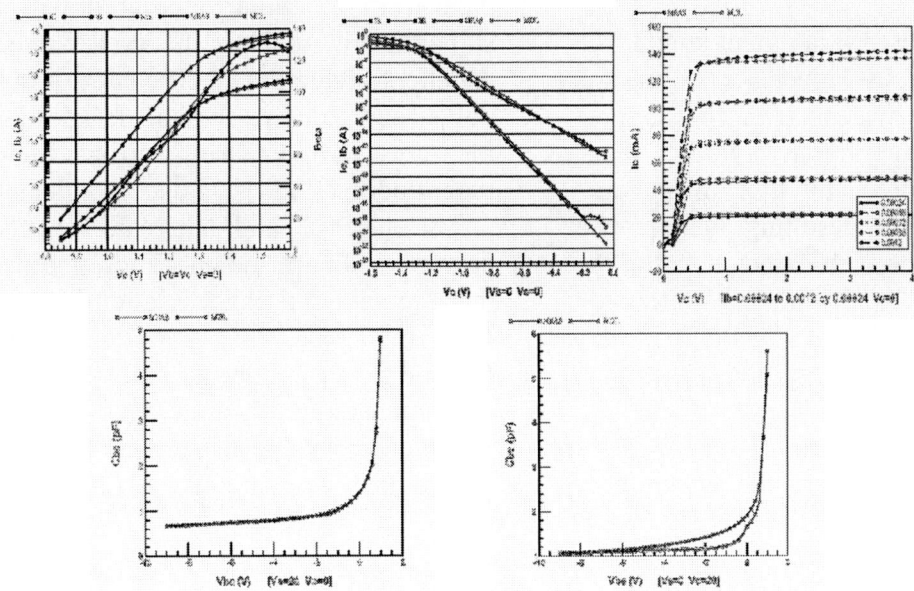

Figure 2: Comparison of Forward, Reverse Gummels, I/Vs, and CV characteristics where TCAD data is in blue and model data is in red.

A comparison of the S-parameter characteristics is shown in Figure 3. Once again the good match indicates that the model accurately reflects the model data.

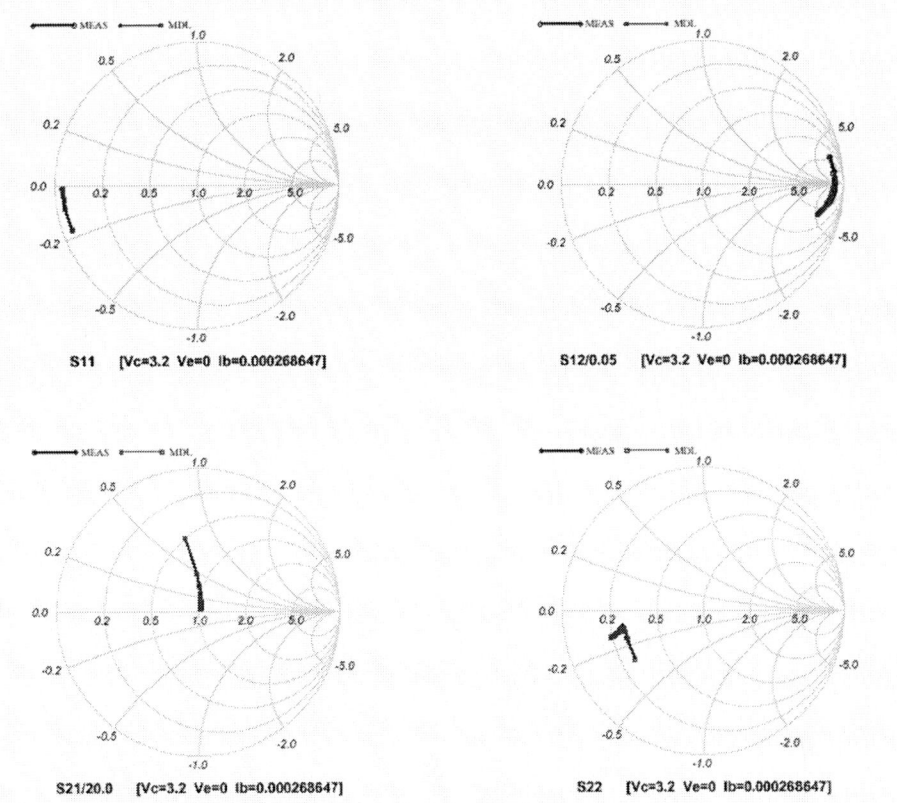

Figure 3: Comparison of S parameter characteristics where TCAD data is in blue and model data is in red.

This model is used in a circuit similar to the one shown in Figure 1. Using an algorithm that sweeps over the source and load plane iteratively the best performing source and load match are selected.[9] The resulting load plane comparison of efficiency between measured data from a similarly designed part is shown in Figure 4. Black is the reference measured data, and red is the data from simulation using this model.

A power sweep from the maximum efficiency point compared to measured data from a similar measured part is shown in Figure 5. In this power sweep excellent prediction of efficiency, output power and gain are shown. In addition, measures of linearity error vector magnitude (EVM), adjacent channel power (ACP), and alternate channel power (ALT) are shown. These indicate that the gain and phase relationships are well modeled. For device designs currently of interest in wireless communications accurate prediction of the linearity characteristics, EVM, ACP, and ALT are very important.[11]

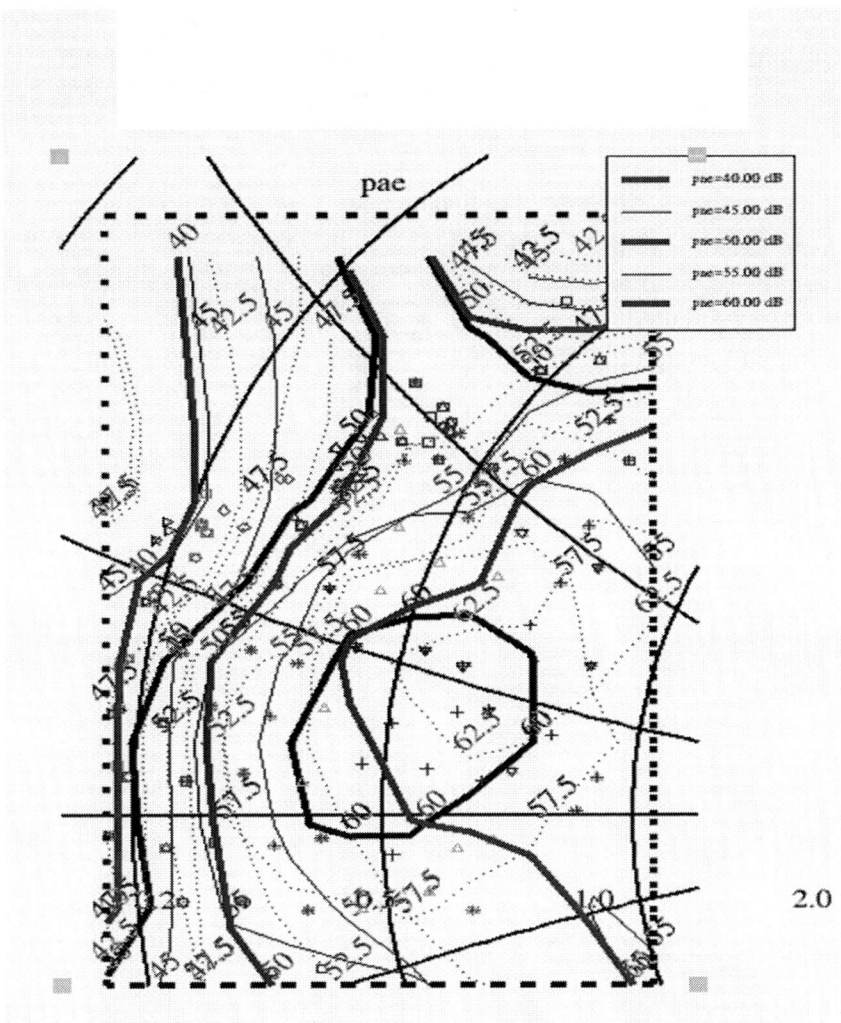

Figure 4: Contours of Efficiency

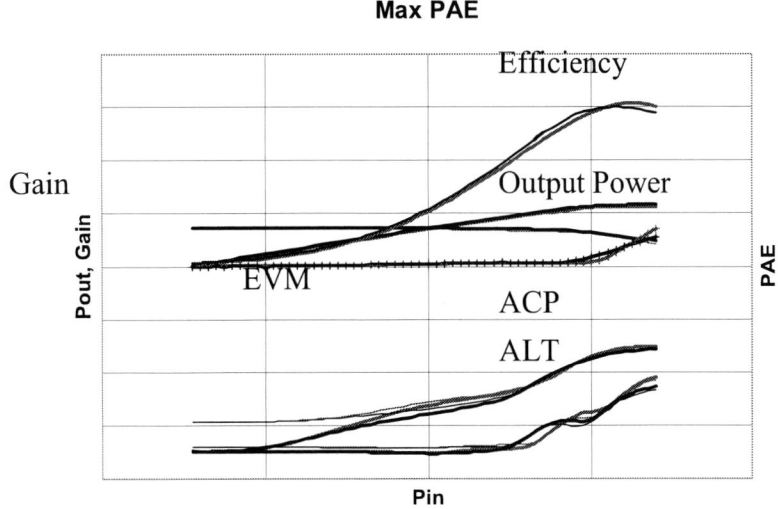

Figure 5: Max Efficiency power sweep

Conclusions

Four methods of determining large signal RF performance have been discussed. Each has been evaluated in terms of advantages and weaknesses. One method, possibly the most powerful at this time, has been used in a comparison to measured data. This demonstrates the ability to use TCAD based modeling to predict large signal performance that is accurate in the source and load plane.

References

[1] Tirumala, S., Y. Mahotin, et al. (2006). Bringing Manufacturing into Design via Process-Dependent SPICE Models. 2006 ISQED 7th international Symposium on Quality Electronic Design: 801-806

[2] Blakey, P. A., M. G. Khazinsky, et al. (1996). Direct numerical simulation of the large-signal RF behavior of power transistors. High Performance Electron Devices for Microwave and Optoelectronic Applications Workshop.

[3] Troyanovsky, B., Z. Yu, et al. (1995). Relaxation-based harmonic balance technique for semiconductor device simulation. Internation Conference on Computer-Aided Design, 1995. ICCAD-95, San Jose.

[4] Tornblad, O., C. Ito, et al. (2005). Linearity Analysis of RF LDMOS Devices Utilizing Harmonic Balance Device Simulation. International Conference on Simulation of Semiconductor Processes and Devices, 2005.

[5] Loechelt, G. H. and P. A. Blakey (2000). "A computational load-pull system for evaluating RF and microwave power amplifier technologies." Microwave Symposium Digest., 2000 IEEE MTT-S International **1**: 465-468.

[6] Pantoja, R. R., M. J. Howes, et al. (1989). "A Large-Signal Physical MESFET Model for Coputer-Aided Design and Its Applications." IEEE Transactions on Microwave Theory and Techniques **37**(12): 2039-2045.

[7] Stefan Jahn web site, qucs.sourceforge.net/tech/node31.html

[8] Root, D. E., M. Pirola, et al. (1993). "Measurement-based large-signal diode modeling system for circuit and device design." IEEE Transactions on Microwave Theory and Techniques 41(12): 2211-2217.

[9] Hartin, O., R. Uscola, et al. (2002). Extraction And Full Load Plane Validation of Large Signal RF VBIC Models. PA Workshop, San Diego.

[10] Hartin, O., M. Ray, et al. (2001). Compound Semiconductor Physical Device Simulation for Technology Development at Motorola. GaAs IC Symposium, Baltimore, MD.

[11] Evan Yu, HBT Validation Summary, Freescale Internal Presentation, 8/23/05

Three-dimensional TCAD Process and Device Simulations

I. Avci, P. Balasingam, K. El Sayed, J. Gharib, M.D. Johnson, , K. Kells,
G. Kiralyfalvi, V. Koltyzhenkov, A. Kucherov, E. Lyumkis, O. Penzin, B. Polsky,
V. Rao, S.D. Simeonov, N. Strecker, Z. Tan, L. Villablanca, and W. Fichtner
Synopsys, Inc., 700 E. Middlefield Road, Mountain View, CA 94043, U.S.A.

Abstract

Shrinking feature sizes, novel device designs as well as stress engineering increase the need for three-dimensional process and device simulations.

We present several application examples for full 3D process and device simulations using Sentaurus TCAD, including a 3D NMOSFET with shallow trench isolations (STI), a PMOSFET device with SiGe pockets for stress engineering (similar to the structure presented in Ref. [1]) and a Ω-FinFET (similar to structures presented in Refs. [2,3]). TCAD simulations of the full process flow as well as of the electrical device characteristics are performed. We also show examples of 3D oxidation simulations with Sentaurus Process.

Introduction

The full-flow process simulations presented here are based on a particularly robust approach to 3D process simulation (which is well suited for many applications). This approach leverages the observation that for the simulation of geometry-altering processing steps (patterning, etching, deposition, etc) it is sufficient to represent the structure as a set of boundaries. For dopant-related processing steps (implantation, annealing), however, it is necessary to represent the entire volume of the structure by a finite-element mesh. For many applications one can gain speed and robustness by separating these two kinds of processing steps as follows:

Step 1: Sentaurus Process (*Sprocess*) simulates the full 1D and 2D initial parts of the process flow up to the point where structure becomes 3D.

Step 2: Sentaurus Structure Editor (*SSE*), which operates on boundaries only, performs all 3D deposition and etching steps, and saves the intermediate structures at all steps where doping implantation and diffusion is to be simulated.

Step 3: Sentaurus Structure Editor creates a composite structure by superimposing the boundaries of all intermediate structures and saves the information on how to restore a particular intermediate structure by changing material properties of regions in the composite structure ("paint-by-numbers" table).

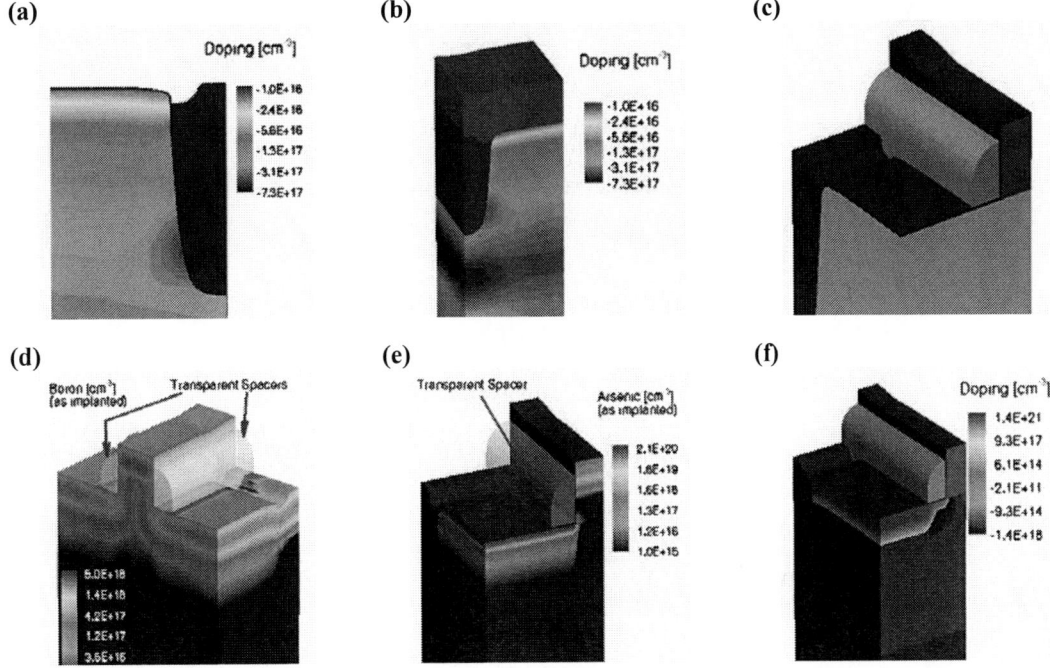

Figure 1: Snapshots of the 3D process simulation of a 3D STI NMOSFET. (a) Trench formation in 2D. (b) Extrusion to 3D. (c) Gate/spacer formation. (d) Halo implant/anneal. (e) Extension implant/anneal. (f) Final structure after source/drain implant/anneal.

Step 4: Sentaurus Process loads the composite structure and creates a volume mesh once. It then simulates dopant implantations and diffusions in the various intermediate structures by changing the material properties of the given regions according to the "paint-by-numbers" table created in Step 3.

This approach is similar to a scheme used in Ref. [4].

3D STI NMOSFET

Figure 1 shows snapshots taken during the process simulation on a 3D STI NMOSFET transistor. The initial processing steps, namely STI creation, V_t–adjust implantation and anneal, gate oxide formation and poly-silicon deposition are simulated with Sentaurus Process in 2D mode (Fig. 1a). After an extrusion to 3D (Fig. 1b) the remaining boundary-altering steps (gate formation and spacer formation) are performed with Sentaurus Structure Editor using process-based commands (Fig. 1c). Finally, the halo (Fig. 1d), source-drain extension (Fig. 1e) and source-drain (Fig. 1f) implantation and diffusion are simulated with Sentaurus Process in 3D mode. In this simulation the following genuine 3D effects were encountered:

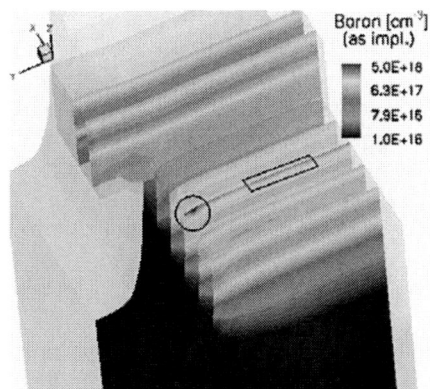 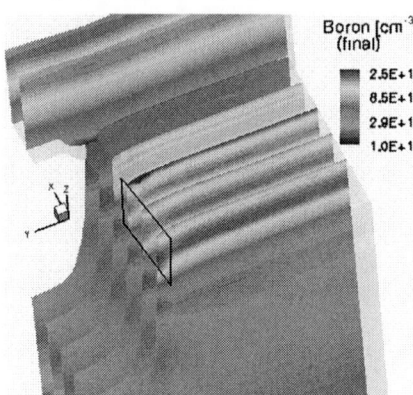

Figure 2: Three-dimensional boron halo implantation including shadowing effects. The rectangle highlights the increased boron concentration due to shadowing from the gate stack. The circle highlights the trench shadowing effect. The boron profile is shown along a series of slices for six different x values.

Figure 3: Three-dimensional boron diffusion. The rectangle highlights the boron pileup at the trench-silicon interface. The boron profile is shown along a series of slices for six different x values.

For shallow implantation angles, typical for halo implants, shadowing effects become important. Shadowing effects occur, for example, when the ions first pass through the edge of the poly-silicon gate and only then hit the silicon surface. The passage through the first material slows down the ions and the range in silicon is greatly reduced. The resulting shallow peak is highlighted by the rectangle in Figure 2 which shows the implanted boron concentration after a quad halo implantation with a tilt of 30°, a total dose of 2×10^{13} cm^{-2} and energy of 20 keV. The top of the final oxide in the trench is actually higher than the silicon in the source–drain area, which leads to shadowing. The circle in Figure 2 highlights the area where the gate and trench shadowing regions cross, resulting in an extra boron peak.

Boron diffuses through interstitials. Silicon-oxide interfaces are interstitial sinks. For certain RTA conditions, this leads to boron pileup at the silicon-oxide interface. The trench surface constitutes an additional interface, which makes boron diffusion a genuine 3D problem. (Similar statements apply to other species.) Figure 3 shows the final boron profile after annealing. The boron pileup at the trench–silicon interface is highlighted by a rectangle.

To understand the influence of decreasing device width on device performance, three different structures with widths W = 0.1, 0.7, and 1.0 µm were simulated. This study revealed that the stresses during the trench liner oxidation can profoundly influence the characteristics of the final device.

Figure 4 shows the pressure distribution after the trench liner oxidation. For the narrowest device (W = 0.1 µm), the oxygen diffuses all the way under the nitride mask and the entire mask is lifted during the oxidation process. Consequently, the nitride mask is relatively stress-free. For the wider

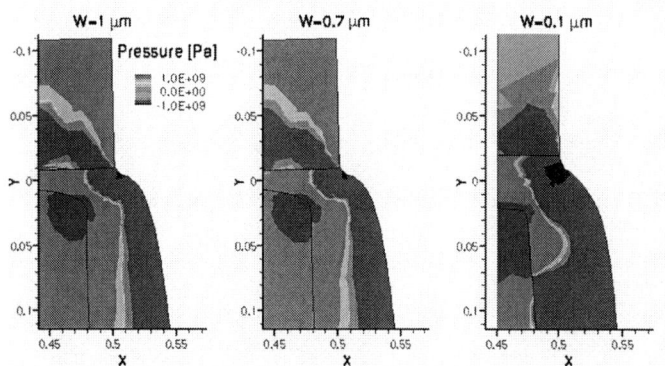

Figure 4: Pressure distribution after trench liner oxidation. The shown layers (top to bottom) are: silicon-nitride masks, trench liner oxide, and silicon substrate.

devices (W = 0.7 and 1.0 µm), the oxidation lifts only part of the mask and bends it. The resulting pressure in the oxide layer retards further oxidation and the liner is thinner at the trench edge for the wider devices. In the end, this results in a very different divot shape: The poly-silicon gate dips deeper into the trench in the wider channel devices compared to the narrowest one.

This results in anomalous behavior of the threshold voltage, defined as the gate voltage at which the (scaled) drain current reaches a certain level. It would be expected that V_t decreases with W because the parasitic edge transistor, which has a lower threshold voltage, becomes more dominant. Indeed, ΔV_t is negative for W = 0.7 µm. However, the narrowest device shows a positive ΔV_t value. The reason is the divot shape. The small poly-silicon dip into the trench strongly diminishes the parasitic edge transistor.

3D stress engineering

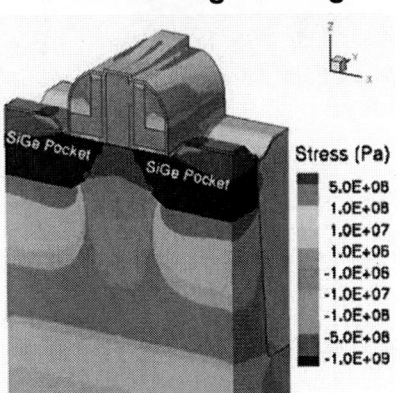

Figure 5: Final structure of the PMOS transistor with SiGe pockets.

Figure 6: Final stress profile.

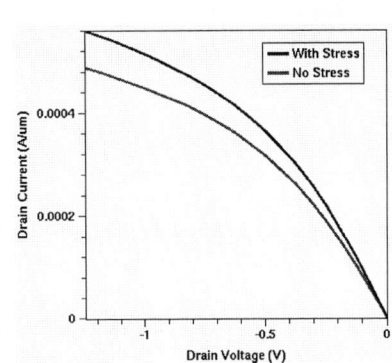

Figure 7: Drain current vs. drain voltage for the PMOSFET with (blue) and without SiGe pockets (red).

Semiconductor companies have successfully adopted strain engineering for 90 nm technology nodes to improve transistor performance [5]. Local strain techniques were found to be a favorable option due to their low cost and easy integration. These techniques utilize stresses induced by shallow trench isolation (STI), strained cap layers, silicidation, and SiGe or SiC pockets.

The effects of stress on transistor performance are usually estimated using a piezoresistance mobility model [6]–[8], and the performance change can be expressed as a function of various local stress components. As device dimensions continue to decrease, the control of local strains becomes increasingly challenging due to the complexity of stress patterns arising in a three-dimensional local geometry. Full 3D modeling is necessary to account for realistic transistor geometry, processing, and layout, and to understand their influence on the final stress distribution. We present here an example of a 3D process simulation for a PMOS transistor including the stress history. Transport I_d–V_{gs} simulation is performed to reveal the stress influence on the device performance. The geometry and process flow of a PMOS transistor similar to the one presented in Ref. [1] is used to set up this simulation. The final transistor structure (half of the device) is shown in Figure 5. The main features include a 3D STI and elevated SiGe source/drain. The dimensions used for the simulation are a gate length of 50 nm and a channel width of 200 nm, a SiGe pocket depth of 100 nm, and a SiGe source/drain elevation of 30 nm. The modeling consists of the following principal steps: structure generation, process simulation, meshing, and device simulation. Figure 6 shows the final stress profile (xx-component) of a PMOS device with SiGe pockets in the source-drain areas.

Device simulations are performed using Sentaurus Device for structures with and without SiGe pockets. Figure 7 presents final I_d–V_{gs} curves. The simulated enhancement in the drain current compared to the unstrained device is about 15%.

Ω-type FinFET

Figures 8-11 show simulation results for FinFET devices. Figure 8a shows the silicon channel fin of an Ω-type FinFET simulated with Sentaurus Process and Sentaurus Structure Editor.

Fig. 8b shows the final device geometry. The activation/annealing is simulated by Sentaurus Process using the pair diffusion model. The silicon layer is assumed to have initially a uniform boron concentration of 5×10^{18} cm^{-3}. However, due to the complex interaction of all dopants (arsenic, phosphorous, and boron) with point defects as well as among themselves via the electric field of the charge impurities, the diffusion process becomes genuinely 3D in nature and results in a the boron distribution which is quite non-uniform after the activation/annealing.

1-4244-0267-0/06/$25.00 ©2006 IEEE

(a) (b)

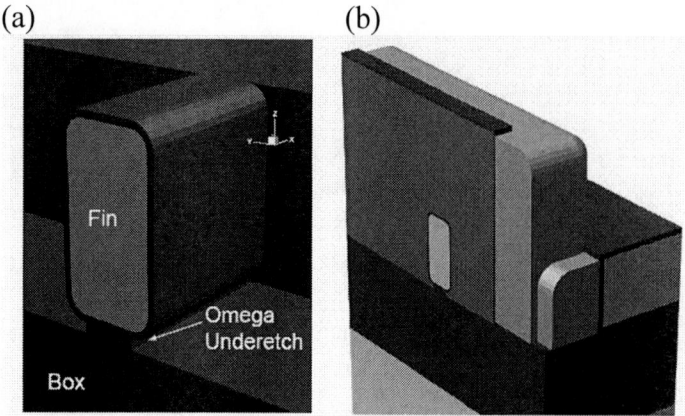

Figure 8: (a) Silicon channel fin of an Ω-FinFET. (b) Final FinFET geometry.

Figure 9 shows the final boron distribution along the XY plane at a Z-coordinate, which corresponds to 75% of the silicon layer height. It can be seen clearly that during the annealing process, boron is redistributed in a complex fashion: The arsenic extension implant introduces a large amount of interstitials, which during the annealing diffuse quickly to the surface, where they recombine. As boron diffuses only as boron–interstitial pairs, boron is transferred to the surface in this process. Due to small volume in the channel fin, a limited amount of boron is available and, therefore, the boron pileup at the surface in the extension area is accompanied by a corresponding depletion of boron in the center of the channel fin in the extension area. Near the p-n junction, the electric field from the arsenic ions slightly depletes the boron concentration on the p-side and, further at the channel–gate interface, boron segregation increases the surface concentration by approximately 7%.

Sentaurus Device (the successor of DESSIS) is well known for its robustness, and robustness is particularly important when simulating modern, deep submicron, MOS-type devices, where a very advanced set of transport models must be used. For example, for the Omega FinFET structure considered here, the body may not be fully depleted. Therefore, the continuity equations for both electrons and holes must be solved simultaneously. The very short gate length of 25 nm mandates the use of the hydrodynamic transport model. Further, the thin oxide thickness (2 nm) and relatively high body doping level ($\sim 5 \times 10^{18} cm^{-3}$) require the consideration of quantization effects. Here, the advanced quantization model (density gradient model) is used. [9]. Within the density gradient model, an additional partial differential equation is solved to determine the effective quantum potential. For the 3D FinFET, Sentaurus Device solve, in a self-consistent manner, five partial differential equations (Poisson equation, electron and hole continuity equations, electron energy balance equation, and the quantum potential equation).

Figure 10 shows the electron concentration at four cross-sections of the fin for the bias point $V_{gs} = V_{ds} = 1$ V. It can be seen that the carrier concentration is not uniform along the perimeter of the fin (channel). The highest concentrations are found near the four corners of the fin. For this reason, an accurate control of the shape of the fin corners is important for the accurate modeling of Omega FinFETs. The slice at the drain side (12.5 nm) shows a reduced electron concentration due to the pinch-off.

Figure 11 shows the low-drain and high-drain bias I_d–V_{gs} as simulated with Sentaurus Device. The dashed lines in Fig. 11 show the corresponding results of a classical transport simulation (that is, omitting the density gradient model). It is clear that quantum effects lead to a threshold voltage shift of 60 mV. For this device, this

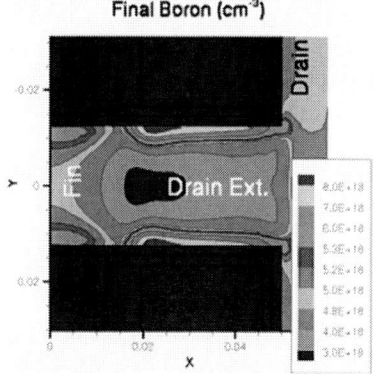

Figure 9: Final boron distribution in the Omega FinFET along the XY-plane at a Z-coordinate, which corresponds to 75% of the silicon fin height.

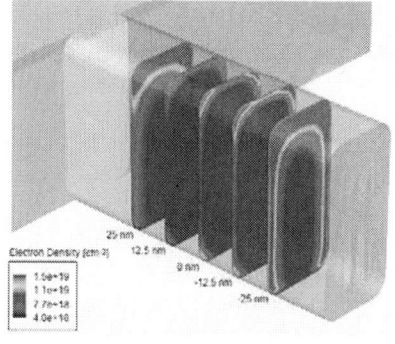

Figure 10: Electron concentration at four cross sections of the fin for the bias point $V_{gs} = V_{ds} = 1$ V.

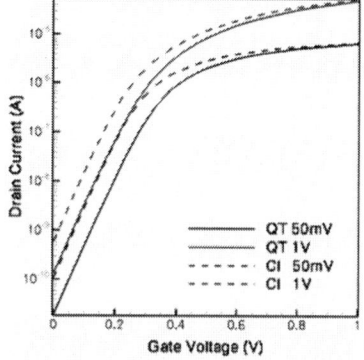

Figure 11: Drain current vs. gate voltage for a drain bias of 50 mV (blue) and 1 V (red). Solid lines are quantum transport (including the density gradient model) and dashed lines are classical transport.

corresponds to 20% of the threshold voltage, underscoring the importance of quantum transport simulations for modern CMOS devices. When using analytic doping profiles, the threshold voltage is 20% lower compared to runs based on simulated profiles, because the effects of boron redistribution as discussed above are neglected.

Oxidation in 3D geometries

Sentaurus Process is also well suited to study oxidation in 3D geometries. An example application is the investigation of the mask dimensions at which lifting of the nitride mask in LOCOS sets in. Fig. 12 shows the thermal oxide and the L-shaped nitride mask after a LOCOS step. Figure 13 shows the results of a poly-buffered LOCOS growth simulation, and a simulation of the poly-silicon gate re-oxidation step in a CMOS flow is shown in Fig. 14.

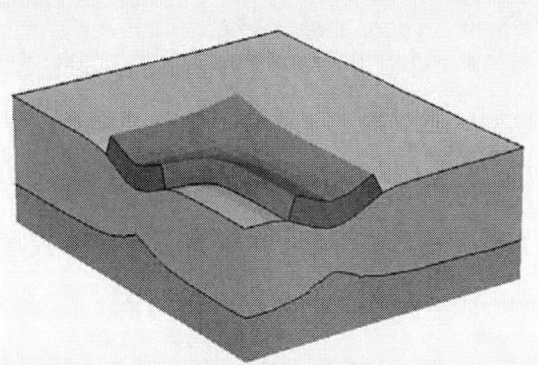

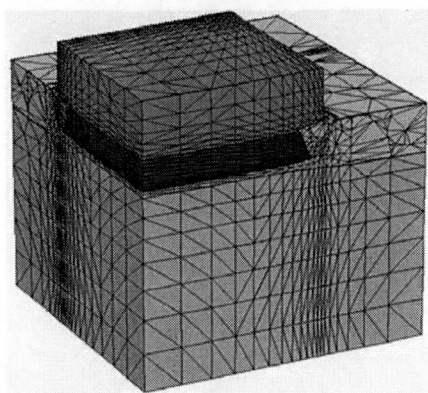

Figure 12: LOCOS oxidation simulated with Sentaurus Process. Oxidation conditions were: 65 min in a wet environment at 1050°C. Top golden layer: nitride. Middle magenta layer: poly-silicon. Middle yellow layer: Thermal oxide. Bottom pink layer: silicon.

Figure 13: Poly-buffered LOCOS oxidation. Oxidation conditions were: 50 min in a wet environment at 1050°C. Top golden layer: nitride. Middle magenta layer: poly-silicon. Middle yellow layer: Thermal oxide. Bottom pink layer: silicon. The simulation mesh is shown by the black lines.

The 3D oxidation simulation is performed, as soon as the anneal conditions and the current layer structure require such simulation. Oxidation simulations starts with the formation of a native oxide layer at all interfaces where this is required. During subsequent anneal time steps first the oxidant diffusion and reaction is simulated, followed by the stress equations. For the given velocity field, the maximum time step is computed that can be performed on the given mesh topology. The velocity is used to displace the mesh points and the time step when the first volume element becomes "flat" is computed. This grid limited time step is used to solve the dopant and point defect equations using time dependent coordinates and time dependent box weights. This approach avoids most of the interpolation error and preserves the dose for the dopant concentrations during the anneal step.

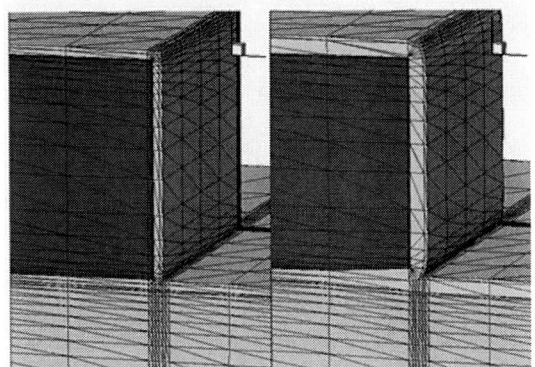

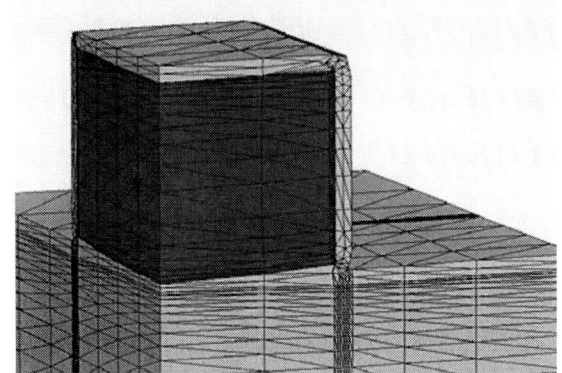

Figure 14 (a): Poly silicon gate re-oxidation. Middle yellow layer: thermal oxide. Bottom pink layer: silicon. The left panel shows the structure with the initial native layer and the right panel gives the structure after the oxidation.

Figure 14 (b): Same a Fig. 14(a) but seen from a different view angle. The simulation mesh is shown by the black lines. Oxidation conditions were: 20 min in an O_2 environment at 1000°C. Top golden layer: nitride

At the end of each such diffusion time step, a full face flipping is performed to optimize the quality of the mesh elements. After the face flipping remaining elements that are flat or nearly flat are selected and removed using a

1-4244-0267-0/06/$25.00 ©2006 IEEE

variety of local mesh clean-up operations. Both during the face flipping and the local mesh cleanup the change of geometry is monitored and the operations are performed only if they increase the quality of the worst element involved. During each time step new mesh elements are generated on the oxide side of the oxidation front to prevent edges, nearly perpendicular to the oxide surface from growing too large.

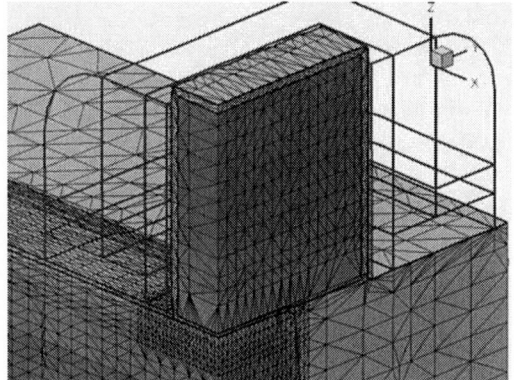

Figure 15: 3D oxidation in the framework of "paint-by-numbers": The green lines show region boundaries. The lines in the ambient area show the region boundaries of what will later be the spacers. During the poly-reoxidations these regions are assigned the material property "gas".

If the face flipping and local mesh cleanup do not succeed to remove all poorly shaped elements, the next subsequent grid limited time step may be quite small. A minimum grid limited time step can be specified by the user, where a global Delaunay triangulation algorithm is used to construct the new mesh from the surface faces and the collection of bulk points that are not contained in volume elements of poor shape. Since all this remeshing is done for a fixed geometry and fixed point locations, the interpolation error is tightly controlled.

The described local algorithm provides a boundary fitted mesh for every time step of the anneal simulation. Using appropriate weight calculations, the dopant and point defect equations and similar the oxidant diffusion/reaction and stress equations have been solved successfully even for meshes of rather poor element quality.

By default, during 3D oxidation simulations, regions of the same material are merged, if they touch along faces. If the user chooses to not allow region merging, the described 3D moving grid algorithm can be combined with the "paint-by-numbers" scheme to handle small changes of the geometry as they occur during a poly-gate reoxidation or a liner oxidation step. During such steps the topology of the structure does not change and just the boundaries of the oxide and mask regions are deformed. In subsequent process steps still some of the regions, temporarily labeled as "gas" can be used, e.g. as nitride spacer. Figure 15 gives an example of a poly-gate reoxidation simulation within the "paint-by-numbers" scheme.

Summary

It was shown that many 3D effects in modern semiconductor processing technologies, as well as in the electrical behavior of modern devices can be successfully simulated with Sentaurus TCAD.

References

[1] T. Ghani *et al.*, "A 90nm High Volume Manufacturing Logic Technology Featuring Novel 45nm Gate Length Strained Silicon CMOS Transistors," in *IEDM Technical Digest*, Washington, DC, USA, pp. 978–980, December 2003.

[2] F.-L. Yang *et al.*, "25 nm CMOS Omega FETs," in *IEDM Technical Digest*, San Francisco, CA, USA, pp. 255–258, December 2002.

[3] F.-L. Yang *et al.*, "35nm CMOS FinFETs," in *Symposium on VLSI Technology*, Honolulu, HI, USA, pp. 104–105, June 2002.

[4] S. Cea et all., Proceedings of IEDM 2004.

[5] D. James, "2004 – The Year of 90-nm: A Review of 90 nm Devices," in *IEEE/SEMI Advanced Semiconductor Manufacturing Conference and Workshop*, Munich Germany, pp. 72–76, April 2005.

[6] C. S. Smith, "Piezoresistance Effect in Germanium and Silicon," Physical Review, vol. 94, no. 1, pp. 42–49, 1954.

[7] S. E. Thompson et al., "A Logic Nanotechnology Featuring Strained-Silicon," IEEE Electron Device Letters, vol. 25, no. 4, pp. 191–193, 2004.

[8] F. Nouri et al., "A Systematic Study of Trade-offs in Engineering a Locally Strained pMOSFET," in IEDM Technical Digest, San Francisco, CA, USA, pp. 1055–1058, December 2004.

[9] M. G. Ancona and H. F. Tiersten, "Macroscopic physics of the silicon inversion layer," Phys. Rev. B, vol. 35, no. 15, pp. 7959–7965, May 1987.

1-4244-0267-0/06/$25.00 ©2006 IEEE

Investigation of the Performance Limits of III-V Double-Gate n-MOSFETs

Abhijit Pethe, Tejas Krishnamohan, Donghyun Kim, Saeroonter Oh, H. -S. Philip Wong, and
Krishna Saraswat
Department of Electrical Engineering
Stanford University
330 Serra Mall, Stanford CA-94305

Abstract—The performance limits of ultra-thin body double-gated (DG) III-V channel MOSFETs are presented in this paper. An analytical ballistic model including all the valleys (Γ-, X- and L-), was used to simulate the source to drain current. The band-to-band tunneling (BTBT) limited off currents, including both the direct and the indirect components, were simulated using TAURUS[TM]. Our results show that at significantly high gate fields, the current in the III-V materials is largely carried in the heavier L-valleys than the lighter Γ- valleys, due to the low density of states (DOS) in the Γ, similar to current conduction in Ge. Moreover, these high mobility materials like InAs, InSb and Ge suffer from excessive BTBT which seriously limits device performance. Large bandgap III-V materials like GaAs exhibit best performance due to an ideal combination of low conductivity effective electron mass and a large

I. INTRODUCTION

Due to their small Γ-valley electron mass, Ge and III-V materials like GaAs, InAs and InSb are being investigated as high mobility channel materials for high performance NMOS [1, 3, 4, 5]. Under ballistic conditions, the main advantage of a semiconductor with a small transport mass is its high injection velocity. However, these materials also have a very low density of states in the Γ-valley, which tends to greatly reduce the inversion charge and hence reduce drive current. Further, the very high mobility III-V materials like InAs and InSb, have a much smaller direct band gap which gives rise to high band to band tunneling (BTBT) leakage. Materials such as InAs and InSb have a high dielectric constant and hence are more prone to short-channel effects (SCE). In this paper we have thoroughly investigated and benchmarked Double-Gate (DG) n-MOSFETs with different channel materials (GaAs, InAs, InSb) under ballistic transport taking into account band structure, quantum effects, BTBT and SCE.

II. DEVICE STRUCTURE AND SIMULATION METHODOLOGY

The device structure simulated is shown in Fig. 1. The effective masses used in this work are listed in Table I. To calculate the electron wavefunction and sub-band energy levels we solve the 1D- Poisson-Schrödinger as described in [6]. Due to the small effective mass in the Γ-valley, the quantization and the lower DOS causes the inversion charge to populate the

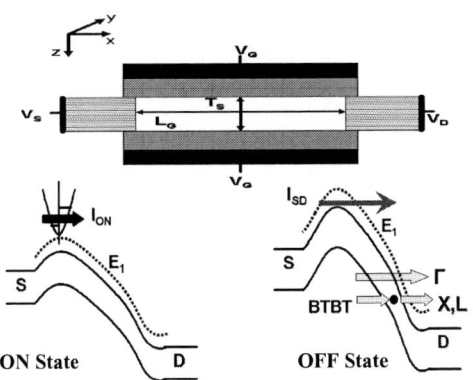

Fig. 1: The device structure used in the simulation. Undoped semiconductor body was assumed with the S/D doping of 1e20 /cm^{-3} which act as perfect absorbers. The off current is assumed to be limited by the BTBT. Both the direct and indirect modes are simulated.

higher (X- and L-) valleys, which cannot be neglected. We include all the valleys, Γ-, X- and the L-, for the III-V materials due their relatively small valley shifts. Parabolic E-k relationship is assumed in all valleys. The drive current for the device is calculated using a ballistic transport model [8]. Due to the higher dielectric constant, the SCE in these high mobility materials must be taken into account [7]. The relative performance of these devices is highly dependent on their sub-threshold characteristics [10]. We use TAURUS[TM] to estimate the sub-threshold characteristics and SCE. The BTBT leakage is calculated taking into account QM effects using the model and parameters from [3, 9]. In this paper, Ge (110) orientation is used, as it has been shown to have the highest drive currents [10]. The (100) orientation was simulated for the III-V materials and Si.

A. Single Ladder Semiconductor

Effective masses in the x, y and z direction are crucial for ballistic current operation as they determine the injection velocity and the charge induced in the channel. To observe the independent effects of changing the effective mass in the x, y and z direction, we consider quantum-ballistic transport through four different constant energy surfaces at varying semiconductor film thicknesses (T_S) (Fig. 2). The electron sub-band energies and injection velocities at $N_{inv}=10^{13}$ cm^{-2} are shown in Fig.. 2. In the thin film regime, the sub-band energy level is strongly dependent on the m_z in the ladder. The injection velocities are proportional to m_x^{-1}. However, even for

1-4244-0267-0/06/$25.00 ©2006 IEEE

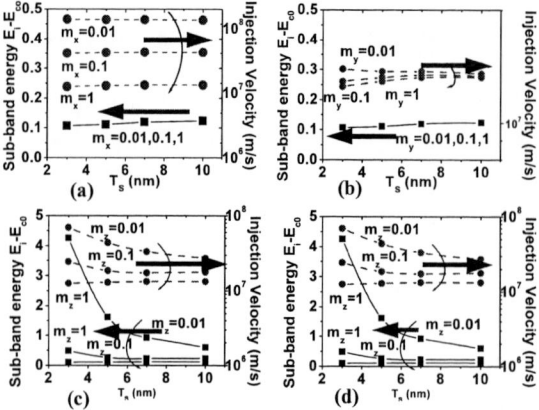

Fig. 2: Single ladder. Effect of individually varying the effective masses in the x-, y- and z- directions on the sub-band energy levels and associated injection velocities. Si-like semiconductor assumed with $N_{inv}=10^{13}$ cm^{-2}. Solid lines represent energy and the dashed lines represent velocities. m_x, m_y and m_z are varied in (a), (b) and (c) keeping the masses in other directions constant. The mass is varied isotropically in (d)

the same m_x, the v_{inj} is 2X when m_z is reduced by 100X. This is due to higher quantization and hence higher carrier velocity. Scaling the effective masses isotropically, sharply increases the injection velocity due to a combination of both the effects discussed above. The gate capacitance (C_{Geff}), which determines the channel charge, is also a strong function of the effective masses. For lower masses, the C_{Geff} reduces drastically - a signature of low density of states (Fig.. 3). Evaluating the ballistic current, we find that for the best performance, current should be carried in a valley with low m_x (strongly increased v_{inj}), high m_y (higher density of states) and low m_z (higher v_{inj}).

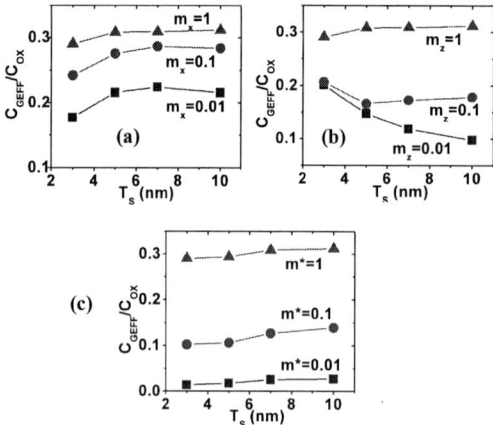

Fig. 3: Effective gate capacitance C_{G-eff}/C_{OX}. m_x and m_y show similar trends and their effects are depicted in (a). m_z is varied in (b) and the mass is varied isotropically in (c).

B. Electronic Quantization and Ballistic Currents

The high carrier velocities in III-V materials like GaAs, InAs and InSb are due to the very low effective mass in the Γ-valley. However, since this is an isotropic valley it also has a very low DOS. Quantization due to space and/or E-field strongly affect the relative occupation of carriers in these valleys. Fig. 4 shows the electron sub-band energies and the

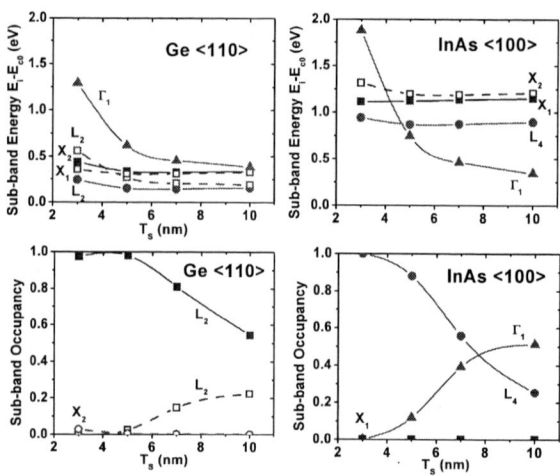

Fig. 4: : Electron sub-band structure and occupancies in the thin films in all valleys. The occupancies in the lowest energy levels in all valleys (Γ-, L- and X-) are plotted. Inversion charge density is $5X10^{12}$ cm^{-2}. Crossover between Γ and L valley is seen in all III-V as T_S decreases. InAs is chosen as a representative III-V material.

occupation of ladders in the different semiconductors at an inversion charge density of $5X10^{12}$cm^{-2}. As we scale T_s or induced more inversion charge, the Γ-valley energies in all the III-V materials rise up rapidly due to very low m_z resulting in most of the inversion charge in the thin films moving to the heavier L-valley. Similar effects are seen in the Ge<110> direction, where charge begins to leak into the heavier X-valley from the L-valley at strong quantization. Fig. 5 shows the injection velocities in the various valleys. Clearly, the advantage of the high injection velocity in the Γ-valley of the III-V material is lost if the charge moves into the L-valley. As seen from Fig. 6 and Fig. 7, at large inversion densities more than 50% of the current is carried in the L-valley. This fraction increases with the extent of quantization. The L-valley velocities in the III-V materials are similar to the L-valleys in Ge. Most of the III-Vs lose the advantage of their smaller Γ-valley electron mass and begin to conduct very similar to Ge under strong quantization (thin T_s and/or high N_{inv}).

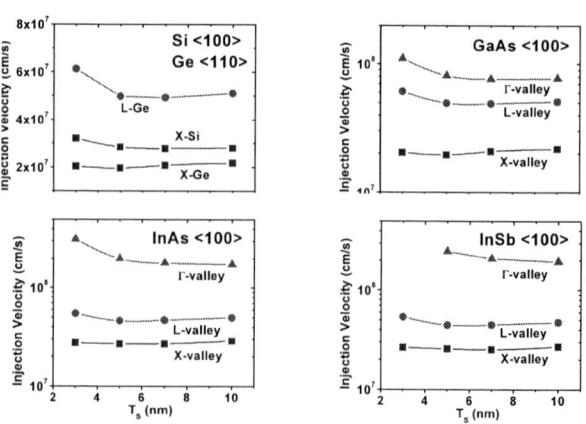

Fig. 5: Injection velocities in the semiconductor materials. The injection velocities in the Γ-valleys in the III-V materials are ~10X than injection velocities in Si $N_{inv}=1e13$ cm-2.

1-4244-0267-0/06/$25.00 ©2006 IEEE

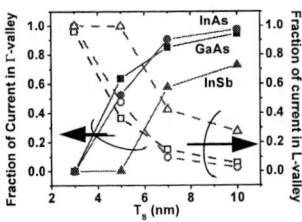
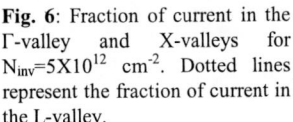

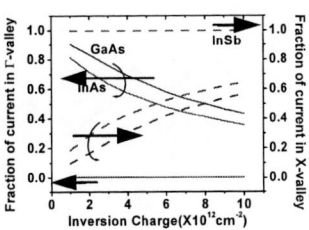

Fig. 6: Fraction of current in the Γ-valley and X-valleys for $N_{inv}=5 \times 10^{12}$ cm^{-2}. Dotted lines represent the fraction of current in the L-valley.

Fig. 7: Fraction of current in the Γ-valley and X-valleys for $T_S=5$nm. The inversion charge is varied from 1×10^{12} to 1×10^{13} cm^{-2}. Dotted lines represent the fraction of current in the L-valleys. The charge occupies L-valley with increase in degree of quantization

III. DRIVE CURRENT SCALING

The high mobility III-V materials typically also have higher dielectric constants than Si and suffer from worse SCE (Fig. 8) InSb ($\varepsilon_r=17$) exhibits the worst characteristics. The poor SCE

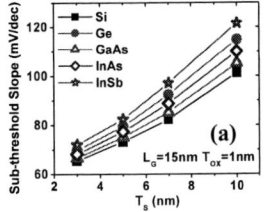
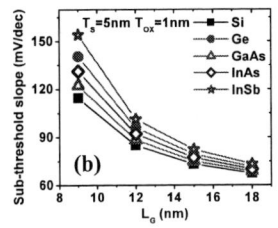

Fig. 8: SCE Nominal device set to $L_G=15$nm, $T_S=5$nm, $T_{OX}=0.1$nm. The Off current is set to $0.1\mu A/\mu m$ by varying the gate workfunction. Fig (a) shows the immunity to variations around T_S while Fig (b) shows the immunity of the device around L_G.

in InSb is one of the main factors for the reduction of its current drivability.

A. Effect of scaling semiconductor film thickness on drive currents

The ballistic drive currents for $T_{OX}=1$nm and 0.1nm are shown in Fig. 9. $T_{OX}=0.1$nm, enables us to evaluate the ultimate intrinsic device performance, where the quantum/semiconductor capacitance dominates and SCE are negligible. At the technological limit when T_{OX} is scaled to 0.1nm, the III-V materials perform better as T_S is scaled from 10nm to 3nm. At thicker T_S, Ge devices have the highest drives because of a higher inversion density due to the larger C_{DOS}- the density of states capacitance. In contrast, most of the charge in the III-V materials at thick T_S is present in the Γ- and L- valleys which have a low density of states (DOS). InSb performs better than InAs and GaAs at thick T_S, since it has the highest DOS in the L-valley among the III-V materials considered. As T_S is reduced, charge fills the heavier X-valley in Ge, causing a reduction in its drive currents. On the other hand, reducing T_S moves the charge from the Γ- valley into the L-valleys in the III-V materials. These L-valleys have lower conductivity effective masses than Ge and hence carry higher currents than Ge. The drive current for Si MOSFETs remains largely unaffected due to reducing T_S. This is because the C_{DOS}

reduces as T_S is reduced as charge moves from the four-fold valley into the low DOS two-fold valley. Fig. 9b illustrates drive current scaling at $L_G=15$nm with an effective $T_{OX}=1$nm. At these dimensions, the effect of C_{DOS} is reduced and the SCE severely impact device performance. At thick T_S, GaAs MOSFETs have the highest drive currents. This is because a significant portion of this drive current in GaAs is carried in the very low mass Γ-valley as compared to the InSb or InAs MOSFETs. As T_S is scaled, in the III-V MOSFETs, most of current is carried in the L-valleys. GaAs has the lightest conductivity effective mass in the L-valley among the III-V materials and hence continues to provide higher drive currents than InAs or InSb. Even though InSb has a larger DOS in the L-valley than InAs, these devices have comparable drive currents are due to a larger contribution to the current in InAs from the Γ-valley. Ge devices have lower current drive than the III-V MOSFETs due to the lower DOS and lower injection velocity. As L_G is scaled, the ON characteristics of the device are completely dominated by the SCE. Si performs the best at $L_G=7$nm and thicker T_S due to better SCE.

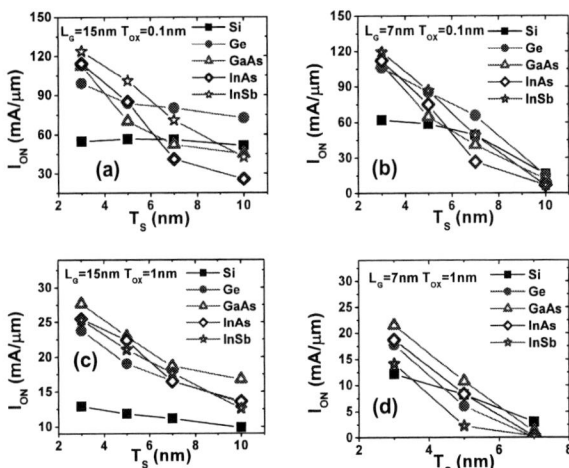

Fig. 9: Ballistic drive currents. (a) and (c) are for $L_G=15$nm, matched to $I_{OFF}=100$nA/μm. (b) and (d) are for devices with $L_G=7$nm and $I_{OFF}=1\mu A/\mu m$. V_{DD} is fixed at 1V. The off current does not include BTBT

B. Effect of supply voltage V_{DD} on drive current scaling

Fig. 10a Fig. 10b illustrate the effect of scaling the supply voltage on the drive current in MOSFETs with $L_G=15$nm and $T_S=5$nm. As V_{DD} is reduced, less charge is induced in the channel reducing the effect of the E-field quantization. At $T_{OX}=0.1$nm, InSb MOSFETs exhibit highest drive currents because of higher DOS in the L-valley which carries most of current in the III-V materials. GaAs MOSFETs however, have lower drive currents at higher V_{DD} due to larger contribution of the heavier X-valley. Ge MOSFETS also perform comparable to the III-V MOSFETs because of their higher DOS which compensates for their lower injection velocities. For thicker T_{OX} and hence lower inversion charge at a given V_{DD}, GaAs exhibit highest drive currents attributed to firstly, a higher component of Γ-valley current, and secondly due to the lower conductivity effective mass in the L-valley. The drive currents

1-4244-0267-0/06/$25.00 ©2006 IEEE

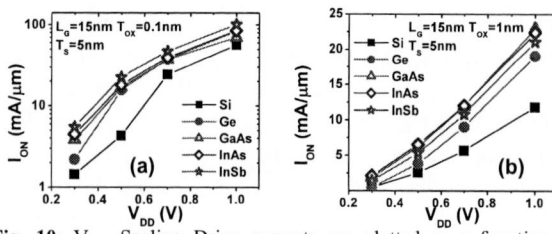

Fig. 10: V_{DD} Scaling Drive currents are plotted as a function of V_{DD} L_G=15nm, T_S=5nm. The characteristics of the T_{OX}-0.1nm are dependent largely on the semiconductor properties and less on the device structure.

in the InAs MOSFETs have a higher Γ-valley component and hence are higher than those in InSb MOSFETs. Ge MOSFETs have lower drive currents than the III-V MOSFETs due the higher conductivity effective mass in the L-valley.

IV. BAND-TO-BAND TUNNELING AND DEVICE DESIGN

BTBT leakage is very important especially in small bandgap (E_G) high mobility materials. We follow the approach in [9] to calculate the I_{BTBT} using TAURUS™ PMEI. All the valleys are included, taking into account band-gap widening and reduction in the DOS due to quantization. Fig. 11 depicts the BTBT limited off state leakage for a nominal device as the supply voltage is scaled. The tunneling currents in the III-V materials are substantially reduced because of marked band-gap widening due to quantization. Fig. 12 illustrates the effect of quantization on Ge and GaAs devices. As the III-V films are quantized, the BTBT current becomes increasingly dominated by the indirect component of the tunneling. Fig. 13 summarizes the relative advantages of III-V materials compared to Si and Ge MOSFETs. The intrinsic delay of the MOSFET $C_{Geff}V_{DD}/I_{ON}$ is used as a performance metric and compared to the BTBT limited OFF state leakage indicative of the static power dissipation in the device for a fixed device dimension. All the III-V MOSFETs except InSb provide lower intrinsic gate delays than Si and Ge. However, even after quantization the InAs and InSb have substantially high off state leakage cannot be successfully turned off even for modest OFF current specs of 0.1µA/µm. Furthermore, GaAs MOSFETs exhibit a lesser increase in the intrinsic gate delay than Si or Ge MOSFETs as V_{DD} is scaled, an ideal candidate

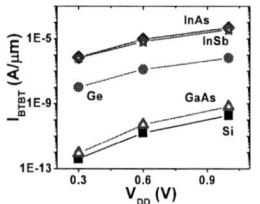

Fig. 11: The BTBT limited Off current in devices with L_G=15nm, T_S=5nm and T_{OX}=0.7nm

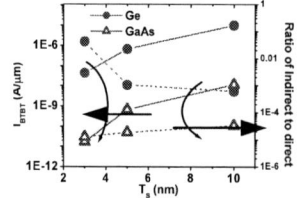

Fig. 12: BTBT limited off current Ge and GaAs as T_S is scaled. V_{DD} =1V. L_G=15nm, T_{OX}=0.7nm

for low V_{DD} applications. This is because, as V_{DD} is scaled, less charge is induced in the channel, but in the case of GaAs and InAs, most of the charge is present in the lighter Γ-valley.

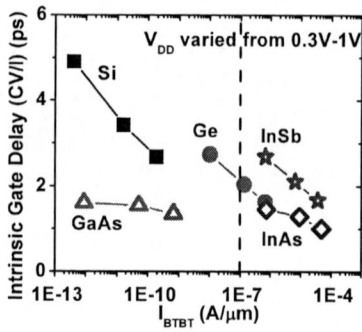

Fig. 13: Performance vs. off state tradeoff for all materials considered. L_G=15nm, T_S=5nm and T_{OX}=0.7nm.

V. CONCLUSIONS

In this paper we have thoroughly investigated and benchmarked Double-Gate (DG) n-MOSFETs with different channel materials (GaAs, InAs, InSb, Ge and Si) under ballistic transport taking into account band structure, quantum effects, BTBT and SCE. Our results show that under normal operation, a significant portion of the I_{ON} in the III-V materials occurs through the heavier L-valley, and hence they perform similar to Ge. However, the high mobility small bandgap materials like InAs, InSb and Ge, suffer from excessive BTBT current and poor SCE, which limits their scalability. GaAs has a significantly higher I_{ON} at a reduced I_{OFF} when compared to Si making it a very promising candidate for future scaled n-MOSFETs. Further material optimization can be made by considering ternary alloys such as GaInAs and by using heterostructures in the channel as in [11]

ACKNOWLEDGMENT

This research was supported by the MARCO MSD center

REFERENCES

[1] S.Datta et.al., *Int. Solid State Integrated Circuits Conference*, Beijing, China, Oct 2004
[2] T.Ando et.al., *Rev. Mod. Phys.*, Vol. 54, No. 2, pp. 437-672, April 1982.
[3] M.V. Fischetti, *IEEE T-ED.*, Vol. 38, No. 3, pp. 634-649, March 1991.
[4] R. Chau et. al., *IEEE Trans. on Nanotech.*, Vol. 4, No. 2, pp. 153-158, March 2005.
[5] M.V. Fischetti and S.Laux, *IEDM* pp. 481-484, Dec. 1989.
[6] L. Ge and J. G. Fossum, *IEEE T-ED.*, Vol. 49, No. 6, pp. 287-294, June 2002.
[7] Q.Chen et.al., *IEEE T-ED.*, Vol. 49, No. 6, pp 1086-1094, June 2002.
[8] K. Natori, *J. App. Phys.*, Vol. 76, No. 8, pp. 4879-4890, Oct. 1994.
[9] E. O. Kane, *J. Appl. Phys.*, Vol. 32, No. 1, pp. 83-91, Jan. 1961.
[10] A. Pethe et. al., *SISPAD*, pp. 359-363. Sep. 2004.
[11] T. Krishnamohan et.al., *VLSI Tech. Symp.*, June 2005.

Simulation of Quasi-stationary and Transient Effects in GaN Based Heterostructure Field Effect Transistors

N. Braga, R. Mickevicius, V. Rao, and W. Fichtner

Synopsys, Inc., Mountain View, California, 95113, USA. Email: nbraga@synopsys.com

1 Introduction

In the last decade, AlGaN/(In)GaN Heterostructure Field Effect Transistors (HFETs) have gained wide recognition as potential devices of choice for ultra high-power microwave systems and power electronics.

The understanding of deleterious transient effects in GaN HFETs, such as current collapse, as well as the ability to predict precise device DC and RF behavior require understanding of physical phenomena involved. Physics-based device simulations complementing experimental measurements are the key for gaining qualitative and quantitative insight into the above mentioned phenomena.

The spontaneous polarization component in commonly grown III-nitride material is nearly zero in transversal directions, with the residual polarization related to a typically small deviation from the growth direction from the c-axis. However, although the primary source of strain comes from lattice mismatch in the layer system, also resulting in nearly zero in-plane piezoelectric polarization [1], stressed overlayers and electrode patterning may introduce non-uniform stresses that lead to in-plane polarization and spatial variation of the vertical polarization due to strong piezo-polarization in III-nitrides.

Also, there is experimental evidence that the presence of SiN surface passivation layers reduce the dispersion between DC and high-frequency operation regimes [2-4]. The proposed explanations for collapse mitigation evolve around the reduction of surface density of states and blocking electrons from reaching the interface states. However, fringing electric fields in the higher dielectric constant passivation layer might also be accountable for collapse reduction.

This paper addresses DC and transient effects of stress dependent 3D polarization fields and of fringing fields for devices with and without a SiN passivation layer.

2 TCAD Simulations

The simulated structure was formed by a SiC substrate, capped by a 2 μm thick GaN buffer layer and a 25 nm $Al_{0.3}Ga_{0.7}N$. Mechanical constants of Titanium were assumed for the source and drain metallization whereas Nickel was used for gate metal. The gate length was fixed at 1.1 μm and source/gate and drain/gate separations were set to 1.0 and 1.5 μm, respectively.

Simulations were performed for overlayers with dielectric constants corresponding to stoichiometric SiN or to that of air. Moreover, for the case of the SiN, simulations were carried out for intrinsically stressed and unstressed films.

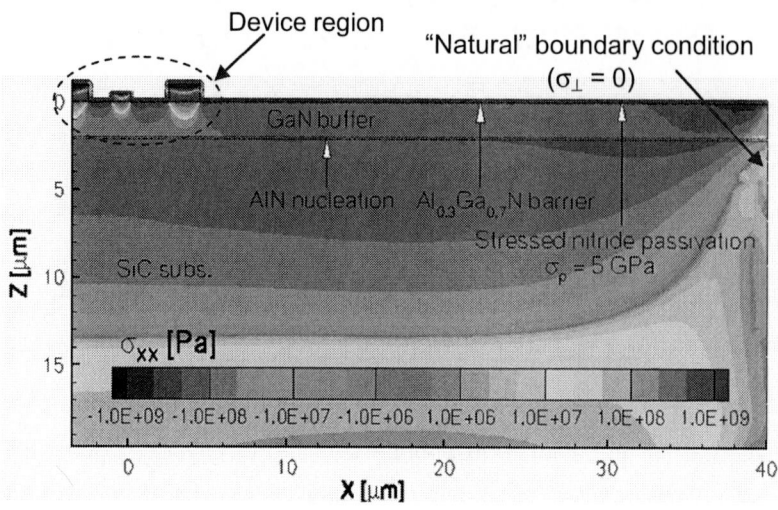

Figure 1 Simulated stress distribution (σ_{xx}) including extended domain

2.1 Simulation of non-uniform stress distribution

The process simulator Sentaurus Process (SProcess), was used to form the layer structure and compute non-uniform stress distributions from lattice mismatch and intrinsically stressed layers. Although the process simulator is capable of handling viscoelastic materials, this work concentrates on the purely elastic problem.

Wurtzite nitrides have hexagonal crystal lattices and transverse isotropic symmetry. To fully describe the stress-strain relation we need five stiffness constants: c_{11}, c_{12}, c_{13}, c_{33}, and c_{44}. However, the dispersion in published data for mechanical constants is still significant. Also, average values indicate that $c_{11} \approx c_{33}$, and $c_{12} \approx c_{13}$ suggesting that the mechanical behavior does not depart drastically from that of a cubic system. Therefore, in this work, we assume cubic symmetry in the computation of stresses, with only three different stiffness constants: c_{11}, c_{12}, and c_{44}.

Converse piezoelectricity is not computed in Sentaurus Process. In other words, Sentaurus Process computes only the mechanical part of constitutive equations of piezoelectricity and hence the resulting tensor is comprised of the sum of the stress and piezoelectric contribution tensors. However, it is possible to show that polarization vectors can be obtained from these fields as long as permittivities measured under constant strain, ε_S, are used in the Poisson equation of the device simulator [5].

The simulated structure was extended laterally 40 μm beyond the device region, and the boundary condition was set to null normal stress applied to the lateral walls, *i. e.* homogeneous-Neumann boundary conditions to allow for a small but finite stress relaxation of the GaN buffer layer. A large portion of the SiC substrate was also included in the simulation domain to minimize the influence of the Dirichlet ($v_y = 0$) boundary condition applied to the bottom plane. A partial strain relaxation might lead to a reduction of the polarization charges. Therefore, we also examine non-uniform stress effects in partially relaxed AlGaN films. The GaN layer is assumed to be initially fully relaxed.

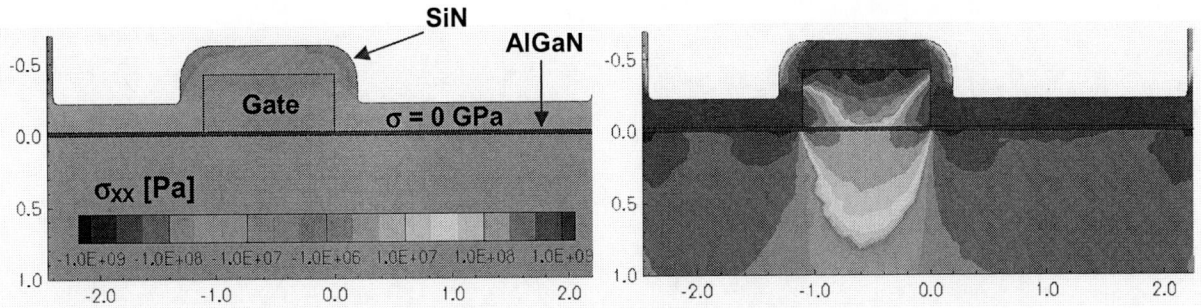

Figure 2 σ_{XX} component of stress distribution for unstrained and strained nitride passivation.

Figure 1 shows the in-plane stress distribution, σ_{XX}, for the case where a 5 GPa intrinsically stressed SiN layer was deposited on top. Positive numbers refer to tensile stress while negative represents compressive stress.

For comparison, simulations of the model structure were also carried out for the case of unstrained nitride passivation. Figure 2 shows contour plots comparing the σ_{XX} stress component distribution (including the piezoelectric contribution) in the device region for the two different scenarios. A 10% AlGaN strain relaxation was assumed in these simulations. As clearly seen, the presence of the stressed nitride layer, combined with the softer metal gate, leads to significant variation in σ_{XX} in the channel region.

Other components of the stress, including shear stresses, are also computed in the simulator. A common assumption while modeling polarization fields in III-nitrides is that the vertical stress component is null. Figure 3(a) shows a contour plot for the computed vertical stress component, σ_{ZZ}, where we can see that the magnitude of the vertical component within semiconductor regions remains below 1 GPa but reaches values large enough to slightly influence the piezoelectric polarization field. Also, non-zero shear components such as σ_{XZ} might lead to in-plane components of the polarization vector as visible in Figure 3(b).

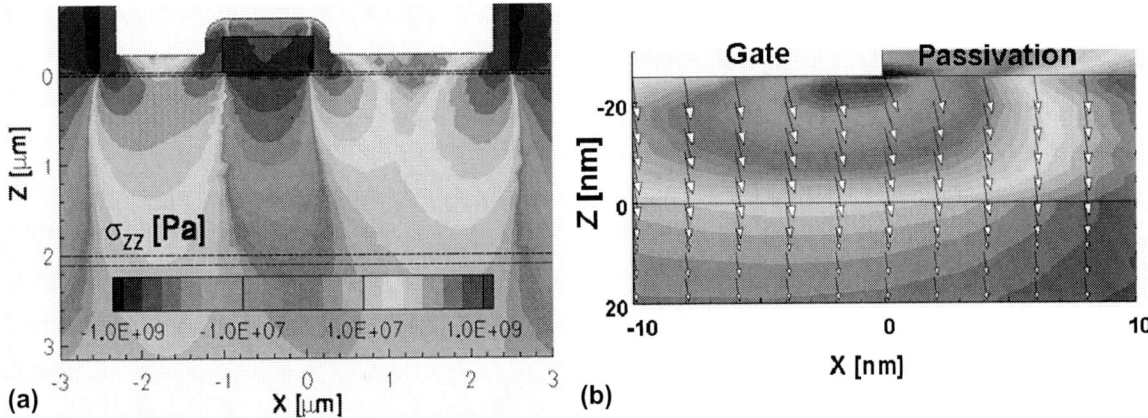

Figure 3 (a) Vertical stress field σ_{ZZ} and (b) polarization vectors showing in-plane component around the gate edge due to shear stresses

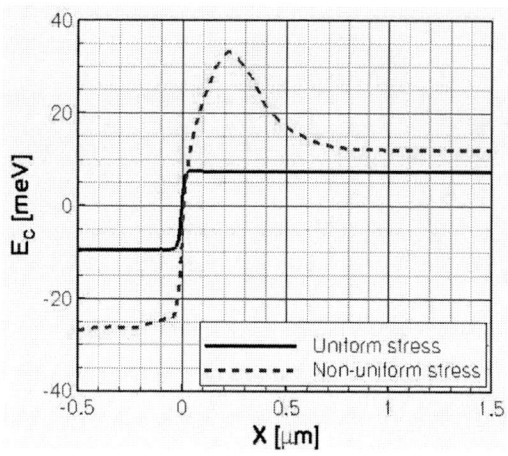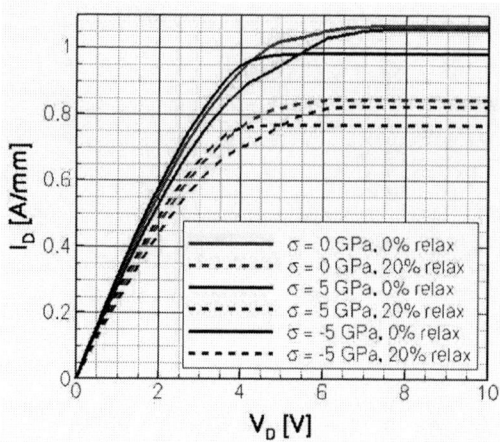

Figure 4 (a) Conduction band edge along the channel under the gate comparing the cases with and without intrinsically stressed nitride and (b) I_D vs. V_D curves showing effects of stressed passivation layers on I_{Dss}.

2.2 Simulation of electrical characteristics

Important model parameters for III-nitrides, such as energy band structure, mobilities, and saturation velocities were based on the book by Levinshtein *et. al.* [6]. Linear interpolations were adopted to compute parameter values as a function of mole fraction in AlGaN. More details on parameter values employed in simulations can be found elsewhere [7,8].

III-nitrides materials still tend to exhibit a significant amount of structural defects, such as threading or misfit dislocations, or point defects, which translate into bulk traps. Therefore, incorporation of traps in the simulations is mandatory. Also, nitride semiconductors with wurtzite crystal structure have different values of the dielectric constant in the direction parallel to the *c* axis and in the plane perpendicular to the c-axis, with the *c* axis component being typically larger. This anisotropy in dielectric constants plays an important role in the potential distribution in HFET devices and hence in the I-V characteristics. Therefore, an anisotropic model for the dielectric constants was accounted for in the simulations.

Finally, hot electrons play important part in the charge transport, even in long channel AlGaN/GaN HFETs, by overcoming potential barriers and spreading to barrier layers and bulk GaN where they are captured by surface and bulk traps. Dynamic capturing of hot electron may then lead to negative output conductance [7] and current collapse [8,9].

The resulting structures from SProcess simulations, including various stress distributions, were subsequently passed on to Sentaurus Device (SDevice) for the simulation of DC and transient behaviors of AlGaN/GaN HFETs with the various stress distributions and passivation layer dielectric constants.

Non-uniform stress fields lead to volume charges via the piezoelectric component of polarization in nitrides and consequently to variation in electrostatic potential via its divergence. For instance, Figure 4(a) compares simulated conduction band edge profiles along a horizontal cut just inside the GaN channel. Clearly stress variations lead to

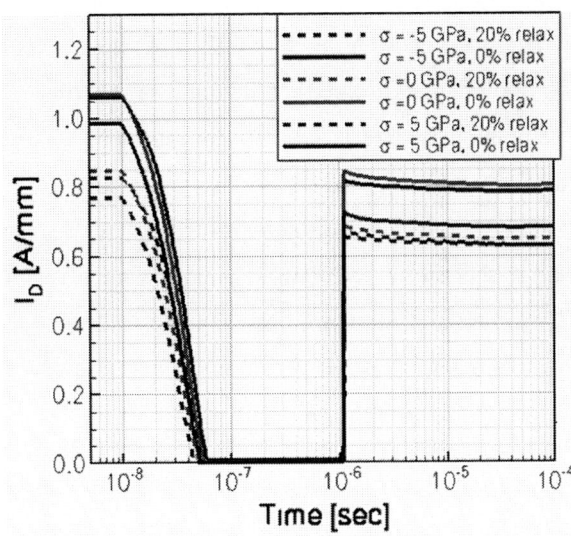

Figure 5 Gate lag simulations with V_{DS} = 10V and a 1 μs pulse from 0V down to below pinch-off and back to 0V

different conduction band profiles around the gate edges and hence to variation in electrical behavior of devices as seen in Figure 4(b) that plots DC I_D vs V_D curves for different passivation layer stress schemes. Notice, however, that even for this case where very large intrinsic stresses were assumed for the passivation layer, combined with a 20% relaxation in the AlGaN film, only a 10% reduction is observed in simulated I_{DSS}.

Non uniform stresses may also result in different transient behavior. For instance Figure 5 compares simulated gate lag experiments for V_{DS} = 10 V and a gate pulse ΔV_{GS} from 0 to -5 V and back to 0 V in 1 μs. The pinch-off voltage of the device is around -4 V. The collapse observed goes from about 20% for the unstressed case to 32% for the simulation with 5 GPa tensile stressed passivation, and down to 17% for the one with -5 GPa (compressive) stress in the passivation layer.

It is important to mention that non-uniform stress effects on bandgap, band offsets, and mobility might also play important roles but were not included due to lack of available data in the literature.

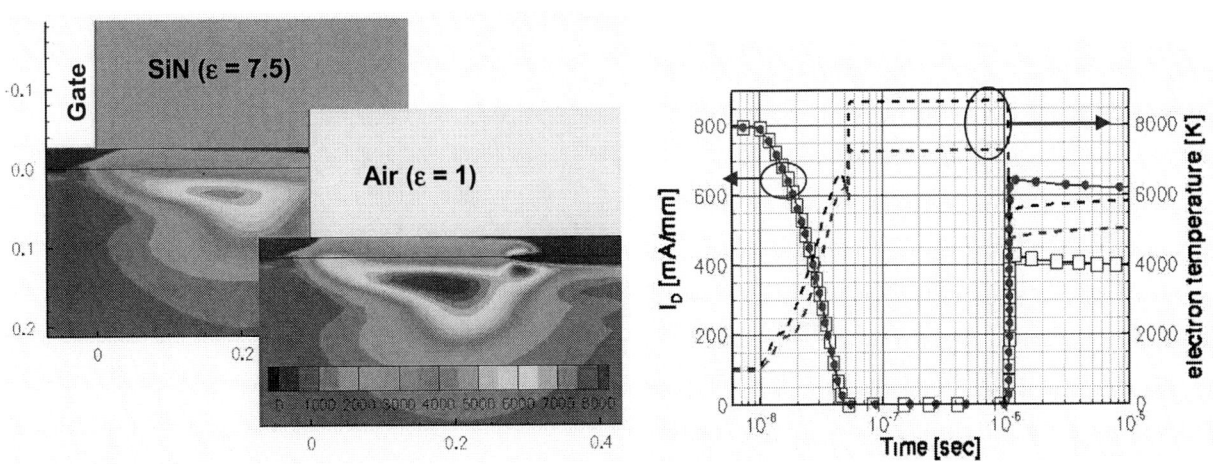

Figure 6 (a) Electron effective temperature and (b) current collapse (gate lag) for different passivation layer dielectric constants (left *y* axis) and maximum electron effective temperature (right *y* axis)

1-4244-0267-0/06/$25.00 ©2006 IEEE

In addition to stress, the higher dielectric constant of a passivation layer compared to that of air may have significant impact in device performance according to simulations. Fringing electric field lines between the gate and the channel in the gate vicinity smoothes down electric fields and leads to less hot electrons. This is clearly visible in Figure 6(a) that compares simulated electron effective temperature in devices with and without passivation layers. It has been proposed that hot electrons near the gate edge can reach far down into the bulk and barrier layers where they are temporarily captured leading to current collapse [5]. Indeed, Figure 6(b) shows simulated gate lag experiments for devices with and without passivation, where collapse is significantly reduced in the presence of the SiN overlayer. Since simulations were carried out for identical interface or surface density of states, the higher coupling due to larger dielectric constant provides an alternative explanation, or at least a contributing factor, for collapse mitigation by the presence of passivation layers.

3   Conclusions

Numerical simulations to compute stress distributions resulting from stressed overlayers reveal that significant stress fields can penetrate deep into device. Piezoelectric polarization effects from stressed overlayers are only mild due to relatively high stiffness in nitrides but can change band profile along the channel, especially under gate edges.

Fringing fields in passivation layers with large dielectric constants can play important role in collapse reduction.

References

[1] O. Ambacher, J. Smart, J. R. Shealy, N. G. Weimann, K. Chu, M. Murphy, W. J. Schaff, L. F. Eastman, R. Dimitrov, L. Wittmer, and M. Stutzmann, W. Rieger, and J. Hilsenbeck, J. Appl. Phys. **85** (1999) 3222

[2] B. M. Green, K. K. Chu, E. M. Chumbes, J. A. Smart, J. R. Shealy, and L. F. Eastman, *IEEE Electron Device Lett.*, **21**, 268 (2000)

[3] A. V. Vertiatchikh, L. F. Eastman, W. J. Schaff, and T. Prunty, Electron. Lett., 38, 388 (2002)

[4] G. Koley, V. Tilak, L. F. Eastman, and M. G. Spencer, IEEE Trans. Electron Devices, 50, 886 (2003).

[5] N. Braga, R. Mickevicius, V. S. Rao, W. Fichtner, R. Gaska, M. S. Shur, , In *IEEE Compound Semiconductor Integrated Circuit Symposium (CSICS) Technical Digest* (2005) 149

[6] M. E. Levinshtein, S. L. Rumyantsev, and M. S. Shur, Editors, **Properties of Advanced Semic. Mat.: GaN, AlN, InN, BN, and SiGe**, John Wiley & Sons, ISBN 0-471-35827-4, New York, 2001.

[7] N. Braga, R. Gaska, R. Mickevicius, M. S. Shur, X.Hu, M. Asif Khan, G. Simin, and J. Yang, J. of Appl. Phys. **95** (2004) 6409

[8] N. Braga, R. Mickevicius, R. Gaska, M. S. Shur, M. Asif Khan, and G. Simin, In *IEEE International Electron Devices Meeting (IEDM) Technical Digest*, (2004) 33.6.1

[9] N. Braga, R. Mickevicius, R. Gaska, M. S. Shur, M. Asif Khan, and G. Simin, In *IEEE Compound Semiconductor Integrated Circuit Symposium (CSICS) Technical Digest* (2004) 287

A 2-Mask NMOS Process Design Fabricate and Test Module for Use In Microelectronics Instruction and Process Development

D. W. Parent
Electrical Engineering Department
San Jose State University
One Washington Square Hall, San Jose CA 95192-0084

Abstract— We have developed a simplified 2-mask n-type Metal Oxide Semiconductor (NMOS) transistor process design and verification module for electrical engineering students enrolled in the Microelectronic Manufacturing Methods class/laboratory at San Jose State University. We have run this module for three years and have found that the simplified process allows the students to learn more because they have the time to design the process fabricate and test in one semester. Student learning is also enhanced because it allows students to make and correct mistakes in the processing the devices. We have also found that the simplified process saves time in process development of more complex processes, by reducing the number of photolithography steps required to fabricate a transistor.

I. INTRODUCTION

To teach semiconductor process development is resource intensive (instructor time, masks, chemicals, etc.)[1] and it is a risk of these resources to allow novices to design, fabricate and verify a semiconductor device on their own. If the devices do not work, then students are left with a feeling that they did not master the material. (They also believe that instructor did not master the material!) One method to reduce this risk is to have students design a process with Technical Computer aided Design (TCAD) tools, but stop short of fabricating the devices. This allows students to be exposed to the constraints of an advanced deep sub-micron process that might not be available at a university. On the other hand, students will not develop a sense of the statistical variations, and modeled vs. fabricated device differences that can only be encountered in actual device fabrication. Another method to reduce the risk of non-functional devices is to have the students fabricate a process that has already been designed and verified previously[5]. Although the device yields can be over 90%, the students do not learn how to truely design a process. We feel that best way to reduce this risk is to reduce the time and money to design and fabricate devices is to reduce the complexity of the process to an absolute minimum. We have developed a 2-mask NMOS process that by using two masks instead of four, reduces the amount of time and raw materials that is the main cost to fabricating devices. (It also reduces the time to simulate (TCAD) a device structure.) Since this process is less resource intensive it is acceptable to allow novices to design and fabricate their own transistor process. If a student makes an incorrect decision in the processing or design, there is usually enough time to fix the problem because the number of processing steps is reduced.

This module is part of a senior/graduate level course on microelectronic manufacturing methods. The 3-4 student teams are multidisciplinary so a student has to have CMOS circuit design, advanced device physics or semiconductor process knowledge to be a successful team member.

The two-mask device is a simplified form of a non-self aligned 4-mask NMOS process[3]. The major features of the device are (Fig. 1):

- The diffusion windows for the source and drain also are the contact windows when the source drain spin on glass is removed before metalization.
- The field oxide and the gate oxide are the same layer. Since we are interested in transistors, diodes and MOS capacitors, we do not have to prevent leakage current between adjoining devices.
- The gate oxide between the source and drain acts as self aligned diffusion mask.
- The amount the junction diffuses under the gate is critical. The metal gate (which has to be notched in from the edge of the source and drain) has to overlap at least the depletion widths of the source and drains to have current flow at small drain voltages. This notching is required because the standard deviation of our registration error is 1.0 micron. If the gate metal were to drift right or left it would short the gate to the source or the drain.
- The mask set consists of four different NMOS width/length pairs with the notch distance (gate to source/drain offset) between the as edge of the drawn source/drain regions and the edge of the metal gate varied from 0.0 microns to 8.0 microns (Fig. 1).
- The width of the transistor is the as drawn vertical distance of the diffusion cut. Since there is no field oxide the actual width is larger due to the fact the diffusion areas move about 2 microns into the surrounding substrate.
- The length is the as drawn distance separating the

two edges of the diffusion cuts. The true length is reduced by the diffusion into the substrate as well.

- The 2 Mask set is made out of quartz and chrome and has been provided by the instructor and consisting of sheet resistance structures, junction depth monitors, MOS caps, different sized MOSFETS as well as inverters, current mirrors and ring oscillators. Each MOSFET structure comes with an x and y Venier scale to measure registration error.

II. Module Overview

The model begins with a semiconductor process review, which consists of a lecture presentation of a simple metal gate NMOS process that has an Athena run deck, linked to a process traveler, with measured device characteristics. (This is the same process that some team members have fabricated in a previous class.) The major processes reviewed are oxidation, diffusion, and ion-implantation. Then an NWELL self-aligned process is reviewed using the Java Applet Site hosted by Suny[4]. The final lecture shows all the process steps of the 2-Mask NMOS process in conjunction with a calibrated run deck, and process traveler.

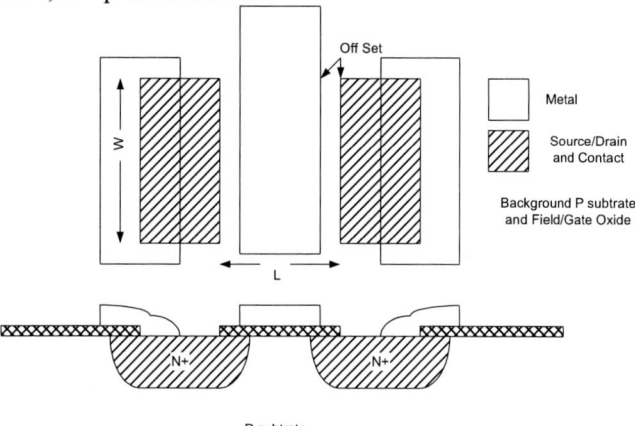

Fig 1. The top view and cross-section of the 2-Mask Process.

In order to design the improve process the students first need to learn the fundamentals of Athena/Atlas by completing a tutorial the guides them though creating their own run deck for a pn junction and learning how and why to refine the mesh[5]. Then the students use the NMOS poly-silicon gate run deck example provided by Silvaco. To prove that they are familiar with TCAD tools the students complete an individual homework assignment to change the channel length and VT of the sample run deck.

The students are then ready to improve the VT of the 2-mask NMOS process. They are given free reign to change any time and temperature of the existing calibrated process run deck (Fig. 2). The students present their improved process to the professor who makes sure that the times and temperatures selected by the students are "reasonable". The students then simulate their design for variations in time and temperatures to

anticipate process variances in the laboratory. The final part of the design is to create a process traveler (Fig. 3.) that will be used to guide then through device fabrication in the SJSU micro fabrication facility. The groups give an oral presentation on their process design in the form of a design review.

```
#SCREENING OXIDE GROWTH STEP 2.0 ASSUMED
method fermi compress grid.ox=.003 gridinit.ox=.003
# STEP 2.0 IS ASSUMED
# STEP 2.1 PUSH
diffus time=15    temp=400 t.final=700  nitro
# STEP 2.2 RAMP UP
diffus time=13.3 temp=900 t.final=1000   nitro
# STEP 2.2 STABILIZE
diffus time=10    temp=900     nitro
# STEP 2.2 SOAK
diffus time=30    temp=900     dryo2
#STEP 2.2 PURGE
diffus time=10    temp=900 nitro press=1.00
#STEP 2.2 RAMP DOWN
#diffus time=10    temp=900 t.final=700    nitro
#STEP 2.3 PULL
diffus time=15    temp=700 t.final=400   nitro
```

Fig. 2 Sample code from a calibrated Athena run deck.

2.0		Oxide Growth 1 - Load	• Load cleaned wafers into quartz boat; put boat 6" into furnace. Lowest number, shiny side in first. Note Wafer order: • N2 flow 10 slm Tube furnace heating for few hours
2.1		2 - Push	• Push in slowly to prevent temperature shock using autoloader • 15 min push at 700 Dry N2 • Target temperature = 700°C • Record oven temp in all zones • , . ,
2.2		3 - Oxidize	• Put on end cap, Leave doors open 15 min ramp to 900 Dry N2 • 10 minute stabilize at 900, Dry N2 • 100 Min soak Dry O2 (Max Flow) (put on flow restrictor after you switch gas flow or the cap will fly off and break!) • Take off restrictor • 10 minute purge N2 • XX min ramp down to 700 Dry N2
2.3		4 - Pull	• 15 min pull at 700 • Pull out boat slowly, remove, cool under hood

Fig. 3 Sample instructions from the process traveler.

After an equipment process review and safety lecture the students fabricate their devices in an effective open lab[1]. The teams operate independently in lab, asking for help only when something with the process is unexpected.

---

[1] There are five groups working o different processes at any given time. Therefore even if the instructor is present the students are not being directly supervised at all times.

1-4244-0267-0/06/$25.00 ©2006 IEEE

III. 2 MASK NMOS PROCESS FLOW

The starting substrates are <100> P-type silicon wafers with a resistivity of 5 Ω-cm. The substrate doping is so low that given the large observed fixed oxide charge in our lab ~10^{11}q/cm^2, the threshold voltage will be negative for all oxide thicknesses.

To change the substrate doping B diffusion through a screen oxide (~200Å) is carried out[6], after SC1 and SC2 cleaning steps. The thickness of the oxide is critical in determining the final substrate doping (Fig. 4.). If the oxide is too thin the boron concentration at the surface of the substrate will be much higher than the 10^{17}cm^{-3} range, but rather will be in the 10^{20}cm^{-3} range, cause a V_T in excess of 40Volts. Diffusion through a screen oxide is done rather than an implant step due to the fact our facility does not have an implanter. Implant services are available from a local vendor or SNF, but we have found students do not like to outsource their processing steps. The boron diffusion time and temperature are usually the steps the students change to increase the substrate doping. The boron source is spin on glass (SOG) from Emulsitone[7]. The SOG is spun of for 60 seconds at 300rpm and then baked for one minute at 100°C.

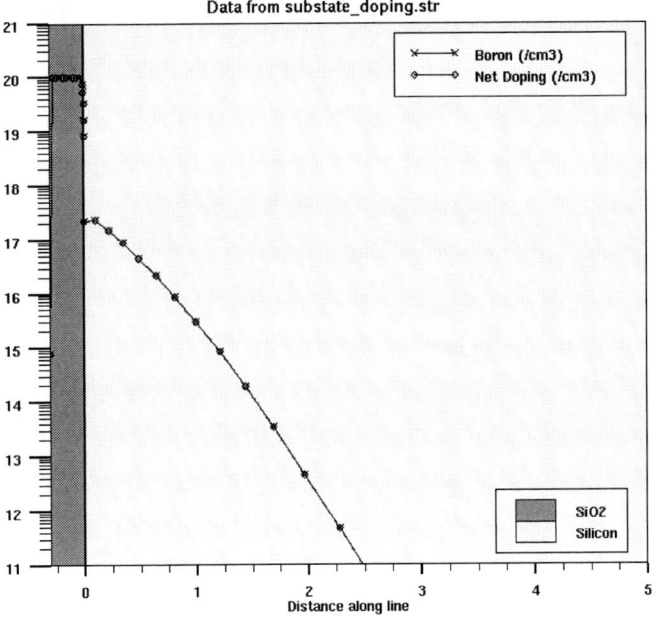

Fig 4. Substrate doping profile after B diffusion step.

Once the substrate-doping step has been carried out the SOG and screen oxide are etched off in buffered hydrofluoric acid (HF). The gate oxide is grown next (Fig. 5.), by means of dry thermal oxidation. The oxidation temperature is 1100°C and a post oxide anneal is done at the oxidation temperature for at least 50minutes. The gate oxide thickness is usually in the range of 3500 Å, as it is over etched in a later processing step. The students adjust the oxidation time but not the post oxide

anneal time or the oxidation temperature to increase the threshold voltage.

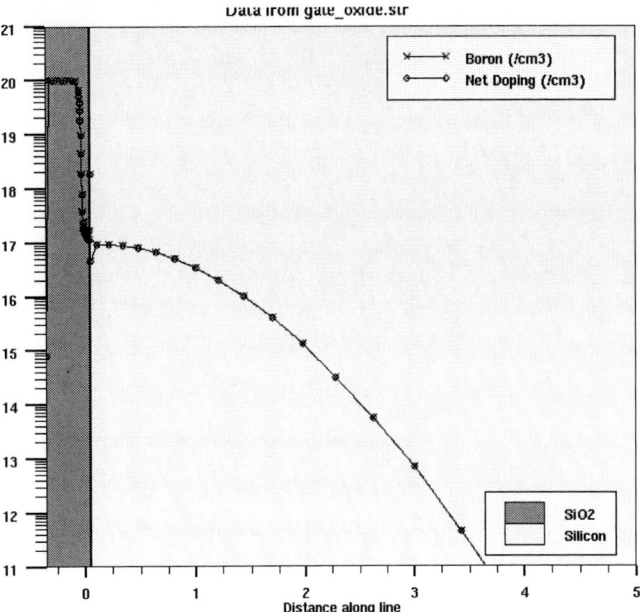

Fig. 5. Substrate doping profile after gate oxidation step.

The first photolithography step (Fig. 6.), source drain diffusion/contact (mask 1) is completed after the gate oxidation step. The source and drains are etched down to below 20Å.

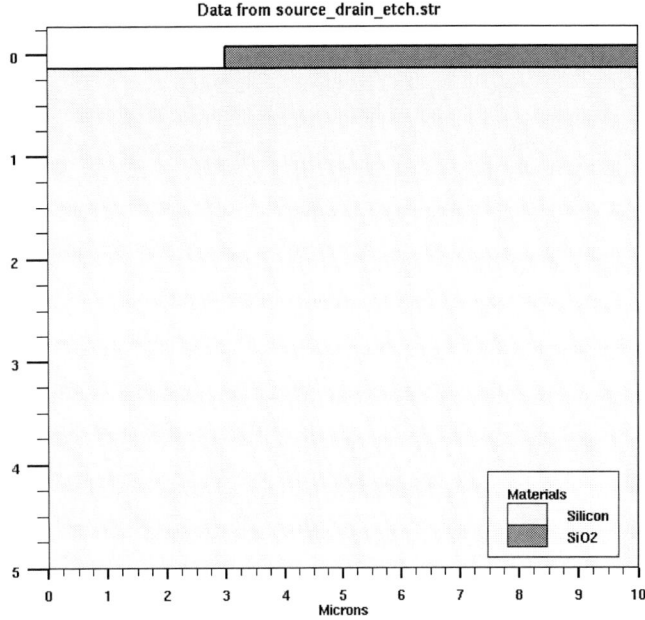

Fig. 6. Source/Drain etch step.

To dope the source and drains, P-SOG is applied to the wafer in a similar manner as the B-SOG. The source and drain diffusion is carried out at 1100°C. The time of the diffusion is critical because the source and drain need to diffuse under the gate to make sure the gate electrode covers at least the depletion widths of each source and drain region (Fig 7.). If the diffusion time is too short, the gate will not overlap the drain and there will be almost zero drain current until the depletion width is extended under the gate due to an increasing reverse bias voltage of the drain. If the diffusion time is too long, the P-SOG can punch through the gate oxide and dope the substrate making normally on MOSFETS. The amount of overlap needed is based on a statical process control (SPC) report on alignment generated by the students earlier in the semester.

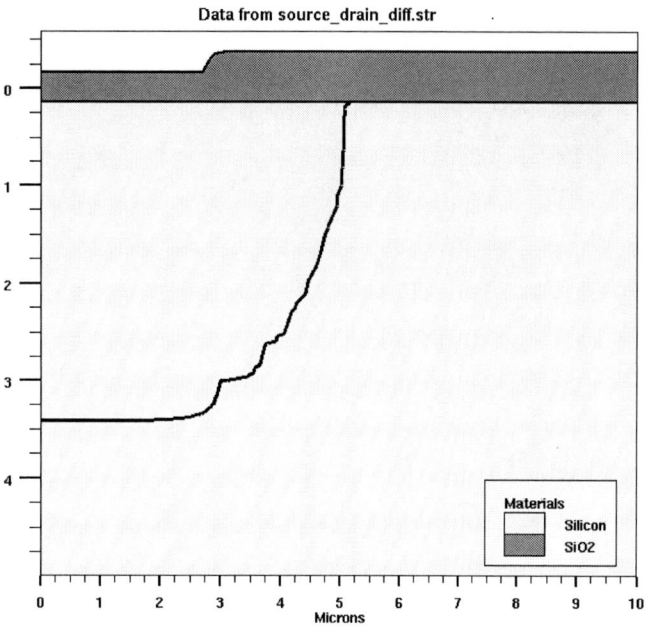

Fig. 7. Source/Drain Diffusion step.

Before the aluminum can be evaporated onto the wafers the P-SOG must be removed from the source and drain regions (Fig. 8.). It was thought at first that the selectivity of the P-SOG would be greater than that of the thermally grown oxide. This is true once the SOG has been removed form the Gate/Field oxide regions, however once the etching of the SOG near the surface of the source and drain regions begins the etch rate of the gate/field oxide remains around 400Å/minute but the SOG etch rate decreases dramatically. Since the SOG has to be removed in order to from a good ohmic contact, 1000Å of oxide maybe removed form the gate/field regions before the source and drain area read 20Å or below as measured with a nano-spec. In some cases the students etch of all the oxide, even after careful monitoring of the etch process.

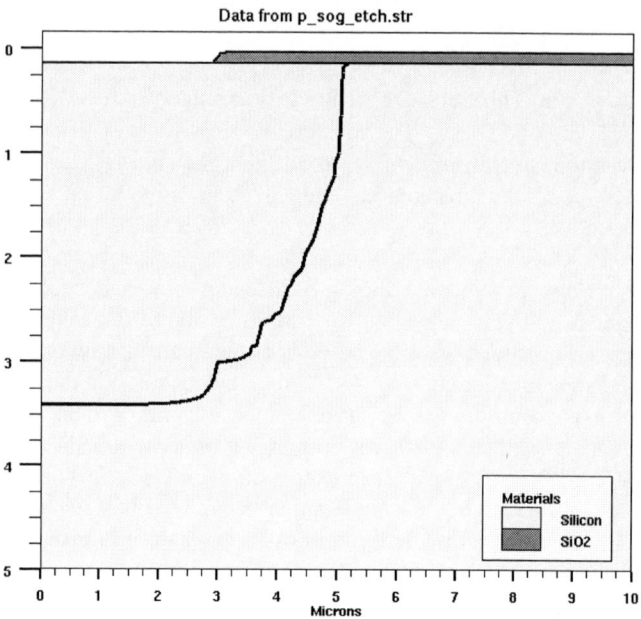

Fig. 8. P-SOG Etch step.

After the metal is evaporated the metal mask photolithography is performed. The metal is etched in Phosphoric acid at 42°C. The wafers are then annealed at 450°C for 30 minutes. The devices are now ready for testing (Fig. 9.). Notice that the junction diffuses almost 2.0 microns under the gate.

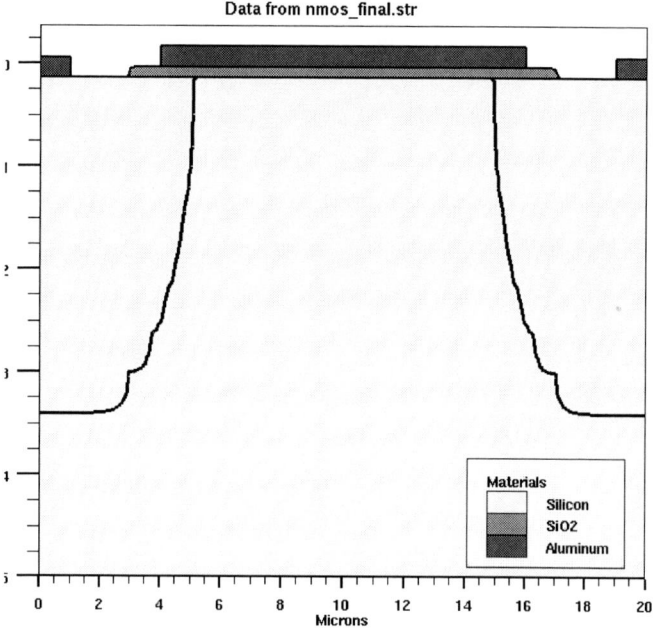

Fig. 9. Completed 2-MASK NMOS Device.

IV. TESTING

The students use IC metrics in conjunction with an HP 4145 or 4156 semiconductor parameter analyzer. The plot set-ups have all been saved previously. This has reduced the training time for testing.

The first test is an I_D vs. V_{DS} to see if transistors were created. Large step sizes are use to speed up the testing. If this test does not work and it can be determined that the gate is not shorted to the source or drain, and then the diodes are checked. If the gate is shorted then devices with a larger offset are chosen to test, as well as those with a minimum registration error. If the diodes work then more MOSFETS are tested with larger voltages. If the diodes are shorted, then usually the wrong dopant was used for the source and drains. (This has happened to me. Since then we have a double check system to make sure we use the correct dopant course.) If the diodes are shorted then a CV test is performed to see if MOS capacitors were formed.

If the process was found to produce working devices, then the students test diodes, capacitors, and try to extract spice level one parameters (Fig. 10.). They also try to find which gate to source/drain offset produces the highest yield of working devices.

Sometimes the process does not produce working devices. The students need extra help to find out exactly why the process did not work. As long as their final report includes a reasoned argument as to why the devices failed and a proposed solution to correct the problem, no points are deducted for the process not working. This is done to promote risk taking in process development. Sometime the students etched of all the gate oxide during the P-SOG removal step. Some times the students did not fully etch the P-SOG form the source and drain regions and there is no ohmic contact to the source and drain.

This past semester the students adopted a very thin screening oxide that was not modeled correctly in their TCAD simulations. Rather than the boron being screened to a substrate surface concentration of $\sim 10^{17} cm^{-3}$, the boron push through the screening oxide and doped the substrate to excess of $10^{20} cm^{-3}$. This resulted in very high threshold voltages. For those students with gate oxides of around 300Å measuring the V_T by either MOS CV or transistor measurement was impossible as the gate oxide broke down at 20V. One group was successful because their gate oxide was etched down only 1000Å and so a V_T of 60V could be measured with out shorting the structure. They did not fully clear the source drain widows for proper ohmic contact formation and one can see the results of a high source drain resistance (Fig. 11.). Since the source resistance due to the metal N+ contact resistance is extremely large the source and body are not shorted. This gives the appearance of an even higher V_T than one would expect due to the body effect.

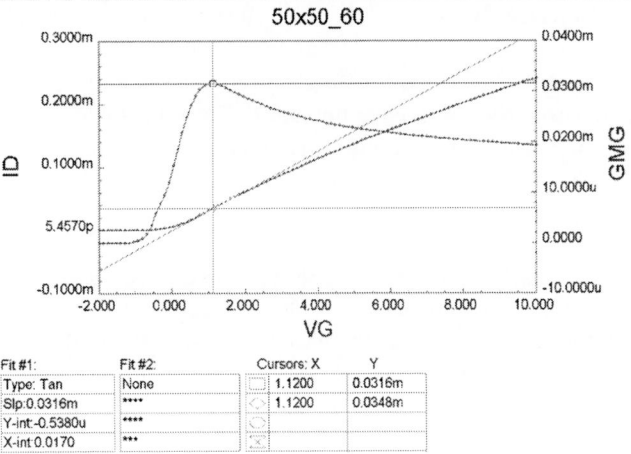

Fig. 10. I_D vs. V_G to extract V_T.

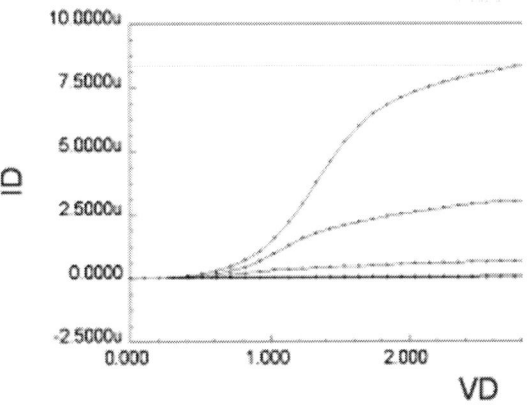

Fig. 11. I_D vs. V_D for a high drain resistance MOSFET.

V. RESULTS/DISCUSSION

The students enjoyed this module even though to complete it required a lot of time spent outside of normal laboratory meeting time to finish. Even though this module was designed as a method to reduce the cost to operate laboratory–based processing instruction, it still is resource intensive to teach. The improvement is that that time is better spent because the students are doing the design, and they have the opportunity to make and correct mistakes.

Students need help interpreting the testing results. The modeling problem of this past spring semester was especially difficult for them to explain as it was the model that was wrong, not the processing. Even when student devices work, they need help to explain how substrate-doping can affects the break down voltage, λ, and even mobility. They do understand the affect of substrate doping on V_T.

After running this module for three semesters, it was observed that the students who made incorrect processing decisions and corrected them seemed to learn more. The students who made a mistake were in general trying to improve the process flow by aggressively scaling back diffusion time, or oxide thickness. The ability to determine what went wrong in a process flow and correct the problem in one semester is quite an achievement. Personally my first BJT's did not work and it took me years to find out what had a happened (Bad P diffusion modeling). It was this kind of experience that made me a better process engineer.

The two-mask process has also been useful in process development for other more complex processes. We were able to reduce the ramp down time for wafers at oxidation/diffusion temperatures by 60%, because of an experiment that we were able to conduct quickly by using only the two-mask process rather than a four-mask process. We probably would not have invested the time to see if we could reduce than ramp time if we had to use a four-mask process in our experimental plan. We were also able to solve a high source/drain resistance problem in our 4-mask NMOS teaching process that was causing the I_D/V_{DS} curves to be concave downwards in the linear regime instead of concave upwards (Fig. 11.). It turned out the surface concentration required for a good ohmic contact was not being maintained. The contact resistance is not modeled well and it was the fact the source drain doping step came after the gate oxide step (unlike our 4-mask NMOS process) that pointed in the direction of how to fix the contact resistance problem. The next studies we will conduct is how much oxide can be left in the source and drain regions and still get a proper source drain region, and a good ohmic contact.

VI. CONCLUSIONS

The 2-mask NMOS process can be used to teach semiconductor design and processing. The reduced the number of processing steps saves time in not only processing the wafers but in the TCAD simulations. The 2-mask process is also helpful in process development. One could use this process to rapidly evaluate high K dielectrics, by simply changing the gate oxide and using implants instead of diffusion.

In addition, this kind of module prepares students for future graduate study in process development because the student will have designed fabricated, and improve a process.

ACKNOWLEDGMENT

We would like to thank Neil Peters for his efforts in managing the MPE laboratory, Kindness Israel for extensive IT support, Linda Shnell of Cadence Design Systems for access to the Virtuoso tool set, and Jerry Kissinger of Intel for the donation of testing equipment and Linux stations.

REFERENCES

[1] D. W. Parent, E. J. Basham, Y. Dessouky, S. Gleixner, G. Young, E. Allen, "Improvements to a Microelectronic Design and Fabrication Course", IEEE Transactions on Education, Vol. 48, No. 3, pp. 497-502 (2005).

[2] D. W. Parent, Y. Dessouky, S. Gleixner, G. Young, E. Allen, " Microelectronics Process Engineering Program at SJSU," *Proceedings of the 15th Biennial IEEE University/Government/Industry Microelectronics Symposium,"* Richmond , VA pp. 128-134 (2001).

[3] D. W. Parent, http://www.engr.sjsu.edu/MatE129/Process%20Handbook.htm Accessed 12 June 2006.

[4] http://jas.eng.buffalo.edu/ Assessed 12 June 2006.

[5] D. W. Parent, "Silvaco Tutorial", http://www.engr.sjsu.edu/dparent/Silvaco/silvaco.pdf, Accessed 12 June 2006.

[6] G. H. Bernstein, R. J. Minniti, X, Huang, "Advanced IC processing laboratory at the University of Notre Dame" IEEE Transactions on Education, Vol. 37, No. 4, pp. 334-340 (1994).

[7] http://www.emulsitone.com/, Accessed 12 June 2006.

Effect of Process and Layout on Strain Enhancement from Dual Stress Liners

Victor Moroz, Munkang Choi, Xi Wie Lin, and Dipu Pramanik

Synopsys

Abstract:

Tensile and compressive stressed nitride liners have been used to increase the carrier mobility in n-channel and p-channel silicon transistors respectively. Simulations indicate how much of the stress in the film is transferred to the channel region and the magnitude of the stress in different directions .A simple bulk piezoresistive model was used to estimate the effect on carrier mobility. It is shown in the case of the n-channel transistors that the enhancement is due to the vertical stress component whereas in the case of p-channel devices the enhancement is due to the in-plane stresses. The effect of different process conditions such as film stress, thickness and method of deposition, on mobility enhancement, was also characterized. It is shown that the enhancement saturates with increasing nitride thickness but scales proportionally with the film stress. Detailed studies of the effect of the circuit layout on the final channel stress allow the critical layout parameters to be identified. The variation of device performance with the layout parameters is quantified and can be used to define design rules as well as equations to modify the device characteristics based on layout.

1-4244-0267-0/06/$25.00 ©2006 IEEE

Modeling Process Impact on Cu/Low k Interconnect Performance and Reliability

Xiaopeng Xu, Greg Rollins, Xiao Lin, and Dipu Pramanik

TCAD DFM Solutions, Synopsys, Inc.,
700 East Middlefield Road, Mountain View, CA 94043, USA

ABSTRACT

This paper studies the impact of layout alteration and structural variation on capacitance and spatial variations of electric and thermal mismatch stress fields. The fabrication process related layout alteration and structural variation include floating dummy fill insertions, silicon nitride cap layers thickness selections, and metal line cross-section shape changes. It is demonstrated that the spatial distributions of electric field and thermal-mechanical stress field have different geometric dependence and process variations have different implications. The layout pattern and interconnect architecture that are optimized for electric performance may be inferior in reliability due to large stress concentrations. The numerical results suggest that in pursuit of manufacturability the tradeoffs between electrical performance and mechanical reliability need to be considered together for future interconnect architecture and process technology developments.

INTRODUCTION

Interconnect performance and reliability is determined by its material composition and geometric architecture. To meet technology requirements new materials are introduced to interconnect structures, and new processes are adapted to integrate the new materials into the ever shrinking structures. While functional design relying on nominal geometries, fabrication process introduces intentional and unintentional structural alterations and variations. These deviations from nominal designs can impact both the electrical performance and mechanical reliability.

To achieve better planarization, dummy fills are inserted in dielectrics between conductor lines prior to chemical mechanical polishing. With floating metal fills altering electric field distributions, coupling capacitances are induced between conductor lines. The total capacitance associated with a conductor line in the proximity is thus affected. The deviation from the nominal design is a function of the local arrangements of conductor lines and dummy fills. Because of the dummy insertions, the local material compositions are changed, so is the surrounding mechanical rigidity of a conductor line in the proximity. These dummy fills also serve as a thermal mismatch stress source in the subsequent thermal steps. The magnitudes and distributions of thermal-mechanical stress fields are also dependent upon local conductor line and dummy fill arrangements. While floating fill impact on interconnect capacitance has been studied [1], its impact on thermal-mechanical stress needs to be investigated.

Dielectric cap layers are important integration components in damascene structures. They are used for two main functions, assisting patterning and preventing diffusion. When the thickness of the higher permittivity cap layer varies, the overall effective dielectric constant of the multilayer stack is changed, so are the local electric field and electric flux. These changes may lead to the unintended change of the total capacitance of a neighboring conductor line, especially if the cap layer makes significant contributions to the overall effective k. The cap layer typically has distinct mechanical properties from other materials. Depending on the layer thickness, the cap layer can have a large impact on the magnitude and gradient of the stress fields within itself and in the adjacent metal and dielectrics regions.

The shape of a conductor line fabricated by copper damascene process is generally trapezoidal with the top width larger than the bottom. The exact shape and dimensions are a function of process control. With process induced changes in shape and dimensions, the actual electric characteristics deviate from the design nominal that is based on regular shape and dimensions. Of particular concerns are the corresponding electric field and thermal-mechanical stress field variations. The former is linked to dielectric breakdown [2], and the later is related to mechanical damage.

In this study, the impact of these process induced geometric variations on coupling capacitance, electric field variations, and thermal-mechanical stress field modulations is investigated using industry standard process simulators and field solvers [3]. The interconnect structures are generated using the process data representative of 65nm technology or taken from [1]. The geometry models include the intentional and unintentional structural alterations. The numerical simulation setup details were discussed in [4] and the stress-free temperature for copper is taken as 250°C degrees for the thermal-mechanical stress modeling.

METAL FILLING PATTERN EFFECT

The layout configurations shown in Figure 1 contain 4 copper conductor lines. Multiple floating dummy fills are inserted to attain a metal density of ~42% in the region adjacent to the terminating wire and the wire with a jog in the middle of the layout section.

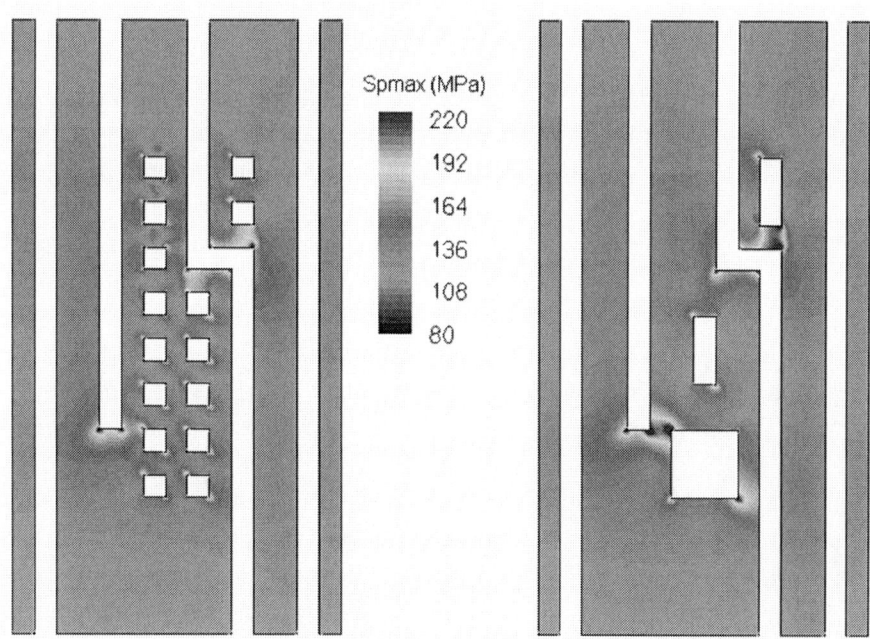

Figure 1. Maximum in-plane principal stress distributions in dielectrics surrounding metals after annealing for the layout configurations studied in [1]. (a) Regular fill pattern. (b) Fill

pattern with recommended guidelines. The metal lines and fills are not shown.

The left configuration has all 15 dummies arranged with a regular fill pattern. The impact of these floating fills on wire capacitance was studied in [1] and the guidelines to reduce coupling capacitance were outlined. The filling pattern shown in the right configuration results from the applications of these guidelines and it represents a 16% reduction in coupling capacitance between the two middle wires from that in Fig.1 (a). Notice that the individual dummies in Fig. 1 (a) are merged into three larger blocks in Fig. 1 (b).

Generally speaking, the geometric dependence of stress field is different from that of electric field. The maximum in-plane principal stress distributions after annealing for the two layout configurations are displayed in Figure 1. Two areas of high stress concentrations are evident in Fig. 1 (b): one between the corner of the terminating wire and the corner of the enlarged dummy block, and the other between the turning corner of the jog and the end of elongated dummy block. These high stress concentrations pose reliability concerns. Therefore, the optimal pattern design for floating fill insertions has to take both the electric performance and the mechanical reliability into consideration. The optimization process often leads to trade-offs between the two requirements.

CAP LAYER THICKNESS EFFECT

Figure 2 shows two cross-sections of damascene process fabricated copper lines with surrounding low k dielectrics and top silicon nitride cap layers. The thickness of the cap layer is 50nm in Fig. 2 (a), and 20nm in Fig. 2 (b). The dielectrics constants used in simulations represent these of spin-on porous polymer and CVD silicon nitride. The spatial distributions of simulated electric field magnitudes under 1 volt bias are mapped on the cross-sections. Because of the symmetry, only halves of the coppers cross-sections are shown.

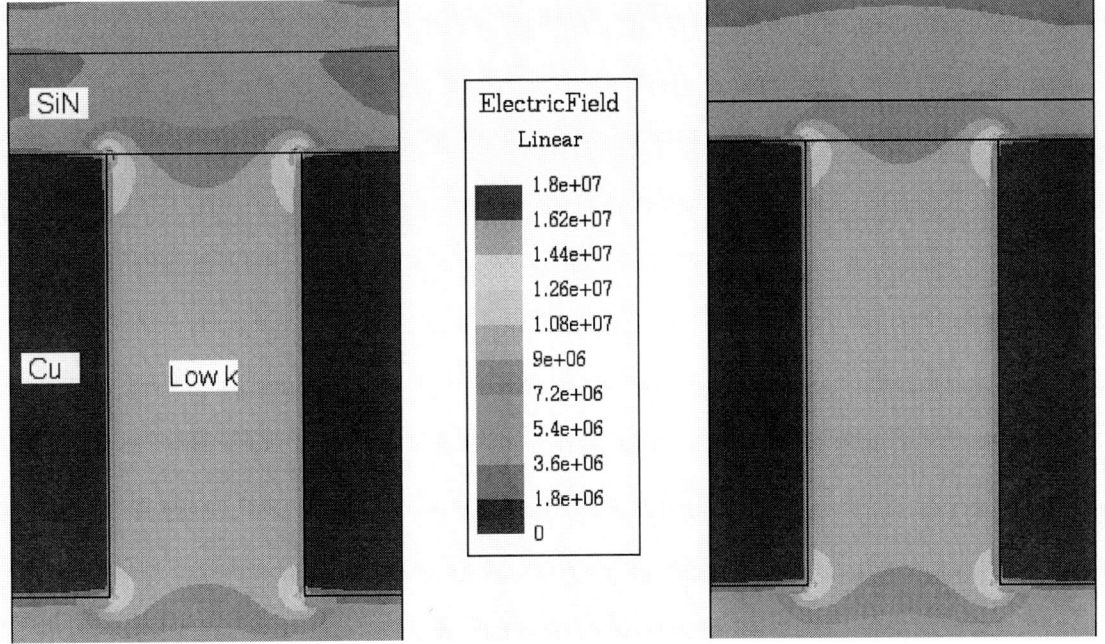

Figure 2. Electric field magnitude distributions under 1 volt bias between two copper lines

for two designs with different cap layer thickness. (a) 50nm cap layer. (b) 20nm cap layer. The legends are in V/m.

As the cap layer thickness reduces from 50nm to 20nm, the coupling capacitance between neighboring copper lines decreases by ~10%. It is interesting to note that the maximum field magnitude is achieved around the location where copper, low k and nitride regions meet, and the maximum field magnitude decreases by ~11.5% with the thickness reduction. This decrease in electric field magnitude may be beneficial for preventing local dielectric break-down [2].

The normal stress Sxx distributions after annealing for the two designs are shown in Figure 3, with Ta barrier layers highlighted. Due to thermal and mechanical property mismatches, the cap layer is under large compression for the sections above copper and large tension above dielectrics. As the layer thickness decreases, the maximum magnitudes of compression and tension increase. Large stresses in cap layers are of mechanical reliability concerns [5]. It is also interesting to note that the stress state in Ta barrier layers around Cu conductors is very different from that of Cu due to the difference in thermal deformations.

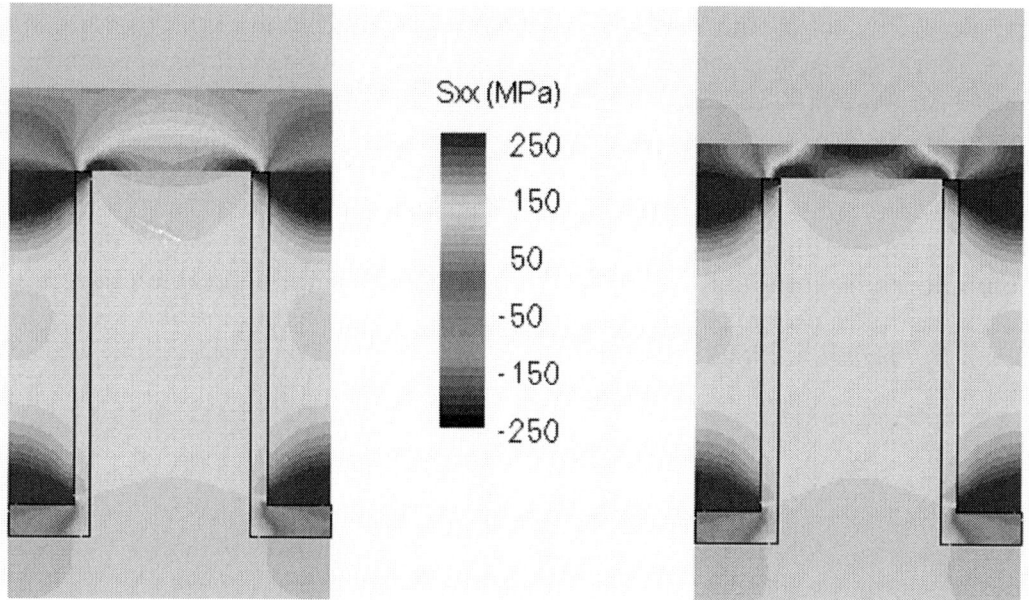

Figure 3. Normal stress Sxx distributions around two copper lines after annealing for the two designs with different silicon nitride cap layer thickness. (a) 50nm cap. (b) 20nm cap. The highlighted polygons represent barrier layer boundaries.

LINE CROSS-SECTION SHAPE EFFECT

When the shape and dimensions of line cross-section vary due to process variations, the actual resistance and associated capacitance can deviate from the design nominal. Depending on the degree of the geometric changes, local stress concentrations can occur. The shape of metal lines formed by copper damascene process is generally trapezoidal rather than rectangular, and the final geometry is a function of fabrication process. The deviation of line cross-sections from the nominal design can have performance and reliability implications.

Figure 4 shows the electric field magnitude distributions around metal lines for two cross-sections with different shapes. In the case with trapezoidal cross-section, the side wall slope of the metal line is about 4 degree and the spacing at the middle is the same as that with a rectangle cross-section. Because of the side wall slope, the line spacing at the top surface or bottom reduces or increases by 7.5%. As a result, the capacitance is only slightly affected, but the maximum electric field magnitude around the upper corners of the metal lines increases by 15%. The impact on capacitance has been studied for variations in metal line width, thickness, slops, and spacing. The maximum impact is found from the variation in spacing where a 10% spacing reduction leads to over 15% coupling capacitance increase.

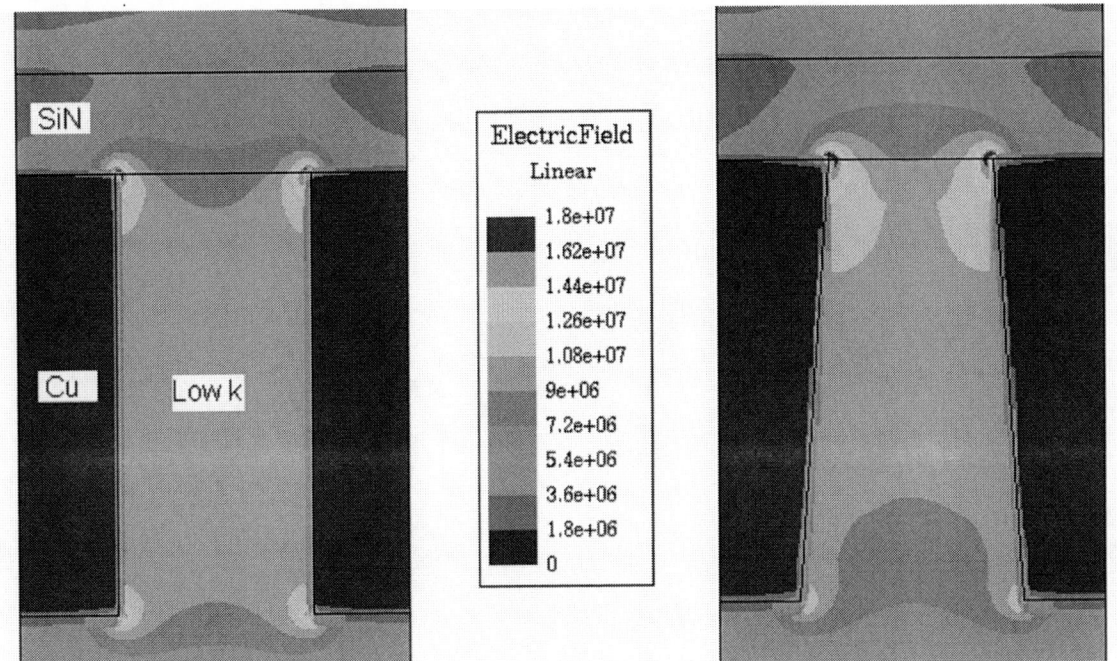

Figure 4. Electric field magnitude distributions under 1 volt bias between two metal lines for two cross-sections with different shapes. (a) Rectangular. (b) Trapezoidal. The thickness of the cap layer is 50nm. The legends are in V/m.

The normal stress Sxx field distributions after annealing for the two cross-sections are shown in Figure 5. The change in cross-section shape and the reduction in spacing at the top surfaces lead to larger stress concentrations around the top interfaces inside both metal lines and cap layers. These elevated stress fields and stress gradients can nucleate voids and grow pre-existing imperfections, and lead to local mechanical damages [6].

SUMMARY

Three case studies are carried out to examine the impact of layout alteration and structural variation on capacitance, electric and thermal-mechanical stress fields. When dummy metal fills are inserted into dielectrics between conductor lines, both the electric field and the thermal-mechanical stress distributions are altered, so is the coupling capacitance between conductor lines. It is found that the layout configuration with the least capacitance coupling is inferior in

1-4244-0267-0/06/$25.00 ©2006 IEEE

mechanical reliability due to large stress concentrations around the merged fills. When the cap layer thickness is reduced, both the coupling capacitance and maximum electric field are decreased but the stress magnitudes and gradients in and around the cap layer are increased. For line cross-sections with side wall slopes, the reduction in top surface spacing leads to increase in electric and thermal-mechanical stress fields. The elevated stress magnitudes and gradients in local regions are responsible for mechanical reliability deteriorations in interconnects, especially with reduced overall dimensions and porous dielectrics. These results demonstrate that both layout and architecture design, and process technology development need to be optimized together with constraints for both better performance and superior reliability.

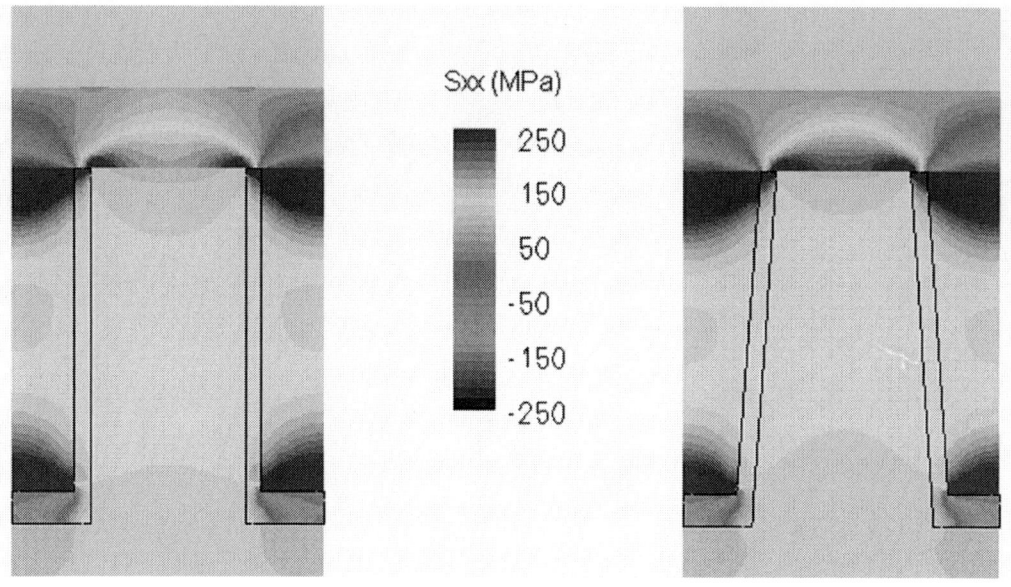

Figure 5. Normal stress Sxx field distributions between two metal lines for two cross-sections with different shapes. (a) Rectangular. (b) Trapezoidal. The highlighted polygons represent barrier layer boundaries.

REFERENCE

1. A. B. Kahng, G. Robins, A. Singh and A. Zelikovsky, Intl. Symp. on Quality Electronic Design, pp 691, (2006).
2. G. S. Haase, E. T. Ogawa and J. W. McPherson, J. Appl. Phys. **98**, pp 034503-1, (2005).
3. Synopsys Process-Aware TCAD tool suite manuals, (2006).
4. X. Xu, D. Pramanik and G. Rollins, MRS 2006 Spring Meeting Proceeding, to appear.
5. Y-L. Shen, MRS 2006 Spring Meeting Proceeding, to appear.
6. Z. Suo, "Reliability of interconnect structures" pp. 265-324 in Volume 8: Interfacial and Nanoscale Failure (W. Gerberich, W. Yang, Editors), Comprehensive Structural Integrity (I. Milne, R.O. Ritchie, B. Karihaloo, Editors-in-Chief), (2003).

Lithography Challenges toward Nano Scaled Device

HeeMok Lee*, JinSoo Kim, YoungKeun Cho, SangCheol Jeon,

KiNam Kim, KwangHee Kim, JaeSub Oh, and HeeChurl Lee

National NanoFab Center
53-3 Eoeun-dong, Yuseong-gu, Daejeon city, 305-806, Korea

ABSTRACT

Lithography has been eagerly explored into nanoscale beyond sub-micrometer in the fields of leading-edge technology applications. NNFC (National NanoFab Center) has specially concentrated on direct electron beam lithography and nanoimprint which are flexible and effective methods to be applicable to sub-100nm patterning. Nanoscaled FinFET and MRAM(Magnetic RAM) were evaluated, using hybrid e-beam lithography (double masking method, mix & matching method), i.e. optical exposure tool used to reduce the total patterning time of direct electron beam. The results from these patterning methods were able to fabricate the world's smallest transistor, 5nm FinFET and to adapt new material and device structure to magnetic device. Also 50nm (Line & Space) fabrication capability of UV imprint template (stamp in UV nanoimprint) is demonstrated in this paper.

Keywords: electron beam lithography, nanoimprint, FinFET, MRAM, template,

1. INTRODUCTION

In semiconductor, the development has been getting accelerated to have valuable functions such as high speed, low power operation and non-volatility as well as high density. Thus FRAM (ferroelectric RAM), MRAM (magnetic RAM), PRAM (phase change RAM), FinFET (fin field effect transistor), etc. are emerging as next generation devices following DRAM, SRAM and Flash memory device.

Referring to the International Roadmap of Semiconductors (ITRS) [1], ArF (193nm) lithography is likely to extend to even sub-45nm following 65nm half-pitch by the immersion technology. Also EUV, EPL and UV nanoimprint are estimated to be next generation

lithography technology which is responsible for sub-32nm technology.

As the reason above, National NanoFab Center (NNFC) has focused on e-beam lithography in the point of device development, especially hybrid e-beam lithography, which uses optical lithography to overcome the low throughput in e-beam process, i.e. double masking method or mix & matching method in between direct e-beam and optical lithography process. The benefit of hybrid e-beam lithography, combining high resolution and fine alignment of direct e-beam with high throughput of optical lithography, enables new devices to be evaluated and integrated effectively. Also NNFC has fabricated the qualified 50nm imprint template which is the key to succeed in UV nano-imprint technology. The results of hybrid e-beam lithography in 5nm FinFET and MRAM process and 50nm level template fabrication are shown and discussed.

2. EXPERIMENTAL RESULTS

2-1. Hybrid Electron Beam Lithography

• Sub 10nm Fin and Gate in FinFET device

Figure1 shows the world's smallest transistor using all-around gate (AAG) FinFET known as one of structures to provide scalability in size and flexibility in device design [2]. In the e-beam lithography of AAG FinFET fabrication, due to large pad area, hybrid e-beam lithography [3] i.e. "double masking with contact aligner" was used to improve the throughput as illustrated in Figure 2.
The nanoscale features such as fin and gate patterns were defined by 100kV direct e-beam system (ELS-7000, Elionix) with negative e-beam resist (HSQ: hydrogen silsequioxane), then large pad areas were formed by contact aligner with g-line photoresist (novolak based resin). During direct e-beam writing (1st masking), exposure dose at specific patterns were compensated for pattern shape and density. 2nd masking was performed

by contact aligner after 1st e-beam masking. Briefly, FinFET process is described in Figure 2. After hybrid e-beam lithography of Fin layer, Silicon-fin etching, sacrificial oxide growing, HfO_2 atomic layer deposition for gate dielectrics and hybrid e-beam lithography of Gate layer were followed. Finally novel transistor was achieved with nanoscale dimension such as sub-5nm gate length (L_G), averaged 3nm fin width (W_{FIN}), and 14nm fin height (H_{Fin}). This was performed on (100) SOI wafers.

In 2nd masking by contact aligner, below 0.5um of resist thickness was used for vertical pad edge shape. Also overlay accuracy between Fin & Gate and pad patterns was controlled within ±1um. Figure 3 shows the alignment results of Gate e-beam lithography according to the alignment mark quality, especially edge-shape of alignment mark. The alignment mark formed by e-beam could get good results of overlay error within ±10nm compared to the results by contact aligner. The alignment key formation seems to be a major role in hybrid e-beam lithography.

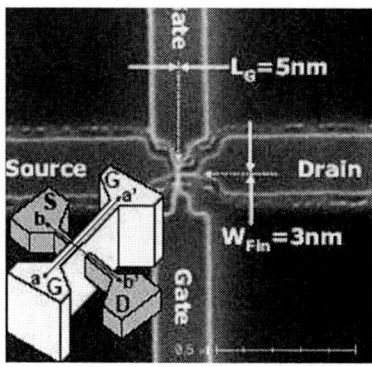

Figure 1: Schematic and SEM top-view of 5nm FinFET

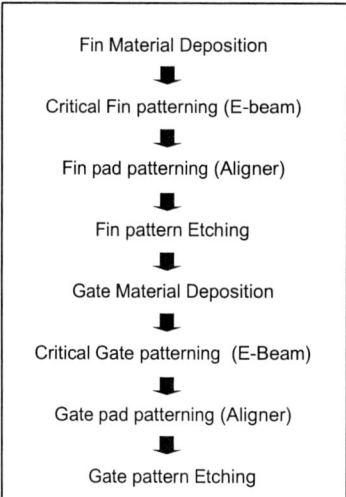

Figure 2: Process Flow of FinFET process

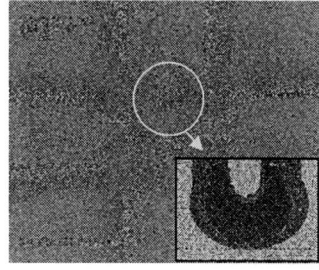
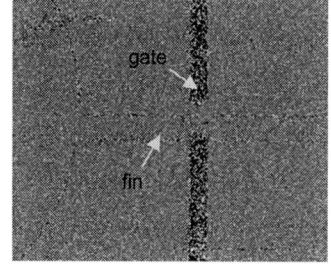

1-4244-0267-0/06/$25.00 ©2006 IEEE

(a)

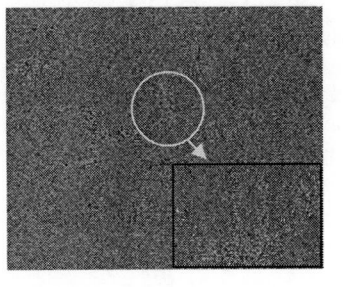 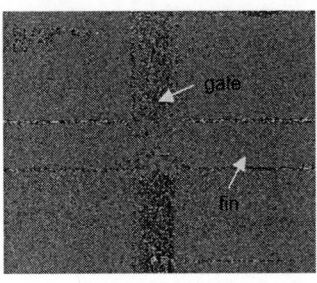

(b)

Figure 3: SEM top-views of alignment marks and overlay results between fin and gate patterns
(a) alignment mark formed by contact aligner, (b) alignment marks formed by e-beam

● **Mix & Matching for MRAM**

MRAM device has become one of potential candidate of high density memory device and lots of device structure and magnetic materials have been under investigated. Figure 4 shows one of MRAM structure which was executed in design verification with hybrid e-beam lithography "mix & matching method" at NNFC, i.e. bottom electrode was formed by optical exposure tool (KrF stepper) and magnetic cell was built by direct e-

beam system (JX00-9300FS, JEOL). The benefit of mix & matching method is to modify the structure design and to adapt magnetic material easily. For hybrid e-beam lithography (mix & matching method), the alignment marks were optimized with its shape, position and depth shown as Figure 5. Also in Figure 6, overlapped cell features i.e. cell by e-beam on bottom electrode by KrF stepper demonstrated that overlay tolerance of 50nm was well controlled in this experiment.

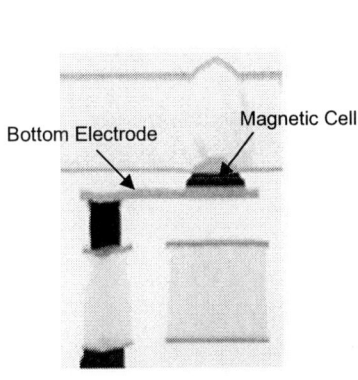

Figure 4: SEM cross-view of MRAM structure
at this time of evaluation

Figure 5: Schematic representation of alignment mark

Figure 6: SEM top-views of cell patterns on bottom electrode using optimized alignment mark

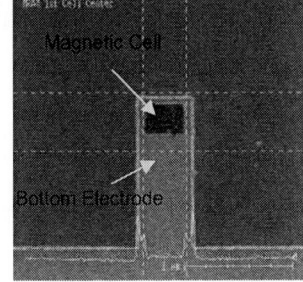

2-2. Template fabrication and Imprint Lithography

Recently, Next Generation Lithography (NGL) as like EUV, EPL and nanoimprint have been emerging as technologies responsible for nano patterning following optical lithography. Among NGL technologies, nano-imprint lithography (NIL) is expected to be potential and low cost of ownership (COO) technology in nano patterning [4]. In this point of view, NNFC has been developing the quartz template for UV imprint lithography. Figure 7 shows the quartz template fabrication method used at NNFC.

Figure 8 shows the results of 50nm line width at 100nm pitch after quartz etching. For this fabrication, 100kV direct e-beam writing and ICP etch system were used. So far NNFC has been supplying high-quality NIL templates to researchers in Korea. Figure 9 shows imprinted patterns with the in-house template. These results were not enough to satisfy sub-30nm technology yet, but we expect to go beyond sub-30nm in the near future by means of applying the e-beam proximity effect correction, improving mask materials and optimizing quartz etch process.

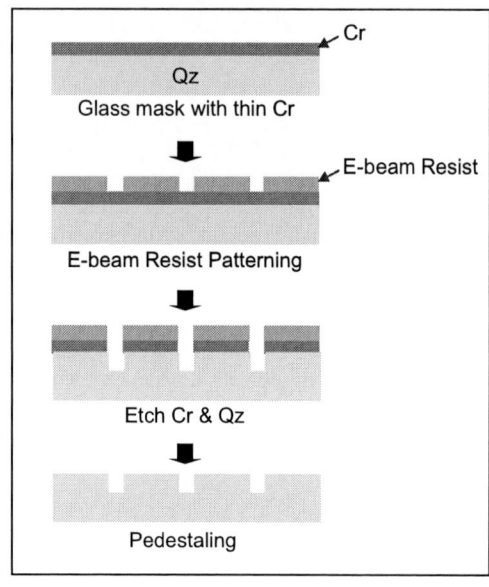

Figure 7: Process Flow of fabricating quartz template

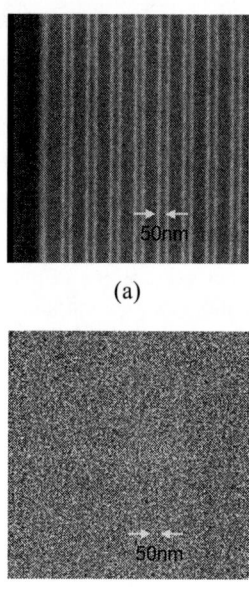

Figure 8: SEM top-views of (a) Before Cr etching and (b) After Cr / Qz etching

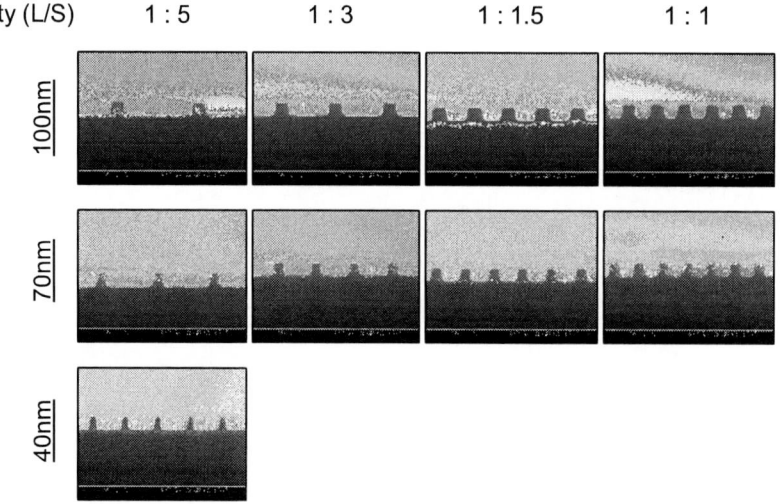

Figure 9: SEM top-views of imprinted patterns (40nm, 70nm, 100nm)
with various duty ratios (1:5, 1:3, 1:1.5, 1:1)

3. SUMMARY AND CONCLUSIONS

NNFC has established hybrid e-beam lithography which
could improve the throughput of e-beam process owing
to optical lithography. Especially NNFC evaluated new
devices such as FinFET and MRAM using hybrid e-beam
lithography, i.e. double masking and mix & matching.
During this hybrid e-beam lithography process, the
alignment mark optimization was one of major factor.
Thus NNFC accomplished world's smallest transistor
(5nm FinFET) and evaluated new structure with new
material for MRAM device. Also NNFC shows 50nm
(L/S) UV template fabrication capability which is
applicable to NIL (Nano Imprint Lithography).

REFERENCES

[1]   ITRS   2005   Edition   for   Lithography:
www.public.itrs.net/

[2]   S. D. Suk *et al., IEDM*, p.735, 2005.

[3]   S. E. Steen et al., Proc SPIE 5751, p.26, 2005

[4]   S. Murthy et al., Proc SPIE 5751, p.964, 2005

1-4244-0267-0/06/$25.00 ©2006 IEEE

Academic Development in Test Engineering

Tamara A. Papalias[1], R. Bryan Gonzalez[2], and Frank Gurtovoy[2]

Abstract - A 2-semester program of lectures and laboratory work in mixed-signal test development engineering has been initiated at San Jose State University. Class one involves bench-top assessment on a specifically-designed printed circuit board to explore basic issues of testing: accuracy, guard bands, repeatability, correlation, and printed circuit board design. Class two revisits many of these issues in off-line and on-line programs for manufacturing testers like Agilent's 93k. The trade-offs in accuracy versus test time are focal. This paper outlines the courses and presents an example of student ATE lab work.

Index Terms – DSP-based test, design for test (DfT), mixed-signal test, test development, ATE.

INTRODUCTION

A series of courses on test development engineering are being developed and implemented at the electrical engineering department of San Jose State University. While an electrical engineering education focuses on issues of design, EE-trained engineers fill an array of jobs in the IC industry. Specifically, as systems-on-chip (SOC) increase in popularity, their complexity drives the cost of testing to become dominant. With typical training programs of 18 months for new test engineers, the industry would benefit from candidates with exposure and aptitude for the specific array of skills needed in the development of IC testing. The training time can be reduced significantly.

LABORATORY

Every bench in the lab has been outfitted with the necessary equipment to model a per-pin tester environment. Students first manipulate the equipment manually and work towards computer automation. A list of equipment is provided in Table 1.

TABLE I
LIST OF EQUIPMENT PER LAB BENCH

Equipment	Description
Agilent 34401	6.5 Digit Digital Multimeter
Agilent 52624A	4-channel 100MHz Digital Oscilloscope
Agilent E3620A	Triple Output Power Supply (quantity = 2)
Agilent 33120A	Function Generator
Weller WTCTP	Soldering Iron
Fluke 187	Handheld Multimeter
-	Grounding Box
-	Log Book of Bench Use

The lab also includes workstations with SmarTest, the off-line tester simulation software for Agilent's 93K testers. Networking is planned to allow the students access to a local tester at specified times.

CLASS ONE

The first class has been designed to introduce students to the techniques and issues of test engineering through a bench-top, hands-on environment. A group of practicing test engineers from Agilent Technologies and National Semiconductor joined San Jose State faculty to generate lecture material [1], laboratory exercises and a test board for lab work (Figure 1).

A bare test board and a pile of components are provided to each student. The soldering of through-hole and surface-mount components creates first-hand experience with the parasitics of non-ideal connections. This exercise naturally leads to the first laboratory on measurements. A set of resistors is given to each lab group—half are carbon and the other half are metal film. Students measure the resistance of every resistor to analyze the distribution. Students then investigate the inaccuracies of their measurements and are introduced to the 4-wire measurement method. Resistance measurements are collected versus temperature, guardbanding is discussed, and histograms are generated.

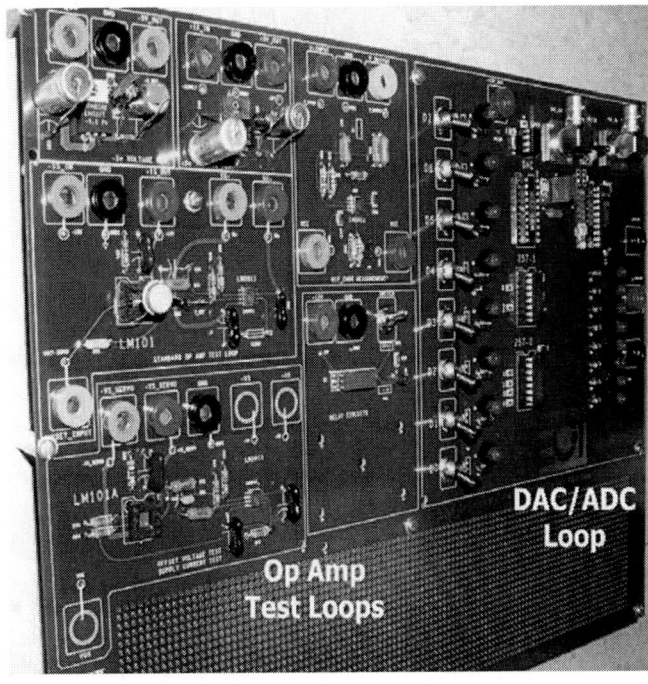

Figure 1. Class Test Board Supplied to Each Student

1-4244-0267-0/06/$25.00 ©2006 IEEE

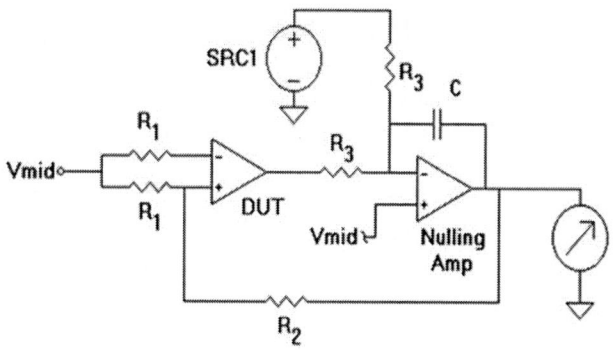

Figure 2. Op Amp Test Loop Circuit

The second lab presents a standard op amp test loop as shown in Figure 2. Students take DC measurements such as gain, offset voltage, and common-mode rejection ratio after tackling the bias and stability of the feedback loop. This is the first introduction of testing circuitry, using additional components to make it easier to take measurements. Students are reminded of accuracy issues, especially in regard to the settings of the test equipment and additional components.

The third lab challenges the students with 8-bit digital-to-analog converters (DACs) and analog-to-digital converters (ADCs). Students determine the size of the least significant bit as well as full-scale. The quality of conversion is found in differential nonlinearity (DNL) and integral nonlinearity (INL) measurement standards. Since DACs and ADCs are the core of digitizers and arbitrary waveform generators within testers, examining this basic block prepares the student for the tester environment.

The fourth lab integrates the previous labs into one. First students measure basic AC characteristics of the op amp under test. Next, they measure the gain and phase effects on a signal traveling through the DAC in series with the ADC. Finally, they repeat the measurement of the AC characteristics of the op amp while it is in a loop with the DAC and ADC, as in Figure 3. This lab most closely represents the manufacturing test environment of AWG (arbitrary waveform generator) and DIG (digitizer) used in test programs generated to evaluate production ICs.

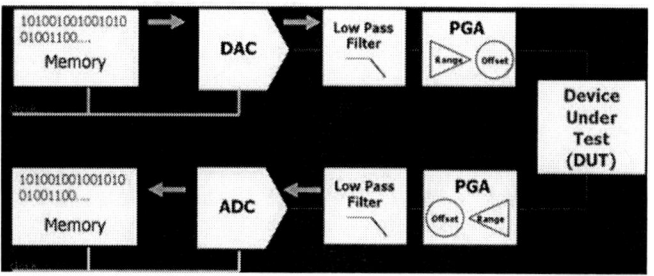

Figure 3. DAC-DUT-ADC Loop to Represent Simple Tester

The final project for the course requires the students to choose a current production part, design and fabricate an evaluation board, characterize the performance of their board, and present the results to a panel of test engineering professionals. These professionals also reviewed the students' lab notebooks.

CLASS TWO

Students navigate off-line tester simulators to characterize voltage regulators, clock chips, op amps, DACs and ADCs. On-line tester work will be accomplished at local companies. Universal tester issues like timing, levels and vectors are presented in generic form along with focused calibrations, correlation, and shmoo plots. Students are encouraged to use the utilities provided with the program and then to write their own test methods.

Students will have exposure to a number of tester platforms. Since many of the students were able to acquire summer internships related to test engineering, their course work will be continued on the platform which they have become accustomed to. Teamwork on laboratory assignments ensures exposure to multiple platforms, such as Agilent, Credence, Eagle, and Teradyne.

In an effort to emphasize the value of minimizing test time, students share their discoveries with each of the exercises. The following is an excerpt from an assignment to test a 555 timer chip.

TEST TIME OPTIMIZATION EXAMPLE

This work describes several ways to reduce the test time of a free-running oscillator using only digital resources [2] on an Agilent 93000 SOC tester. There are a total of five different methods presented for optimization. These methods provide over 10% improvement in test time.

We choose not to use any mixed-signal (analog) components to lower the cost of test. The use of mixed-signal components significantly increases the cost while only marginally affecting the test time. It is to be noted that not all methods are specific to the oscillator or Agilent's 93K tester; they could be used for many devices and/or testers.

Instead of specialized (and more expensive) mixed-signal components, five test-time-reduction techniques using digital hardware are discussed: turning off the datalog to report window, error map range, loop and test flow conditions, variables and print statements, and test methods.

A. Turning off the Datalog to Report Window

The data-log to report window is a 93K function that prints test values to the screen as a log. This should not be confused with normal data-logs which print to a file. The data-log to report window is used to debug the test-flow or observe all pass or fail values. It is not a function that needs to be on, especially when running high volume production. Yet, by default, it is left on. The problem with the data to report window is that it takes away computing power from the tester; and in this case, the 93K shares the resources with the tester.

B. Error Map (EMAP)

When testing a free running oscillator, one considerable

1-4244-0267-0/06/$25.00 ©2006 IEEE

problem is syncing the tester to the oscillator. The 93K is dependent on time to sync the two together; it is time-based and not event-based. One way around this is to use the Error Map to calculate frequency and jitter.

By selecting a vector or pattern of all highs or lows, the Error Map will compare the device output signal with the pattern. Each time the vector does not match with the signal it will mark an error on its map and track the time it occurred. This can be seen in Figure 4.

The frequency is then calculated over an EMAP range. In its simplest form the error map acts as an averaging mechanism to calculate frequency and jitter. The resolution to calculate jitter is dependent on the frequency at which the vectors run.

To reduce the test time, we changed the set EMAP range between 100 and 9192 to 100 to 3192. The frequency and jitter maintained the same value, yet, time of test was reduced by 75μs.

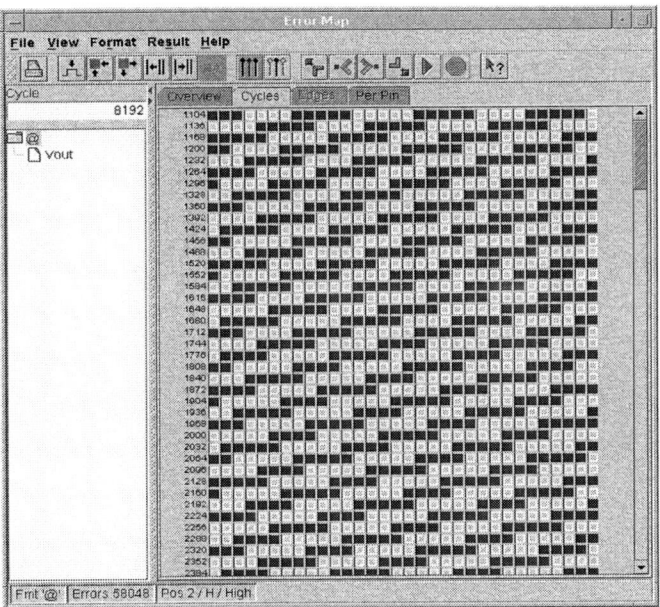

Figure 4. Error Map Used to Calculate Frequency

C. Exiting the Test-Flow

A crucial part of writing a test flow is knowing when to bin a device. By properly setting the loop conditions and order of tests in the test-flow, one can reduce the amount of redundancies and lower the test time.

We implement this scheme throughout the test-flow. One notable case occurs within the output voltage level test. Since we were using I/O digital pins to test the oscillator, we had to improvise a measuring technique. The 93K cannot see the output of the DUT unless the digital capture tool is in sync with the device to see an output. This is due to the architecture of the ATE. The 93K does not trigger on an edge like an oscilloscope. Rather, it has a time-based trigger. If not synced properly, the digital capture would give junk data as seen in

Fig. 5.

Therefore, we had to devise a method to indirectly measure the voltage high and voltage low. To do this, we varied the level specifications while utilizing the frequency test. We incremented the level set from 0.0V to 5.0V. Specifically, when the level is set to 5.0V, anything below 5.0V is a low and anything above 5.0V is a high. This eventually affects the frequency, causing it to vary from the nominal measurement. The first sign of a change higher or lower in the frequency marks Voh or Vol, respectively. This breaks the test loop more quickly, thus saving tester time.

Figure 5. Digital Capture of an Oscillator Without Proper Triggering

There are two trade-offs with this methodology and set-up. The first is that we are not taking any direct measurements and the second is that it yields measurements with an accuracy of approximately 20%. This was not a problem in our simple verification tests.

D. Declaring Variables

Reducing the number of variables and print statements may seem like an obvious way to reduce test time. However, the actual impact on test time is significant and warrants discussion.

We created an experiment in which we declare 30 different variables and print statements for different operations involving math and measurement functions to observe the effect on test time.

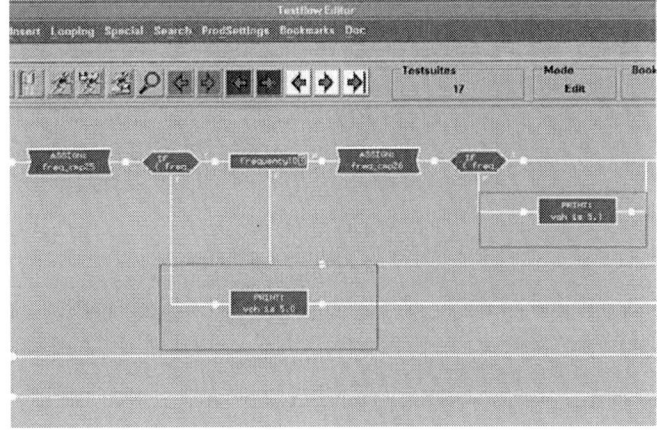

Figure 6. Embedded Print Statements

Our findings have showed that declaring 30 different variables to a test flow adds 3.14μs to test time. While it does not affect the tester measurement speed, it affects the amount of processor resources taken away from the test. This, of course, attributes to an increase in test time. For print statements the test time increased dramatically, an average of 780µs or every 30 print statements.

E. Test Methods vs. Yellow Windows

SmarTest [3] is the software designed to interface with the 93K tester. SmarTest provides a library of standard tests held in an area referred to as "Yellow Windows." They are made to work for any device under all conditions and therefore are not optimized for one particular application. The built-in API commands help the user reduce the amount of syntax and reduce the time needed to develop the test flow. On the other hand, the software also allows the use to write their own tests, or test methods. Test Methods are the single most powerful tool that the SmarTest software offers.

We compared the test time for a sample test using the provided syntax behind the Yellow Window alongside a test using a Test Method. A common test (operating current) is chosen to highlight the improvement. The results show that the Test Method reduces the test time by 200µs, or approximately 10%. The time could even be further reduced by compressing the code.

The improvement from using a Test Method will, of course, vary from test to test and author to author. Yet, ultimately, Test Methods are the truest reducers of test time. Trade-offs between test time and device coverage are often controlled by the author of the test program and the capabilities of the tester, the amount of test time reduction is only limited by the test engineer's time to write and optimize his or her test methods.

CONCLUSION

Two classes are presented that form a new test development engineering program. The first is focused on bench-top analysis and debugging skills, while the second involves programming testers for standard mixed-signal blocks. Students learn about optimization of test time while working on actual testers.

Other courses are planned to broaden the program. One class will explore the issues that arise in RF testing. Since many of the package and board parasitics degrade IC evaluation at RF frequencies, built-in self test (BIST) techniques are the major component. Another class will present graduate-level issues of Design-for-Test (DfT) systems. A project course is also planned to allow student groups to address current topics provided by local industry.

This unique program strives to bring the developing area of test engineering into the university. Economics of test, precision of instruments, and correlation issues of off-shore testing are now available in the university setting and at the undergraduate level. Former students are asked to return to the program as mentors for labs and student projects. Success will be measured through continued contact with the students who have participated in the program. Feedback will be solicited from both the students and their managers to gauge the effectiveness of the program.

ACKNOWLEDGMENT

The authors wish to that Bill DeWilkins for his optimism, tenacity, and involvement. He created opportunity where none had previously existed. The authors would also like to extend gratitude to James Freeman and Ty Papalias for their support.

REFERENCES

[1] Burns, Mark and Roberts, Gordon W., An Introduction to Mixed-Signal IC Test and Measurement, Oxford University Press, 2001.

[2] Mahoney, Matthew, DSP-Based Testing of Analog and Mixed-Signal Circuits, IEEE Computer Society Press, 1987.

[3] Agilent User I and II training manual.

AUTHORS

(1) Tamara Papalias is a professor of Electrical Engineering at San Jose State University in charge of the test development program.

(2) R. Bryan Gonzalez and Frank Gurtovoy are field applications engineers for Agilent Technologies. They are also former students of the test development program.

1-4244-0267-0/06/$25.00 ©2006 IEEE

Low Budget Undergraduate Microelectronics Laboratory

UNIVERSITY GOVERNMENT INDUSTRY MICROELECTRONICS SYMPOSIUM

June 2006

David J. Hunt, *Member, IEEE*

Abstract— Equipment costs for semiconductor fabrication can be millions of dollars just to get started, not to mention the specialized facility expenses required to house such intricate equipment. Specialized facilities such as clean-rooms, special exhaust systems, cryo-pumps, and gas cylinder storage are a huge investment and sometimes hard to justify for undergraduate education. Chemical usage and disposal can also be a considerable expense. Very few institutions can even afford an undergraduate microelectronics laboratory due to the high start-up costs, high equipment costs and continuous and constant maintenance of such equipment. Many colleges have acquired used equipment through donations from industry. However, donated equipment from industry is not always the most feasible option for small colleges with limited budgets and resources. The technical costs of keeping the equipment going can be expensive, not to mention the difficulty of obtaining parts on used industrial equipment. Through a recent New York State Science, Technology, and Academic Research (NYSTAR®) grant opportunity, obtained in collaborations with Alfred University and Rochester Institute of Technology, Alfred State College, a small technical college in rural western New York, has started its own low budget undergraduate microelectronics laboratory facility at a fraction of the cost of comparable industrial equipment. This new undergraduate microelectronics laboratory at Alfred State College has been equipped with Modu-Lab™ semiconductor device manufacturing equipment, which gives students realistic exposure to the semiconductor planer processes. Oxidation, diffusion, photolithography, etch, and vapor deposition stations allow the students the opportunity to design, fabricate, and test their own simple diffused resistors and PMOS devices while gaining experience in microelectronic fabrication techniques. The Microelectronics Laboratory at Alfred State College gives students a realistic experience in semiconductor manufacturing processes. An important concept of this laboratory is hands-on training. Although it is unlikely that the students will work with this type of equipment in industry, the understanding of general processes gained through laboratory experiences will prepare them to either continue their education in the microelectronics field or work in a modern industrial laboratory. Mask layers can be designed on just about any good quality CAD program. Individual layers can be printed out on transparent paper using a good quality laser printer. Diffused resistor and PMOS devices can be designed, fabricated and tested without spending millions of dollars. Functional resistors and aluminum gate

PMOSFET transistors have been successfully fabricated, even though the device sizes are 100-1000 times larger than typical devices fabricated in a clean-room facility. This grant has also led to collaboration with other institutions such as Alfred University and Rochester Institute of Technology who have provided technical expertise and faculty training. Student field trips have enhanced the students' overall experience. With Modu-Lab microelectronics laboratory equipment, IC design and fabrication at the undergraduate level is now a feasible proposition for small undergraduate technical colleges.

Index Terms— Low Budget, Modu-Lab®, PMOS, Fabrication NYSTAR®, Collaboration

I. INTRODUCTION

IN 2001, Alfred University proposed a unique, innovative initiative in the field of photonics that integrates educational, R&D, engineering, manufacturing, and testing resources in order to streamline and advance the photonics industry, specifically as an economic presence in New York State [1]. The project has networked engineers, technologists, and researchers from Alfred University, Alfred State College, and Rochester Institute of Technology and their colleagues in industry. The idea is to have students, faculty, and industrial partners integrated into teams that will work on industrially-relevant problems of photonics R&D, design, and manufacturing in a pilot plant facility. This new concept on collaboration was good enough to secure a grant through NYSTAR, a state agency that funds world-class university/industry partnerships and promotes outstanding high-technology research and commercialization efforts that contribute to New York State's economy [2]. The agency's internationally recognized programs spur the development, design, and manufacture of new technologies in a wide range of areas including nanotechnology, electronics, life sciences, information technology, materials processing, and many others. Even though the photonics industry has since encountered many challenges and even some degree of fall out, the primary goal of the project, which was to produce trained, skilled workers at all levels for immediate integration into the high-tech workforce, has remained intact.

At the time of the NYSTAR grant proposal, the Electrical Engineering Technology Department (EET) at Alfred State College, was teaching a sophomore level course, ELET 4154

David J. Hunt is an Associate Professor at Alfred State College, Alfred, NY 14802 USA (phone: 607-587-4618; fax: 607-587-4615; e-mail: huntdj@alfredstate.edu).

1-4244-0267-0/06/$25.00 ©2006 IEEE

Microelectronics, to undergraduate students in an Electrical Engineering Technology program at Alfred State College. Over recent years, this course has evolved from a thick-film hybrid technology course into a semiconductor technology course.

At Alfred State College, we wanted to give students a realistic experience to enhance an existing microelectronics course. However, the existing laboratory equipment was not suitable for semiconductor fabrication in any way. The NYSTAR grant initiative was an excellent opportunity for the EET department at Alfred State College to obtain modern semiconductor manufacturing equipment, so the department decided to invest about half of the NYSTAR grant money into a new semiconductor manufacturing laboratory facility and Modu-Lab equipment made by EMS (Electro-Mechanical Systems) in Albuquerque, NM [3].

II. SEMICONDUCTOR EDUCATION

Semiconductor Manufacturing is a planar process and is the foundation of the information age. The planar silicon transistor is the very heart and soul of modern electronics. Modern IC devices have a multitude of transistors, capacitors, and resistors. The complexity and scale of these devices is mind boggling. Billions of these devices in the submicron dimensions are routinely fabricated in today's modern IC devices. Fabrication of these devices is rather simple in theory -- deposit, pattern, etch, and repeat. However, the actual fabrication process is unbelievably detailed at every step. For very complex devices, there can be 500 or more individual process steps where the slightest mistake at any of these steps can render the entire device useless [3].

The semiconductor manufacturing process demands the highest level of training and diligence for its technically-trained workforce [4]. Fierce global competition in the manufacturing sector makes it necessary to have effective training programs that enable the flexibility necessary to stay in the game. In order for the United States to compete in the global marketplace, it is essential to bring a large number of high quality manufacturing technicians to the industry in a very short period of time.

The nature of the integrated circuit (IC) fabrication process is highly complex and absolutely intolerant of mistakes. The training is equally complex and is built on basic blocks requiring highly specialized equipment that, until Modu-Lab, cost a fortune to acquire, install, and maintain. Industrial microelectronics equipment can easily cost millions of dollars. The housing and facilities for the systems are equally maintenance-intensive and expensive. Used equipment donated from industry is not always a feasible option either, especially at small technical colleges with very limited budgets. A full-blown clean-room filled with intricate production tools is a superb research resource and can be effectively used for teaching. Unfortunately, this type of facility can be cost prohibitive. Just the installation of some production tools can cost more than an entire set of Modu-Lab modules -- delivered, installed, and running! Another problem with used industrial equipment is that it is typically designed for production not for education. Modern industrial

equipment is at the point where it is virtually hands-free; many of the process steps are automated. With the Modu-Lab system, the user does not have to spend millions of dollars to get into the business of educating tomorrow's high-tech workforce. Modu-Lab allows students the opportunity to design, build, and test their own simple solid state devices in a non clean-room environment. The simple, rugged design ensures long life and allows students with minimal experience to perform lab exercises without the risk of bankrupting the department. The lab experience strengthens the learning curve and, perhaps more importantly, enhances the students' interest level. The Modu-Lab system is designed around a lab size of 12 students that can each process 1 layer in about 1.5 hours. This kind of laboratory equipment is ideal for a small technical college like Alfred State. Modu-Lab processes have been custom engineered to meet the constraints of the teaching lab environment. With maximum safety, minimum waste generation, and the lowest cost of ownership, Modu-Lab was the only affordable option for our microelectronics laboratory at Alfred State College.

III. UNDERGRADUATE MICROELECTRONICS LABORATORY EQUIPMENT

With grant money from NYSTAR, Alfred State College invested nearly $130K into its microelectronics laboratory facility in the fall of 2004. As shown in Table 1, the system purchased in 2003 included a wet process module, lithography module, physical vapor deposition module, oxidation furnace, and a diffusion furnace. This is the bare-bones minimum amount of equipment needed for the PMOS process fabrication [5,6,7]. Not included in the pricing was a system installation package of $5,000 which covered freight, equipment setup, detailed operating instructions, and the basic chemicals and supplies. Due to budgetary constraints at the time, it was not possible to purchase the complete set of equipment. Future acquisitions will include the N-type diffusion furnace, the wet chemical etch bench, and the device characterization module.

Modu-Lab Equipment Costs		
Description	Purchased in 2003	Future Acquisitions
Wet Process Module	X	
Photolithography Module	X	
Physical Vapor Deposition Module	X	
Oxidation Furnace Module	X	
P-Type Diffusion Furnace Module	X	
N-Type Diffusion Furnace Module		X
Wet Chemical Etch Module		X
Device Characterization Module		X

Table 1 Modu-Lab Equipment

Figure 1 - Wet Chemical Process Bench

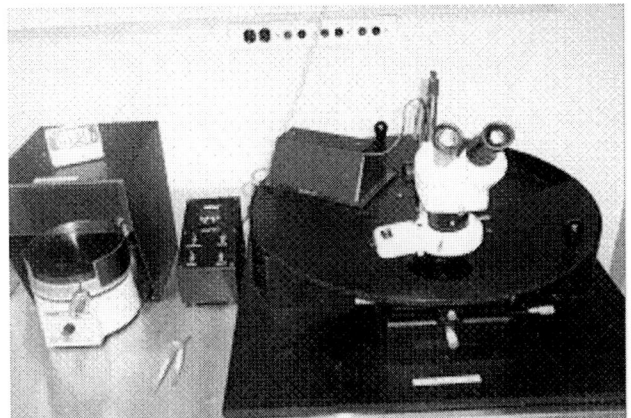

Figure 2 - Soft Bake and Lithography Station

Figure 3 - Oxidation/Diffusion Furnace

Figure 4 - Bubbler System Used for Wet Oxidation

The wet chemical process bench, as illustrated in Figure 1, was designed for single wafer processing, which minimizes hazardous waste generation and exhaust requirements. The wet chemical bench can by used for all the process steps involving chemicals; this includes lithography developing, oxide etching, and metal etching. With the addition of a hot plate, it can also be used for the RCA clean and piranha clean steps [4]. All that is required in terms of facility are 110 VAC power and ventilation through a 6 in. diameter exhaust duct. The photolithography module, illustrated in Figure 2, comes with a wafer spinner, a hotplate designed to soft-bake the wafer resist, and the align and expose tower with capabilities that will support 100 μm minimum feature size. The Modu-Lab align and expose tower is designed to accommodate masks made from acetate transparencies. This feature allows the use of a laser or inkjet printer and readily available presentation transparencies for mask making. The spinner and bake plate are exhausted to minimize exposure to resist vapors. The spinner also features a waste capture bottle in the exhausted housing. All electrical is interlocked 110 VAC.

The oxidation and diffusion furnaces, as illustrated in Figure 3, are very similar. The base unit is a 3 zone, PID-controlled furnace. The diffusion unit will process wafers at temperatures up to 1200°C while the oxidation furnace will only operate to a maximum temperature of 1100 °C. Both furnaces have a programmable controller allowing overnight runs of up to 25 wafers. The furnace features a stainless steel scavenger, quartz-ware, and gas delivery system. Since the lab does not have nitrogen or oxygen gas systems as of yet, dried

compressed air is used instead. A water bubbler system, as illustrated in Figure 4, can be added for wet oxidations. Use of a variac allows precise heating of the de-ionized water to maximize steam production. The diffusion module requires 208 VAC 3ph @ 40 A or 220 VAC 1ph @ 60A as well as exhaust capabilities.

The physical vapor deposition module, as illustrated in Figure 5, is used to evaporate aluminum on the wafers. A mechanical roughing pump is used to pump the chamber down to approximately 10^{-1} Torr. A turbo pump is used for high vacuum pumping which can achieve 10^{-5} Torr in less than 30 minutes. The large view-port and mirror assembly allow users to observe the ion gauge, the turbo, and (of course) the source. The electro-pneumatic system has interlocked controls that prevent equipment damage should the wrong switch be actuated. Currently, the system is used to evaporate aluminum. However, aluminum, copper, and aluminum-copper alloy have both been evaporated successfully. As illustrated in Figure 6, up to 4 wafers can be processed at a time.

The laboratory equipment was first used in the spring 2005 semester. In the first year of use, the class was successful in fabricating diffused resistors on a silicon substrate. Since p-

1-4244-0267-0/06/$25.00 ©2006 IEEE

type wafers were included in the start-up package, the class used phosphorsilicafilm spin-on dopant source to do n-type

Figure 5 - Physical Vapor Deposition Module

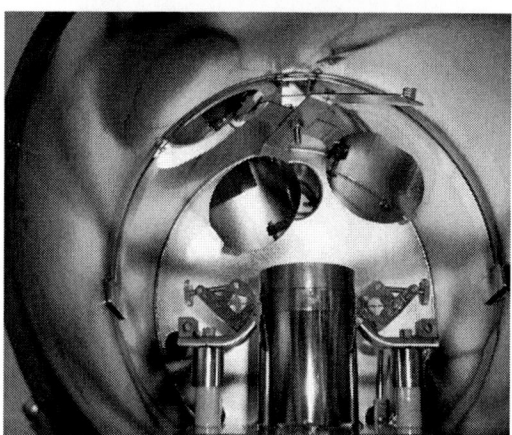

Figure 6 - Inside the Vacuum Chamber of the Physical Vapor Deposition Module

diffusions. In its second year of use, the new microelectronics facility and has switched to n-type starting wafers and Borofilm 100 spin-on dopant to do p-type diffusions. This works perfect for the PMOS fabrication process where only one p-type diffusion is necessary [5,6,7].

IV. MASK DESIGN

Any good quality CAD program could be used to make the masks. However, Microsoft Visio® was chosen because its availability on the computers and familiarity with the program.. There are 4 masks required for the PMOS process [5, 6,7]:

PMOS Design Masks
Layer 1: Diffusion Mask (Green)
Layer 2: Gate Oxide Mask (Red)
Layer 3: Contact Cut Mask (Gray)
Layer 4: Metal Mask (Blue)

Since positive photoresist is used, the diffusion, gate oxide, and contact cut masks need to first be inverted in order to get the proper image etched in the wafer.

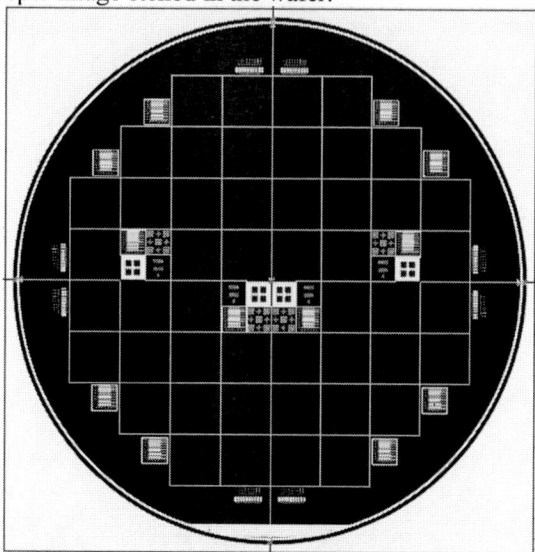

Figure 7 - Wafer Layout Mask for 100mm wafer

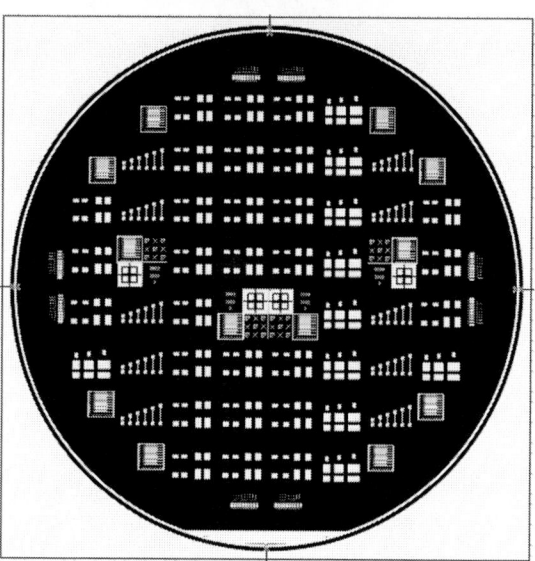

Figure 8 - P-Type Diffusion Mask for 100mm wafer

The metal mask does not have to be inverted. To get started, the mask was set up for a 100mm wafer. The wafer was split up into a working layout that will accept die sizes of 10mm x 10mm. As illustrated in Figure 7, up to 4 dies are reserved as alignment structures, which still leaves room for 48 device dies on each wafer. Device dies are placed around the wafer. Figure 8 illustrates the mask layout with the individual device dies in place for the diffusion mask. Devices are individually designed as illustrated in Figure 9. Individual layers are printed out on clear acetate transparencies which are readily available and cut down to approximately 150mm x 150mm to fit in the mask holder on the lithography module.

It is absolutely critical that Layer 2 align properly with Layer 1. Unfortunately, it was very difficult to see through the transparency paper. It was nearly impossible to see any of the diffusion layer topology when looking through the masks. To overcome this problem, the 4 alignment squares were cut out.

1-4244-0267-0/06/$25.00 ©2006 IEEE

After doing this, it was very easy to through subsequent layers.

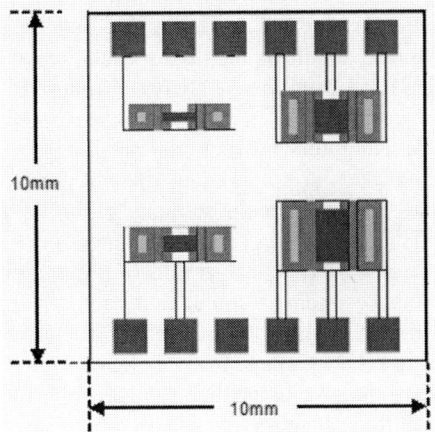

Figure 9 - PMOSFET Design Layout

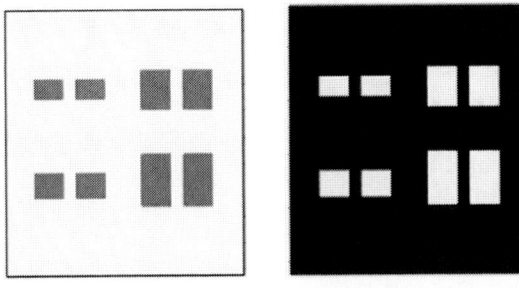

Figure 10 - Layer 1: Diffusion Mask

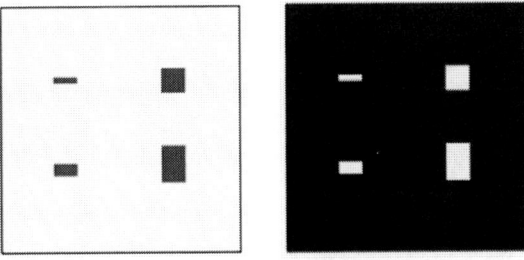

Figure 11 - Layer 2: Gate Oxide Mask

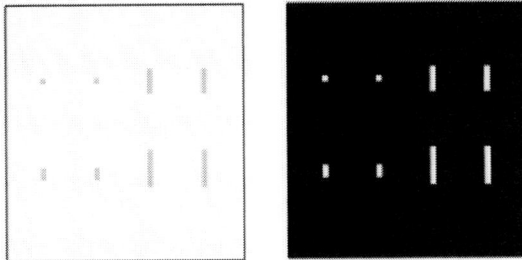

Figure 12 - Layer 3: Contact Cut Mask

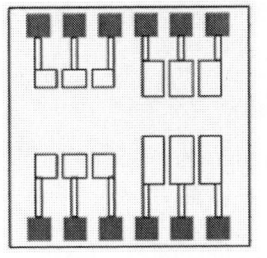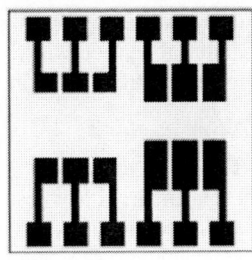

Figure 13 - Layer 4: Metal Mask

V. PMOS DESIGN PROJECT

In the spring 2006 semester, there were a total of 24 students enrolled in ELET 4154 Microelectronics course at Alfred State College. This is a 4 credit hour course with 3 hour long lectures per week and one 3 hour laboratory per week. There were two lab sections with 12 students per section. Most students that take this class are sophomores in the Electrical Engineering Technology program. However, there were some Electromechanical and Computer Engineering Technology majors taking the class as an elective.

For the laboratory, about half the semester was devoted to a PMOS design project. Students in each lab section were initially split into 4 teams. Members of each team were responsible for designing two individual dies, which included diffused resistors and PMOS transistors with varying layout dimensions. Figure 9 illustrates the die layout for 4 PMOSFET transistors with all 4 layers overlaid together. Each team was responsible for compiling all individual designs to complete the 4-layer mask set.

The actual PMOS fabrication process spanned over 5 weeks. Figures 10-13 illustrates the 4 layer mask set for PMOSFET design. Both the positive and negative images are shown. Ultimately, teams were responsible for processing one device wafer, but they worked together on coordinating batch processes such as oxidation, diffusion, and physical vapor deposition runs. There were 3 control wafers for each lab section, and each team was responsible for recording their own data on the control wafers and testing their own device wafer.

VI. OUTREACH PROGRAM

There have been several opportunities for collaboration between Alfred State College, Alfred University, and Rochester Institute of Technology.

A. Field Trips to Alfred University

Students visited an Alfred University lab to use their ellipsometer to measure oxide thickness. The ellipsometer used at Alfred University is illustrated in Figure 14.

B. Instructor Training

In June 2005, Prof. David Hunt from Alfred State College took an IC Processing Short Course at Rochester Institute of Technology. This course provided a hands-on educational experience in the design of ICs using the PMOS fabrication process. The material learned was then applied to the microelectronics laboratory at Alfred State College.

C. Technical Advice

The faculty and staff from the Microelectronics Department at Rochester Institute of Technology have provided technical advice on countless occasions. They have answered questions on the PMOS process steps and they have etched the quartz tubes from the diffusion and oxidation furnaces when the dopant source was switched from n-type to p-type.

Figure 14 - Ellipsometer at Alfred University

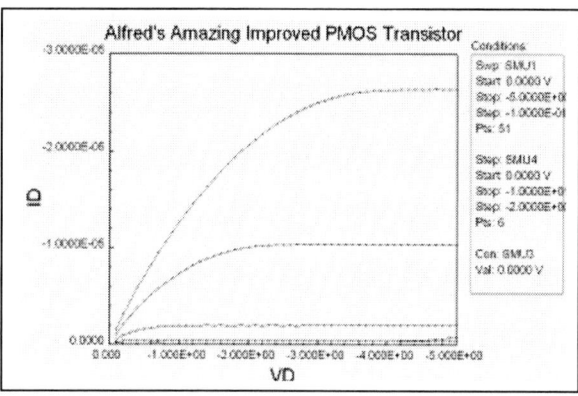

Figure 15 - First Functional PMOSFET I-V Characteristics Verified at RIT

D. Field Trips to Rochester Institute of Technology

All students taking the ELET 4154 Microelectronics class at Alfred State College had the opportunity to visit the Microelectronics Department at Rochester Institute of Technology. The field trip included a complete facility tour, actual testing of devices fabricated at Alfred State College, and hands-on use of the ion implanter and asher. Alfred State College students were able to suit up and go into a fully functioning clean-room. The first functional PMOSFET transistor fabricated at Alfred State College was actually verified during the field trip to Rochester Institute of Technology on April 13th, 2006. Figures 15 illustrates the I-V characteristics of the first functional transistor fabricated at Alfred State College.

VII. CONCLUSION

The Modu-Lab laboratory equipment was designed from the ground up to be safe and economical to own, operate, and maintain. Modu-Lab allows faculty to teach the basics of the planar process safely and inexpensively. Modu-Lab was designed to meet the needs of a growing semiconductor manufacturing laboratory instructional program.

The goal has been to encourage a sense of practicality and improved performance once the students' enter the technology workforce. This has reinforced the weakened link between basic research and the practical aspects of production issues in the United States economy through student involvement. Collaboration with industry and academia has made this possible. Alfred State College may not be building the next Pentium microprocessor in its undergraduate microelectronics laboratory, but it is educating the technically-trained workforce that will be building them.

ACKNOWLEDGEMENT

Funding for the Microelectronics Laboratory came from NYSTAR Capital Facilities Grant C000070 Advanced Research Center for Photonics Education and Training. The author would like to thank all of the students who have taken ELET 4154 Microelectronics course over the past two years. The author would like to especially thank Dr. Lynn Fuller from Rochester Institute of Technology for all the help that he has provided and encouraging the author to write this paper. The author would like to thank Dr. Alexis Clare from Alfred University. The author would also like to thank the faculty and staff in the Microelectronics Department at Rochester Institute of Technology, especially Scott Blondell, Dr. Karl Hirchman, Dr. Robert Pearson, and Dr. Santosh Kurinec. The author would also like to thank Prof. Calista McBride and Mary Hoffman at Alfred State College for helping review this paper.

REFERENCES

[1] Clare, A.G., "Photonics training initiative and pilot plant: A proposal to Corning Incorporated." Alfred, NY: Alfred University, 2001.

[2] *NYSTAR*, http://www.nystar.state.ny.us.

[3] Winder, R., *Electro-Mechanical Services*, http://www.emsi-usa.com.

[4] Wolf, S., *Microchip Manufacturing*. Sunset Beach, CA: Lattice Press, 2004.

[5] Fuller, L.F., Hoomkwap, K., Shakya, S., and Yenrudee, S., "Testing of metal gate PMOS digital integrated circuits," *Proceedings of the IEEE UGIM Conference*, Boise State University, June 2003.

[6] Pearson, R.E., Fuller, L.F., Turkman, I.R., and Ramanan, S., "PMOS metal gate process, ideal for undergraduate integrated circuit fabrication projects," *Proceedings of the 7th IEEE/ISHM/University/Industry/Government Microelectronics Symposium*, June 1987.

[7] Lemmond, T.C., Fuller, L.F., and Pearson, R.E., "An introductory microelectronics laboratory project based upon a NOR logic PMOS gate array," *Proceedings of the 7th IEEE/ISHM/University/Industry/Government Microelectronics Symposium*, June 1987.

1-4244-0267-0/06/$25.00 ©2006 IEEE

 **David J. Hunt** (M'97) was born and raised in rural western New York. He attended Alfred State College from 1989-93, receiving his B.S. in electrical engineering technology in 1993. He did his graduate work at Alfred University from 1996-98, receiving his M.S. in electrical engineering in 1998. His graduate thesis was "RF Plasma Deposition of Magnesium-Aluminum Oxide." Prof. Hunt has worked at Alfred State College in the Electrical Engineering Technology Department since 1998 and has served as Department Chair the past 3 years.

1-4244-0267-0/06/$25.00 ©2006 IEEE

Survey of University Micro/Nanotechnology Cleanroom Facilities as the First Phase in the Development of a U of L Business Model

Kevin Walsh[1], Mark Crain[1], Robert Keynton[2], Lisa Itamura[3], Scott Smith[3], and Bruce Kemelgor[3]

University of Louisville
[1]Electrical and Computer Engineering Department
[2]Bioengineering Department
[3]College of Business
Louisville, Kentucky 40292 USA

Contact: Dr. Kevin Walsh
walsh@louisville.edu
(502) 852-0826

INTRODUCTION

Approximately ten years ago, the University of Louisville made a serious research commitment in the emerging fields of micro/nano/biotechnology by allocating funds towards the construction of an effective, but relatively-small, state-of-the-art, class 100 cleanroom. In 1997 that facility (the Lutz MicroTechnology Cleanroom) was unveiled, placing UofL in the select group of US academic institutions with such facilities and the only university in the state of Kentucky with a cleanroom for general micro/nano-fabrication research and educational training. With the addition of new faculty in recent years and with the unprecedented success of those faculty, the university effectively out-grew the original 1,500 square foot cleanroom facility. To address this critical need, the University committed state funds to the construction of a new state-of-the-art $50M 120,000 sq. ft. Belknap Campus Research Building (the BRB as shown in Fig. 1) which houses a greatly-expanded, true multi-user cleanroom "core facility" to support our escalating research/educational programs in micro/nano/biotechnology.

The April 2006 grand opening of the BRB launched a new era of micro/nano/bio-technology research at the University of Louisville. The 120,000 square foot research building is a unique design in that it intentionally combines complementary and coordinated interdisciplinary micro/nano/bio research efforts from both the School of Engineering (ECE, ME, BE and ChE) and the College of Arts and Science (Chemistry, Physics and Biology). The showcase facility in the BRB is the centrally-located 8,000 sq. ft. core cleanroom positioned on the ground floor (see Fig. 2). Designed by nationally-renown AGI of Phoenix AZ, the large class 100/1000 facility should place UofL among the top universities in the nation in terms of cleanroom size and capacity. The $8.5M micro-manufacturing facility houses a cadre of state-of-the-art processing equipment for prototyping next generation micro and nano-devices for applications such as microelectronics, homeland security, optoelectronics, biotechnology, sensing,

MEMS and nanotechnology. Open to both academic and outside industrial users, the BRB cleanroom promises to enhance the University's visibility and research productivity in the micro/nano fields for many years to come.

RATIONALE FOR "PHASE 1"

As a first phase in the development of a business model for the operation of our new cleanroom facility, colleagues from the UofL Business School teamed with colleagues from the Engineering School to design and conduct a survey among 20-some universities with comparable cleanroom facilities ranging in size from 2,020 sq. ft. to 25,000 sq. ft. (average = 9,026 sq. ft.). Operational metrics included: size, certification (class 10, 100, 1000), personnel (full-time, part-time), number of users (internal faculty, internal students, external industry, external industry, departments), number of projects, areas of concentration, user fee structure, and teaching. Budgetary metrics included: revenues (internal users, external users, university subsidy, state subsidy, other), expenses (salaries/benefits, supplies/expenses, maintenance/repairs/replacement, and other), and value of equipment. The findings were then normalized for our new 8,000 sq. ft. cleanroom to provide a foundation for the development of a specific business plan for our facility, which will be the focus of "phase two" of our business model development plan.

AREAS OF ANALYSIS

In order to capture data that would be relevant to a business model, a short list of questions were generated that would produce a general understanding of each center's operations (see Appendix A). The 13 questions (or metrics) were categorized into two basic areas of analysis: budgetary and operational. Budgetary questions focused on revenues, mix of users, subsidies, expenses, and value of equipment. Operation questions targeted cleanroom size, cleanliness, staffing, number of users, number of projects, areas of concentration, user fee structure, and use for formal academic classes. In summary, the main objectives of this survey were to determine how these other facilities operated, generated revenue and balanced those revenues with expenses.

METHODOLOGIES

A list of approximately 30 universities with cleanroom facilities was identified and subsequently contacted by email and/or telephone for participation in the survey. Not all facilities were willing or able to share their budgetary information, and some were either too busy or unwilling to provide us with any data at all. In the end, the team gathered information from 20 of the 25 centers contacted and in some cases, only the totals were available (i.e., total revenues, total expenses, total users). Only 2 of the 20 facilities produced public annual reports from which our data/statistics could be extracted. The remaining facilities needed to be contacted individually by email and/or telephone, a very time-consuming and laborious task. We agreed to share the data/findings directly with all those that participated in the cleanroom survey, and also agreed that the participants would remain anonymous in any publicly-disseminated documents/papers resulting from the survey.

RESULTS

From these data, a set of ranges and averages for each data point was developed to build a descriptive understanding of the metrics. In addition, the Spearman Rank Correlation Coefficient formula was used to see if relationships existed, or if any pairs of variables were strongly correlated among the data. The results of this analysis guided our recommendations. Appendix B provides a listing of the data obtained from the 20 participants, including range values, number of data points and averages. The average for total revenues was $2.18M (n=9 data points), with the minimum being $600K and the maximum being $6M. The average for annual expenses was $2.08M (n=8), with the minimum and maximum the same as for revenues. Another question asked was the approximate value of the equipment in the cleanrooms, a hard number to estimate and highly subjective. The average value was $42.8M (n=17), with the range being $10M to $100M.

The other focus of this study was operational planning. This includes issues such as number of users (and mix between internal and external), size of facility (broken down by class), and number of personnel. How these issues interact would be useful to our analysis. The average total square footage of the cleanrooms in our survey was 9,026 (n=20), with the smallest facility being 2,020 and the largest being 25,000. Nine of the cleanrooms reported having class 10 space, 16 had class 100 space, 15 had class 1,000 space, and one cleanroom reported having class 10,000 space. The number of users, both totals and subcategories, varied greatly from cleanroom to cleanroom. The total average number of users was 180 (n=16 data points), with the fewest users being 27 and the most being 600. The largest

group of users was internal students, followed by faculty users, external industry users, and external academic users. On average, there were 11 departments using each cleanroom facility (n=18), with the fewest being 5 and the most being 29. In terms of full-time personnel, the average was 8 (n=18), while the range was 1 to 27. Part-time workers were fewer in number overall, with an average of 2 and a range of 0 to 5 workers.

Other issues that were explored, but not included in our correlation analysis because the answers were too varied, included fee structure, areas of concentration and use of the cleanroom specifically for teaching. User fee structures varied among the cleanrooms, with some charging annual fees, monthly fees, supply fees, and/or equipment fees (by the hour, minute or per use). In most cases, the fees were higher for industry users than for internal users. Since each facility had their own mix and cost structure for fees, it would be useful to view each in the "User Fee Structure" in Appendix B.

CONCLUSIONS AND RECOMMENDATIONS

For budgetary metrics, we determined what relationships existed, if any, among the variables. There were a number of strong, positively correlated pairings ($0.8 \leq R^2 \leq 1.0$), including relationships between total revenues and total expenses, total revenues and total cleanroom space, total revenue and total number of users and total revenues and full-time personnel. We next developed scale factors for these strong, positively-correlated pairings. Based on the scale factors and the size of our new 8,000 sq. ft. multi-user, core facility, the study suggests that $1.67 million be the target revenue for our facility with predicted expenses of $1.66 million and a total of 231 users. In addition, since there is a relationship between facility size and full-time personnel, the recommended number of full-time personnel for an 8,000 square foot facility is 10. There was no relationship between size and part-time personnel, so no recommendation is made for this metric.

These targets are based upon the historical operations of other academic cleanrooms and will not necessarily be the same for our new facility. However, they do provide a useful guide for what, on average, an 8,000 square feet cleanroom statistically generates in terms of revenues and expenses, and how many users and full-time personnel it requires.

Fig. 1. Univeristy of Louisville's new $50M 120,000 sq. ft. Belknap Research Building.

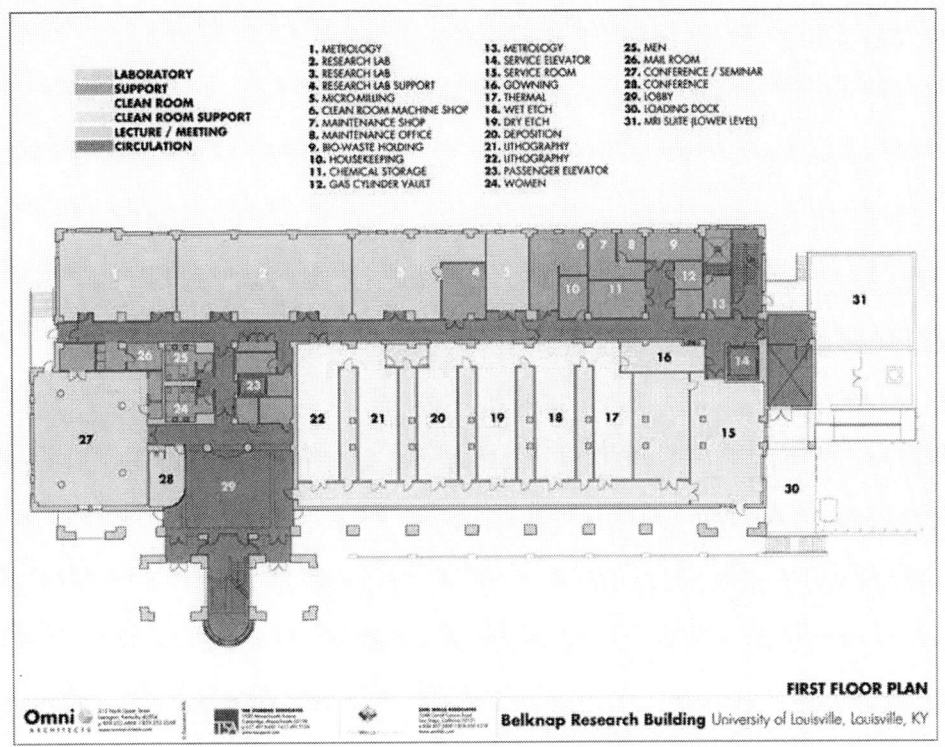

Fig. 2. First floor layout showing the $8.5M 8,000 sq. ft. cleanroom core facility.

APPENDIX A
Questionnaire

1. What were your total revenues in 2004, broken down by those generated by:
    a. Internal users
    b. External academic users
    c. External industry users
    d. University subsidy
    e. State subsidy
    f. Other (grants, etc.)
2. What were your total expenses in 2004, broken down by:
    a. Salaries and benefits
    b. Supplies and expenses
    c. Maintenance, repairs and/or new equipment
    d. Other
3. What is the total approximate value of equipment?
4. How many projects did you undertake in 2004?
5. How many square feet of cleanroom space do you have of the following:
    a. Class 10
    b. Class 100
    c. Class 1000
6. Number of fulltime personnel?
7. Number of part-time personnel?
8. How many of the following does your facility have:
    a. Internal faculty users
    b. Internal student users
    c. Departments using facility
    d. External academic users
    e. External industry users
9. What is the total number of users that you have?
10. What is your user fee structure?
    a. Annual "membership" fee
    b. Monthly user fee
    c. Per entry fee
    d. Per day equipment fee
    e. Per hour equipment fee
    f. Per minute equipment fee
    g. Other fee categories
11. Do your internal user fees differ for external users?
12. Does your facility have an area of concentration in which most projects tend to fall?
13. Is the facility used for teaching (and not just as a research facility)?

APPENDIX B
Survey Data

Metric	Average	Min	Max	Data Points
Total Revenues	$2,180,342	$600,000	$6,000,000	9
(Internal Users)	$814,295	$190,750	$2,114,313	8
(External Academic Users)	$34,167	$0	$100,000	5
(External Industry Users)	$316,551	$54,500	$4,000,000	9
(University Subsidy)	$237,630	$0	$1,250,000	10
(State Subsidy)	$62,500	$0	$500,000	8
(Other)	$115,431	$0	$700,000	8
Total Expenses	$2,075,635	$600,000	$6,000,000	8
(Salaries & Benefits)	$616,988	$183,000	$1,493,694	6
(Supplies & Expenses)	$472,449	$150,000	$1,181,756	6
(Maintenance, repairs and/or new equipment)	$166,209	$0	$300,000	7
(Other)	$25,000	$0	$150,000	6
Equipment value	$42,764,706	$10,000,000	$100,000,000	17
Number of projects annually	178	40	550	17
Total square footage:	9,026	2,020	25,000	20
Class 10	516	0	2,125	9
Class 100	4,164	0	10,500	15
Class 1,000	4,450	0	16,000	15
Class 10,000	809	0	8,000	9
Full-time personnel	8	1	27	18
Part-time personnel	2	0	5	8
Number of Users:	225	50	600	16
Internal Faculty	31	7	59	15
Internal Student	121	14	345	15
External Industry	8	4	15	4
External External	28	15	44	5
Total External	17	1	44	14
Departments Using Facility	11	5	29	18

User Fee Structure (examples):
- $25 per entry fee plus equipment user fees ($35-$105 per run on most equipment, $55-155 per hour on other equipment) and some supply fees
- $10 entry fee, special materials and precious metals, at cost
- equipment hourly rates (different rate for internal external users)
- Access fee is $88.75 month, lab fees are $34.80/hr (capped at $1,200 for academics, $1,600 for industry), special equipment is $22.40/hr, exceptional equipment is $36.60/hr (cap of $1,400 for academics, no cap for industry) and staff services is $67.20/hr.
- Hourly general lab charge ($18 internal/academic and $50 industry), equipment-specific charges by the hour, internal/academic is $25-60/hour and industry is $50-180/hour.
- Equipment use is $100/hour for industry, $50 for external academic, and $30 for internal monthly) and per academic users, staff support minute is $50/hour. Ebeam equipment fees is $50/hour. Lith equipment is $342/hour for internal/academic industry and is $114/hour for academics
- Per entry fee (capped monthly) and per minute equipment fees for major equipment (rates vary for internal and external users)
- Monthly rates ($165/hr capped at $825 for academic users and $165/hr for industry users) or contract rates ($195/hr internal and $210/hr external)

Area(s) of concentration/core facility uses: Sputtering Systems; NEMS and devices are largest groups; opto-electric compound semiconductor processing; None

Used for teaching? Yes: 15; No: 3

1-4244-0267-0/06/$25.00 ©2006 IEEE

National NanoFab Center (NNFC): Nanofabrication Facility

Jong Wan Park, Jeoung Woo Kim
Technical Application Division
National NanoFab Center
53-3 Eoeun-dong, Yuseong-gu, Daejeon 305-806, Korea

Abstract—**Nanotechnology predicts revolutionary changes in human civilization for its applicability to all science, engineering and technical fields, including electronics, materials, medicine, energy etc. It has recently emerged as a new strategic field following information technology and biotechnology. The Korean government has invested on the various national programs, such as nanoFab centers, tera-level nanodevices etc. since 2001 according to a basic plan it formulated for promoting nanotechnology development efforts. It has employed a focused investment strategy on such selected fields as nano electronic devices, especially for infrastructure. The National NanoFab Center was established to encourage and support nanotechnology R&D activities in the academic, research institutes and industry as a centralized public facility for nanofabrication service. In this paper, the overview of representative facilities in Korea, and equipments and activities of NanoFab centers will be introduced.**

I. INTRODUCTION

NANOTECHNOLOGY means the technology of the following paragraphs; science and technology to create material, device or system representing new or improved physical, chemical, biological properties by manipulating, analyzing and controlling materials within the range of nanometer. Moreover, science and technology to ultra-finely fabricate devices, materials etc. within the range of nanometer.

The impact that nanotechnology is currently having on new and existing industries is significant, but the potential for the future is enormous. The National Science Foundation estimates that nanotechnology will have a one trillion dollar impact on the global economy in the next decade. Existing industries, including those not typically characterized as "high tech", are likely to see their product lines and the way they manufacture them influenced by our growing knowledge in nanotechnology.

Research activities on nanotechnology in Korea originated from 'National development plan for nanotechnology' in 2001, to realize the full potential of nanotechnology and achieve the benefits promised by nanotechnology, fueling national competitiveness in three main areas: advanced research and develop concentration on nanotechnologies with comparative advantage, human resources development to satisfy short- and long-term demand, and facilities expansion and infrastructure construction for common use by industry, universities and research labs. The activities related to the progresses in

nanotechnology are leveraged with the enactment 'Law for Promotion of Nanotechnology Development' and new investments funded by Korean government.

Korea's nanotechnology development plan, which is a 10-year, 3-stage plan from 2001 to 2010, is aimed at completing the construction of primary infrastructure for nanotechnology development by 2005 and securing the technological competitiveness on a par with 5 major strong countries in nanotechnology by 2010. For this a total of 1.485 trillion Korean Wons will be invested. As for the operational structure, the Ministry of Science and Technology (MOST) is the primary organ for overall management of the plan and other related departments such as Ministry of Education and Human Resources Development (MOE), Ministry of National Defense, Ministry of Commerce, Industry and Energy (MOCIE), Ministry of Information and Communication (MIC), Ministry of Health and Welfare (MOHW), Ministry of Environment (ME), Ministry of Agriculture and Forestry (MAF) and the Office for Government Policy Coordination jointly execute the plan. Moreover, 'Nanotechnology Subcommittee' established under the Steering Committee of National Science and Technology Council, examines and evaluates the plan and performance of each department. The government policy aims to fuse the 3 main fields information rechnology (IT), biotechnology (BT) and nanotechnology (NT) to upgrade cutting-edge technology and develop basic technology in the furture.

II. NATIONAL NANOFAB CENTER FACILITY

NanoFab means any and all equipments and facilities associated with analysis, fabrication, process or property evaluation, etc. that are required to perform R&D of nanotechnology. The nanofab completely controls and creates physical structure of matter. Nano equipment is very expensive and difficult for individuals to invest in. This maximizes investment efficiency while providing superior equipment and support services. Nanofabrication facility infra in Korea is composed of four core nanofabrication centers: National NanoFab Center (NNFC), Korea Advanced Nano Fab Center (KANC), National Nanotechnology Integration Center for nanoprocesses & Equipments (NICE) and National Center for Nanomaterials & Technology (NCNT). The nanotechnology infrastructures are summarized in the following table 1.

National NanoFab Center (NNFC) as a key national facility

Table 1 Status of nanofabrication facilities in Korea.

Department	Name	Location (Agency)	Emphasis	Budget (Gov.)
MOST	National NanoFab Center	Daejeon (KAIST)	Silicon-Based Device	290 (118)
	Korea Advanced NanoFab Center	Suwon (KIST)	Compound Semiconductor Device	155.6 (50)
MOCIE	National Center for Nanomaterials Technology	Pohang (POSTECH)	Nanomaterials	110.4 (40)
	National Nanotechnology Integration Center	Jeonju (Chonbuk Nat. Univ.)	Nano process, Equipments (Semiconductor)	77.8 (25)
	National Nanotechnology Integration Center	Gwangju (KITECH)	Nano process, Equipments (Display)	81.7 (25)

Fig. 1 Cleanroom layout in NNFC.

infrastructure program was established in Daejeon city in July 2002, to encourage and support nanotechnology R&D activities in the academic, research institutes and industry as a centralized public facility for nanofabrication service, linking with the existing 'Daedeok Science Town' and nearby industrial area named 'Daedeok R&D Special Zone'.

The NNFC has been placed a mission upon as follows. First of all, supporting composite R&D of nanotechnology to main body from industry, school and research institute as a stronghold of nanotechnology. Secondly, significantly reducing the development period as well as promoting joint utilization of research facilities and equipment through providing one-stop service from idea creation to commercialization. Thirdly, training nano-related human power through hands-on experience while utilizing NNFC facilities. Lastly, rearing technology-intensive venture firms and activating business commencement by leading commercialization of nanotechnology based on small-scale investment.

NNFC is open to all of the users from the abroad as well as in the domestic, and would like to make collaborations together with overseas nanotechnology groups. NNFC also provides a variety of support such as IP introduction, technological expansion, commercialization of technology and start-ups. And then we have started for the nanofabrication service from March 2005.

NNFC consists of three major Fabs on the same floor: silicon CMOS (Complementary Metal Oxide Semiconductor) processing, manufacturing of MEMS/NEMS (Micro/Nano Electro Mechanical Systems) and biochip production, thereby providing the nanotechnology commercialization, offering one-stop service from initial idea creation, to verification as well as pilot product manufacturing.

We are striving to do our best for supporting industrialization and offering nanotechnology including united module-process services accumulated, via constructing a global network with outstanding institutes world-wide. The NNFC which already brought a total of 146 different services into being, will play a leading role in the future national economy by offering high quality service in a speedy and stable manner, and by contributing to nanotechnology`s industrialization.

Total spending of NNFC is about 290 M USD including equipment investment, Fab construction, operation and maintenance, land and others. Building capacity has a 5,067 m^2 clean building, 4,588 m^2 support building and 8,354 m^2 research building. The clean building maintains purity from class 1 to class 100. Also various substrates, from piece to 8 inch wafer, are possible to be processed. The main equipment include lithography (e-beam, stepper, mask aligner, imprint, microscope etc), etching (oxide/poly etcher, metal etcher, PR striper, deep silicon etcher etc), diffusion (LPCVD, RTP, wet station, part cleaner, furnace etc), thin film (sputter, ALD, ion implanter, CMP etc), biochemical and new materials (bond aligner, chip aligner, laser micro machine, fusion bonder, nano cluster generator, chemical vapor condensation, nano indentor

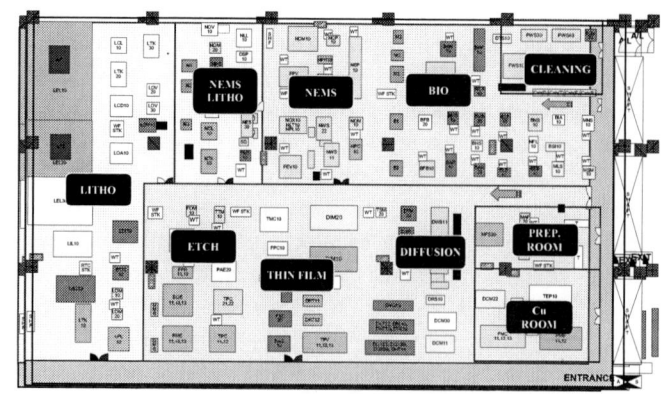

1-4244-0267-0/06/$25.00 ©2006 IEEE

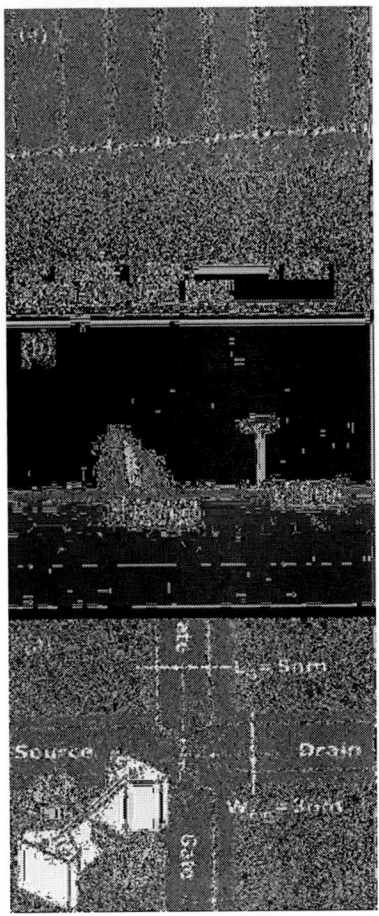

nano-level, ultra-small machines such as minute sensor or actuator including RF-MEMS and optical MEMS. It also support fusion research of nanotechnology and biotechnology i.e., R&D of DNA device, self-assembled molecular device and biosensor (Figure 3 (a) and (b)). The end of the road is lead the technical civilization to prepare the ubiquitous society bring to combine and join with sensor network on CMOS technology.

In addition, Measurement & Analysis lab supports evaluating characteristics of nano devices, materials and structures analysis in real time. Figure 3 (c) shows the dopant profiling of P-N junction in N-MOS device obtained by electron Holography technique.

Nanotechnology is a futuristic technology. Talented human resources are competitiveness. For that reasons, the NNFC provides graduate and doctorate students opportunities to experiment. It also offers custom training on nanotechnology. To effectively deal with everyone's needs, the NNFC has the 'User Association' to make suggestions or inquiries regarding use of the center.

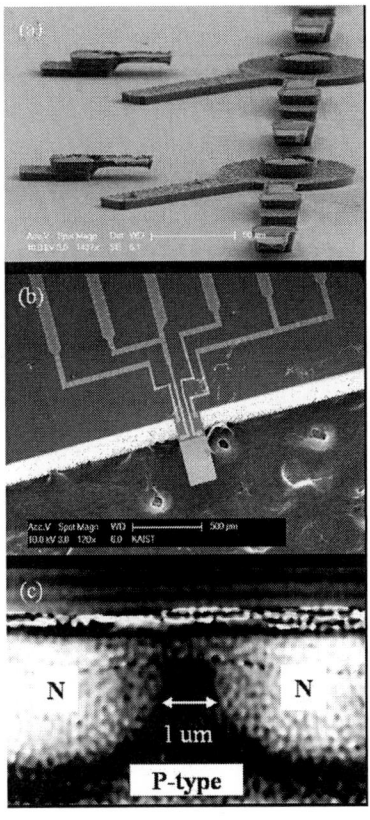

etc) and metrology (FE-TEM, FE-SEM, FIB, AFM etc).

Silicon CMOS lab supports R&D of nano electronic devices and support nano-related fundamental technology including formation of nano structure, nano magnetic material and nano tube and study of nano property, quantum wire and quantum dot. A goal of this lab is supporting advanced nano CMOS technology in the future including standard CMOS technology with litho, etch, diffusion, and thin film teams. Our skill about nano patterning had provided a guarantee against ultra-fine nano patterning with 5nm by e-beam lithography (Hitachi) and fin-gate etch profiling with 10nm. (Figure 1 (a) and (b)) Most recently, the NNFC announced that they successfully developed a silicon-based three-dimensional 3nm Fin Field Effect Transistor (FinFET) in collaboration with professor Yang-gyu Choi of the Department of Electrical Engineering and Computer Science at the Korea Advanced Institute of Science and Technology (KAIST) (Figure 2 (c)). It is the world's smallest transistor that is expected to be applicable to next-generation tera-level semiconductors such as processors, tera-level DRAMs, SRAMs and Flash memory.

The M/NEMS & Biotechnology lab supports R&D of

III. CONCLUSION

The Nano Fab Center will create new industries and become the Mecca and driving force of Korean nanotechnology. It will make the nation even more powerful by sharing information, technology and human resources. The NNFC will become the core facility for conducting nanotech R&D and provide stable and speed services to the research community. The NNFC aims to become the world leading nanotech facility center and to build global research network attracting the world's best researchers to visit and conduct research programs in the center.

ACKNOWLEDGMENT

This work was supported by the Ministry of Science and Technology.

A University-Technical College Nanoscience Training Program

Greg Cibuzar, Steve Campbell, Greg Haugstad, and Michael Flickinger, University of Minnesota, Minneapolis, MN

Deb Newberry, Karen Halverson, and Michael Opp, Dakota County Technical College, Rosemount, MN

Abstract--The projected growth in successful commercialization of nanoscience products and technologies will be slowed without an adequately trained force of technician-level skilled workers. To this end, Dakota County Technical College (DCTC), a two year school in Rosemount, MN received in 2004 an NSF ATE grant to develop a nanoscience training program. The program consists of 3 semesters of coursework at DCTC in various aspects of nanotechnology, followed by a capstone semester at the University of Minnesota (UM). The capstone semester consists of 4 lecture-based courses and 3 lab courses. One of the lab courses is a fabrication course taking place in the UM Nanofabrication Center cleanroom facility. At the conclusion of the capstone semester the students graduate from DCTC with a AAS degree in Nanoscience Technology, and also have 16 credits from the University of MN.

I Introduction

By most accounts, nanotechnology is a rapidly growing area of science and technology that will require large numbers of trained workers at all levels, from Ph.D. researchers down to bench level technicians. Due to the broad applications of nanotechnology, training these workers will require new educational programs with highly integrated curricula in the physical and biological sciences. Designing such courses will involve educators at all levels working across traditional departmental boundaries to create successful training programs.

The federal government, through its National Nanotechnology Initiative (NNI), has recognized this need for training programs to develop a nanoscience workforce, and in conjunction with the National Science Foundation (NSF) has worked to promote the development of nanotechnology educational programs at all levels. The NSF Advanced Technological Education (ATE) program is one example. The ATE program synopsis from the NSF www site [1] states:

With an emphasis on two-year colleges, the Advanced Technological Education (ATE) program focuses on the education of technicians for the high-technology fields that drive our nation's economy. The program involves partnerships between academic institutions and employers to promote improvement in the education of science and engineering technicians at the undergraduate and secondary school levels. The ATE program supports curriculum development; professional development of college faculty and secondary school teachers; career pathways to two-year colleges from secondary schools and from two-year colleges to four-year institutions; and other activities.

There are many ATE programs in a broad range of technical areas such as biotechnology, information technology, and aerospace [2]. The most advanced ATE initiative in the area of nanotechnology is Pennsylvania's Regional Center for Nanofabrication Manufacturing Education. This regional program involves the Penn State University Nanofabrication Facility working with 2 year technical schools in Pennsylvania on a training program for technicians in the area of nanofabrication [3]. Students take 3 semesters of courses at the their home technical college, then spend their final 'capstone' semester at Penn State, working in the Nanofabrication Facility on a series of hands-on courses as well as traditional classroom courses focused on fabrication methods. The program has been very successful for Pennsylvania, as measured by the numbers of graduates successfully finding employment.

In 2001, discussions began to explore the needs for a Minnesota-based nanoscience training program for technicians. Meetings with educators, state government officials, and high technology companies working in Minnesota were held, and employers indicated a need for technicians trained in a variety of areas related to fields commonly included in the nanotechnology umbrella, since the current training programs in areas such as electronics technician were insufficiently diverse in their content. Out of these discussions, a team emerged, led by educators at the Dakota County Technical College (DCTC), committed to the establishment of a two year degree program for technician training in a nanoscience. Unlike the Pennsylvania program that focused on fabrication at the nanoscale, this program is materials-based and would include applications in other areas of nanoscience, such as biological, agricultural, and particle science, in addition to the more standard fabrication focus. To this end, in the fall of 2003 a nanoscience training program grant proposal to the NSF ATE program was prepared, entitled 'Minnesota Nanotechnology Workforce Initiative (MnNano)'. The proposed technician training program was based upon the Pennsylvania model of a capstone semester to be done at the University of Minnesota (UM). The grant was funded, and in the fall of 2004, the first classes began at DCTC with 30 students.

II Minnesota Nanotechnology Workforce Initiative (MnNano)

There are four funded objectives of the program:

- Attract youth to Nanoscience Careers
- Prepare technical employees for nanoscience careers
- Prepare professional employees and scientists for nanoscience careers
- Expand entry into nanoscience careers through dissemination of nanotechnology kits and modules developed by MnNano

This paper is primarily concerned with objective concerning technical employees, and will focus on the curriculum developed for this purpose.

A. *Curriculum*
The broad breadth of nanotechnology is reflected in the diversity of the courses in the program. During their first year, the students will take standard courses in biology, chemistry, physics and math, along with two courses focusing on nanotechnology. The third semester has courses on nanobiology, nanoelectronics, and nanomaterials. The 4 capstone lecture courses at UM focus on microfabrication, thin film deposition, thin film characterization techniques, and bionanotechnology. The three capstone

laboratory courses offer the students the unique opportunity to be exposed to facilities, equipment and processes that are normally found only in companies and at research universities. The capstone experience also includes internships with growing nanotechnology industries in biotechnology, medical devices, agriculture, materials manufacturing and electronics. The course credits earned for the capstone semester are also valid for any students wishing to subsequently enroll in the bachelor's of applied science program at UM.

Nanoscience Technology A.A.S. Degree
Courses Cr
NANO1100 Fundamentals of Nanoscience I 3
NANO1200 Fundamentals of Nanoscience II 3
NANO1210 Computer Simulation 1
NANO2101 Nanoelectronics 3
NANO2111 Nanobiotechnology/Agriculture 3
NANO2121 Nanomaterials 3
NANO2131 Manufacturing Quality Assurance 2
NANO2140 Interdisciplinary Lab 3
NANO2151 Career Planning and Industry Tours 1
NANO2970 Industry Internship & Observation 2
Total 24
Capstone at the University of Minnesota
MT 3111 Elements of Microelectronic Manuf. 3
MT 3112 Elements of Micro & Nano Manuf. Laboratory 1
MT 3121 Thin Films Deposition 3
MT 3131 Introduction to Materials Characterization 3
MT 3132 Materials Characterization Laboratory 1
MT 3141 Principles & Applications of Bionanotechnology 4
MT 3142 Nanoparticles and Biotechnology Laboratory 1
Total 16
General Education
BIOL1500 General Biology 4
COML1400 Intro to Computers 3
ENGL1100 Writing and Research Skills 3
CHEM1500 Introduction to Chemistry 4
MATS1250 Principles of Statistical Analysis 4
MATS1300 College Algebra 4
PHYS1100 College Physics I 4
PHYS1200 College Physics II 4
SPEE1020 Interpersonal Communication 3
Total 33
TOTAL Program Requirements 73 credits

B. *Enrollment*
The first class started fall semester 2004 with over 30 participants. Students ranged in age from 18 to over 40, with many of the older students making a significant career change. 17 of these made it to the capstone semester. Of these 15 graduated. Most have also completed

an internship with a local company. This internship experience, along with the academic coursework and laboratories, should prepare these students to step into a technician-level position with a unique combination of skills not currently found in graduates of two year programs.

III University of Minnesota Participation

The UM participation in the program is to host the students for the capstone semester lecture and laboratory courses. From an academic standpoint, these courses are part of the UM College of Continuing Education (CCE) and not the Institute of Technology (which houses the science and engineering departments). CCE offers a bachelor's of applied science (BAS) degree programs in areas such as Manufacturing Technology, Construction Management, and Radiation Therapy. The courses in the capstone semester are part of the Manufacturing Technology program. Students pay the normal UM tuition for the courses, but laboratory fees are covered by the ATE grant.

The fifteen-week capstone will consist of 12 hours of lecture per week for 15 weeks, as well as three labs per week for the first 11 weeks, to be held in the UM laboratories. Four University of Minnesota faculty members with combined expertise in nanomaterials, nanobioscience, nanofabrication, nanoelectronics, and nanoparticles led the development and delivery of the lecture courses. Faculty members are teaching these courses on an 'overload' basis. One of the courses is being taught by an adjunct faculty. The courses are

- Micro and Nanofabrication
- Thin Film Technology
- Introduction to Materials Characterization
- Principles and Applications of Nanobiotechnology

Each course is designed to be at a pre-calculus math level, comparable to a typical sophomore level, non-major class. Courses are 3 or 4 credits.

Micro and Nanofabrication Course
This course is based upon a standard advanced undergraduate semiconductor microfabrication course offered as part of the normal electrical engineering curriculum at UM. Topics covered include fabrication steps such as lithography, etching, PVD, CVD, oxidation and MEMS. The primary application area is integrated circuit manufacturing.

Thin Film Technology
This course focuses on HV and UHV methods for depositing thin films, vacuum science concepts, and methods of characterizing thin films, primarily process characterization such as film stress, thickness, and index of refraction. Applications areas include optical and magnetic recording.

Introduction to Materials Characterization
Four basic characterization methods are introduced: electron beam microscopy, optical microscopy and FTIR, proximal probe techniques, and x-ray/ion beam scattering.

Principles and Applications of Nanobiotechnology
An introduction to nanobiotechnology particularly focusing on science and engineering miniaturization technologies applied to discrete and multiplexed biochemical analysis.

In addition to the lecture courses, there are three laboratory courses. Each student has 12 hours per week of labs for the first 11 weeks of the semester. The laboratory courses are:

- Micro and Nano Fabrication Laboratory
- Materials Characterization Laboratory
- Particle Technology and Biotech Laboratory

Each lab is developed and run by UM staff members (not student TA's). The costs to the departments for running the labs is currently borne by the ATE grant, and amounts to approximately $2500 per student. If the program continues beyond the life of the ATE grant, the lab costs will have to be paid by the students. Each lab course is 1 credit.

Micro and Nano Fabrication Laboratory
This course is a hands-on processing lab in the UM Nanofabrication Center. Standard labs include: lithography, oxidation, wet and dry etching, PVD, electron beam lithography and focused ion beam techniques.

Materials Characterization Laboratory
Hands-on instruction in the characterization of engineering materials by electron/optical microscopy, atomic force microscopy, x-ray diffraction, spectroscopic methods, and specimen preparation.

Particle Technology and Biotech Laboratory
The use of practical equipment for the detecting particle formation and performing size measurements and aerosol sampling using optical and condensation counters. Also includes a basic introduction to biotechnology laboratory equipment and approaches.

IV Results of Capstone Semester

Generally speaking, the capstone semester received favorable comments from the students. The biggest issue was work load—the students were taking a much larger class and lab load than previous semesters at DCTC, and many had internships at the same time. Survival (passing grades) was the goal for many of them. During the semester, two of the 15 students dropped out, one for this reason and one due to financial problems. Nevertheless, the remaining students were more interactive and responsive than the typical undergraduate student. This probably related to their strong motivation: many were retraining for a new career, and were genuinely interested in learning the material. For the lecture courses, some adjustments will be made in content to prevent overlap, especially with the fabrication and thin films classes. Some of the laboratory courses will need some adjustments in terms of the amount of time outside of class required to do laboratory reports and data analysis.

This summer semester the class sequence is being run again, this time for 14 students from the Chippewa Valley Technical College in Eau Claire, Wisconsin. The next group of students from DCTC is now finished with their first year in the program, and will be coming to UM for their capstone semester in January, 2007. There are currently around 20 students enrolled.

V Conclusions and Future Plans

Nanotechnology needs workers with cross-disciplinary training to level not previously known, thus requiring new educational paradigms for educational and training programs. Through an ATE grant from NFS, a collaborative team of a technical college (DCTC) and a major research university (UM) has developed a unique two year AAS program in the area of nanoscience. The final semester of the program is held at the UM, where students take both lecture and laboratory courses. The laboratory courses give the students exposure to tools and techniques that most two year technical colleges would not be able to provide due to the cost of the facilities and equipment. The first capstone semester students finished this spring semester with generally favorable views of the program. The next group from DCTC will arrive in January 2007.

Recently DCTC successfully solicited planning funds from NSF for the development of a proposal for an ATE regional center in nanoscience workforce training. This would allow the program to be expanded to two year technical colleges in surrounding states.

REFERENCES

1. http://www.nsf.gov/funding/pgm_summ.sp?pims_id=5464
2. http://www.nctt.org/pages/about/ate.php
3. http://www.cneu.psu.edu/default.htm

Low-Noise Amplifier Circuit for Embedded Electrophysiological Recording with Adjustable Gain and High-Pass Filtering

Shahin Farshchi, *Student Member, IEEE*, and Jack W. Judy, *Senior Member, IEEE*

Abstract—This paper describes a fully-integrated, differential, dual-channel, gain-adjustable and bandwidth-adjustable neural preamplifier circuit. This chip has been designed to enable commercial-off-the-shelf (COTS) embedded-networked sensors (ENS) to acquire electrophysiological signals from mobile test subjects, while allowing for the user to remotely adjust amplifier gain and high-pass filtering characteristics for dynamically switching between local field potential and spike acquisition. The preliminary data presented in this paper has been derived from circuit simulations in Spectre, as the chip has been submitted for fabrication in a standard dual-poly dual-metal 1.5-μm process. The 2.2×2.2-mm^2 chip will consume 127 μA of standby current per channel while operating from a single 3-V supply. Gain and high-pass corner frequency are voltage controlled, and can vary between 40 to 90 dB, and 0.1 to 1000 Hz, respectively, while rejecting the large DC offsets that occur at the interface between the tissue and electrode surface. The low-pass cutoff frequency is set to approximately 10 kHz. Simulated input-referred noise is 4.4 μVrms between 0.5 and 5000 Hz.

Index Terms—Embedded neural recording, Low-noise amplifier design, adjustable high-pass filtering, subthreshold design.

I. INTRODUCTION

ELECTROPHYSIOLOGICAL recording is a powerful tool used by neuroscientists to understand the brain function that underlies neurological disease. Traditionally, neuroscientists have performed neural recording on mice by attaching fine wires from the implanted depth electrodes to an external signal acquisition device. The tethering caused by this method of recording prohibits experiments in enriched environments that include tubes, tunnels, and other test subjects with whom the animal can interact. Wireless neural recording can also enable researchers to study fear and aggression, as well as other behavioral phenomena. A wireless neural recording device must be compact and low-power, such that it could be non-intrusively mounted on a

Manuscript received April 3, 2006. This material is based upon work supported by the National Science Foundation under Grant No. 0456125.

S. Farshchi is with the Electrical Engineering Department, University of California, Los Angeles, CA 90095 USA (phone: 925-323-2784, fax: 509-463-5485, e-mail: shahin@ee.ucla.edu).

J. W. Judy is with the Electrical Engineering Department, University of California, Los Angeles, CA 90095 USA (e-mail: jjudy@ee.ucla.edu).

rodent without requiring recharge for several hours at a time. The system must be capable of acquiring the aggregate activity of a population of neurons (known as local field potentials, or LFP), which lie in a frequency band of a fraction of a Hz to a few hundred Hz with an amplitude of several milivolts, while rejecting DC offsets which occur at the interface between the tissue and recording electrode. The system must also be able to acquire spikes, which result from the ion discharge of a single neuron. Spikes generally have a period of several miliseconds, with amplitudes of several 10s of microvolts. Spikes can be obtained by high-pass filtering the LFP signal, and amplifying the filtered signal. Bi-directional communications abilities can also enable the user to switch between different signals (e.g., spikes or LFP) and modes of operation (e.g., continuous recording and transmission or event detection).

A thorough review of existing approaches towards developing wireless biological sensors has been covered in [1]. A novel method for realizing a wireless neural recording system is the overlay of a biological-recording system upon an embedded wireless sensing and communications platform [10]. Such a system has been reported in [1] and [11]. A major limitation of the system was (1) the inability of the user to select between different signals of interest (e.g., spikes vs. LFP), and (2) the amount of noise being coupled into the system via the wires that attach the depth electrodes to the embedded sensors, which rendered spike acquisition impossible. A solution would be a controllable custom IC which can be packaged within (or perhaps integrated with) the recording electrode itself. The gain and bandwidth parameters of the neural amplifier chip (for selecting between LFP and spike acquisition) will be adjusted by the digital-to-analog converters (DACs) on the embedded sensors, while the direct interfacing of the amplifier with the recording electrode will minimize the amount of noise being coupled into the system. In addition, the low amplitude of the spike waveforms requires the amplifier to have an input-referred noise of less than 10 μV$_{RMS}$ over the frequency band of interest (a few hundred mHz to a few kHz) to capture the spikes with a sufficient signal-to-noise ratio.

Existing neural recording devices use neural preamplifier circuits to filter and amplify the signal for subsequent transmission. The low-noise, preamplifier circuits used in [2]-

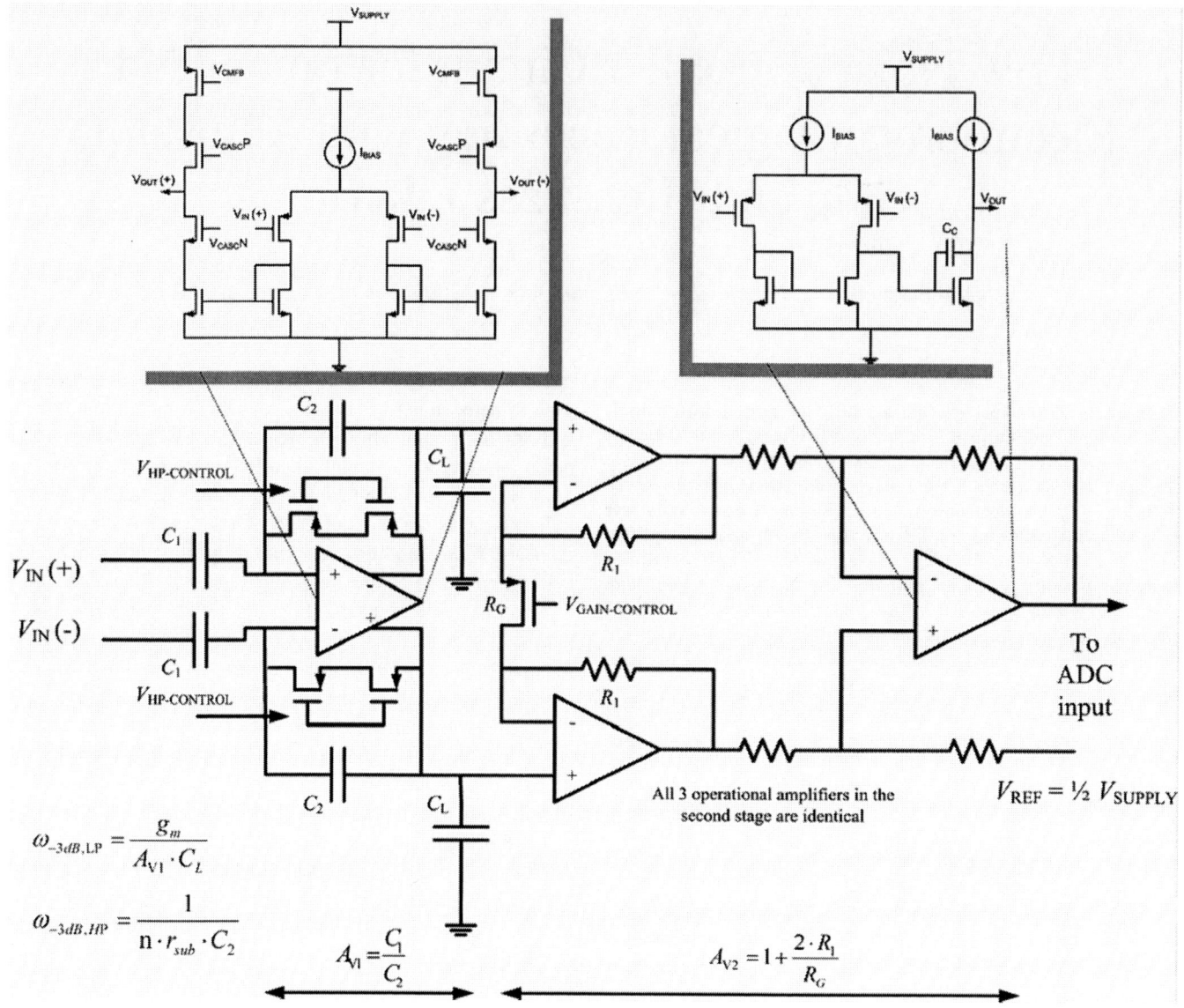

Fig. 1. Complete neural amplifier circuit. The first stage is a high-pass-adjustable low-noise preamplifier stage (adapted from [13]), followed by a gain-adjustable instrumentation amplifier circuit. "n" represents the number of series subthreshold resistors placed in feedback around the first-stage OTA

[5] do not provide sufficient gain for acceptable signal resolution when coupling to an off-chip mote ADC, which is required for this application. For example, the preamplifiers used in [3]-[5] are used to modulate the center frequency of a voltage-controlled oscillator to yield a frequency-modulated version of the neural signal, which is filtered at the receiving end for extracting spikes from the LFP. Simply adding a gain stage to the aforementioned preamplifiers would not suffice, as they do not provide means for adjusting high-pass corner frequency for rejecting LFP when acquiring spikes to avoid saturating the following amplifier stage. Examples of discrete neural amplifier circuits have been demonstrated in [6]-[9].

Charles et al [12] have introduced an ultra-low-noise, low-power preamplifier circuit design that could be modified to provide adjustable high-pass filtering, and hence followed by a high-gain stage. This architecture has thus been chosen as the starting point for this work.

II. NEURAL AMPLIFIER DESIGN

The circuit uses a fully-differential, high-pass-adjustable, low-noise preamplifier stage, followed by a traditional instrumentation amplifier circuit. A supply-independent, wide-swing cascode bias circuit [13] is used to bias the cascade current sources, and a Sooch bias circuit [14] is used for biasing the cascade outputs of the preamplifier circuit. A top-level schematic of the complete biological preamplifier circuit is depicted in Figure 1.

A. Low-Noise, High-Pass-Adjustable Preamplifier Circuit

The preamplifier circuit is an adaptation of the circuit originally introduced by Charles et al [13]. The original design has been modified for use with a single 3-V supply, the addition of voltage-controlled high-pass filtering (to automatically filter spikes from LFP), and fully-differential operation. A very low high-pass pole is achieved by using subthreshold MOS transistors. The operational transconductance amplifier (OTA) that lies at the heart of the preamplifier stage is a differential-pair with cascoded differential outputs mirrored out from the input branches. The common-mode level of the output nodes is set by a continuous-time common-mode-feedback circuit [15].

The preamplifier circuit provides variable high-pass filtering by adjusting the gate voltage of the subthreshold MOS devices to dynamically set the high-pass pole of the front-end gain stage. In [13], the gates of the transistors are tied to their drains, which are biased at roughly the same voltage as their respective sources (by feedback around the high-gain OTA). This ensures that the feedback devices are off, hence providing a very large resistance ($\sim 10^{14}\ \Omega$), for a high-pass pole near DC, which is desirable for LFP recording. By applying a control voltage to the gates of the MOS devices, one can bias them in subthreshold. The I-V characteristic of a subthreshold MOS device is described in (1), where n denotes the subthreshold slope factor.

$$I_D = \frac{W}{L} \cdot \mu \cdot C \cdot V_T^2 \cdot e^{\left(\frac{V_{GS}}{n \cdot V_T} + \frac{\phi_0}{V_T}\right)} \cdot \left(1 - e^{-\left(\frac{V_{DS}}{n \cdot V_T}\right)}\right) \quad (1)$$

Therefore, the resistance across a device biased in subthreshold can simply be expressed as:

$$R = \left(\frac{\partial V_{DS}}{\partial I_D}\right) = \left(\frac{W}{L} \cdot \mu \cdot C \cdot \frac{V_T}{n} \cdot e^{\left(\frac{V_{GS}}{n \cdot V_T} + \frac{\phi_0}{V_T}\right)} \cdot e^{-\left(\frac{V_{DS}}{n \cdot V_T}\right)}\right)^{-1} \quad (2)$$

One can immediately notice that the resistance of a subthreshold device collapses exponentially with rising drain-source voltage, as it does with rising gate-source (or in the case of the PMOS device, source-gate) voltage. Using several devices in series reduces the voltage drop across each device, hence alleviating the effect of output swing on the high-pass frequency. Figure 2 plots resistance as a function of applied gate voltage (or high-pass control voltage) across eight subthreshold devices whose bulks have been tied to their sources to eliminate variation in threshold voltage across the subthreshold transistor chain. The source-drain voltage has been set to 10 mV, which will be the worst-case scenario if the gain of the first stage is set to 100, and a 100-µV spike is sensed at the input.

The exponential characteristic in the relationship between gate voltage and incremental resistance is immediately apparent for control voltages between approximately 0.4 and 0.8 V, which indicates that the devices are indeed in the subthreshold region of operation. The resistance varies over several orders of magnitude in the subthreshold region; however, so long as the resulting high-pass pole is set in the kHz range, LFPs will be attenuated sufficiently to avoid saturating the subsequent gain stage.

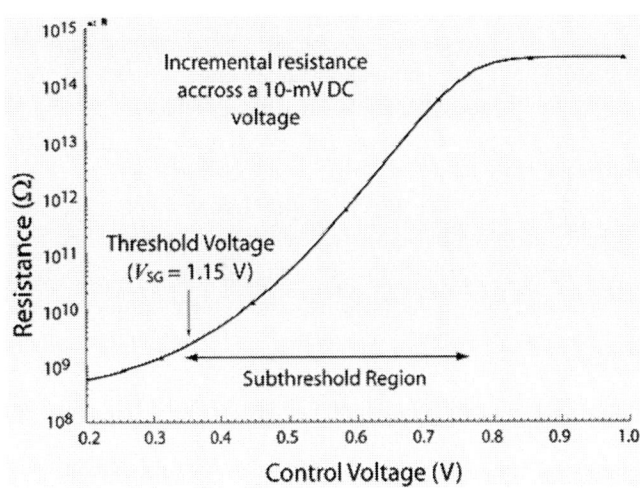

Fig. 2. Resistance of 8 subthreshold PMOS transistors in series as a function of applied gate voltage. The voltage across the devices has been set to 10 mV to simulate the amplification of a 100-µV spike.

B. Gain-Adjustable Instrumentation Amplifier

The instrumentation amplifier circuit provides (1) differential to single-ended conversion with reference to a DC voltage and (2) adjustable added gain. The operational amplifiers used in the instrumentation amplifier are standard compensated two-stage gain stages driving resistive loads. The gain of this stage is controlled by adjusting the gate voltage of a triode PMOS transistor acting as a voltage-controlled resistor. The submitted chip layout has been depicted in Figure 3.

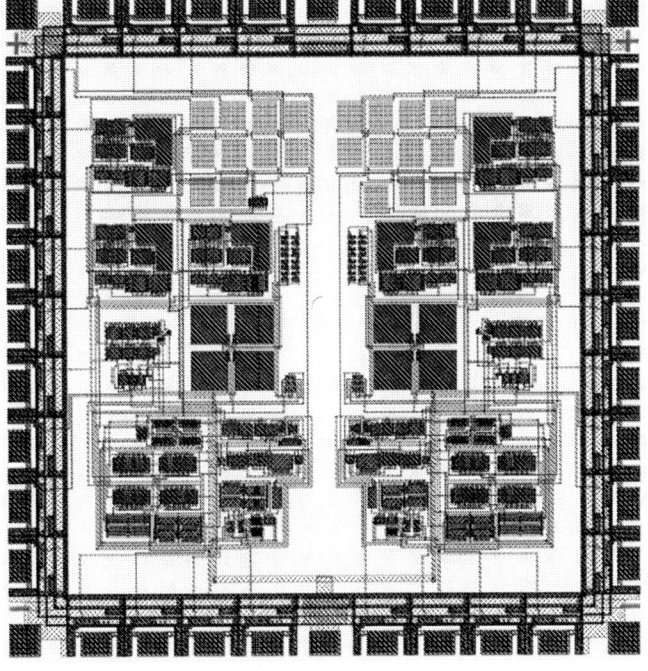

Fig. 3. Neural amplifier chip layout.

III. SIMULATED RESULTS

Spectre simulations were performed to verify circuit performance. The circuit consumes approximately 127 μA from a 3-V supply. Gain is adjustable between 40 and 90 dB, and the input-referred noise is 4.4 μV$_{RMS}$ between 0.5 Hz and 5 kHz. All simulations were performed with typical models provided by MOSIS. First, we applied a previously-recorded neural dataset to the input of the circuit. The data was originally acquired in vivo from freely moving rats using five four-channel MOSFET input operational amplifiers mounted in the cable connector to remove movement artifacts. Data were recorded wide band (0.1 Hz to 5 kHz) and sampled at 10 kHz/channel (16 channels) with 12-bit precision. Figure 4 illustrates the output of the first stage circuit when the raw signal is applied at the input. For LFP acquisition, the high-pass-control voltage is set to half the supply for obtaining a high-pass corner frequency of a fraction of a Hz. The gain control voltage is set to 0 V to provide a total gain of approximately 60 dB to avoid saturating the amplifier.

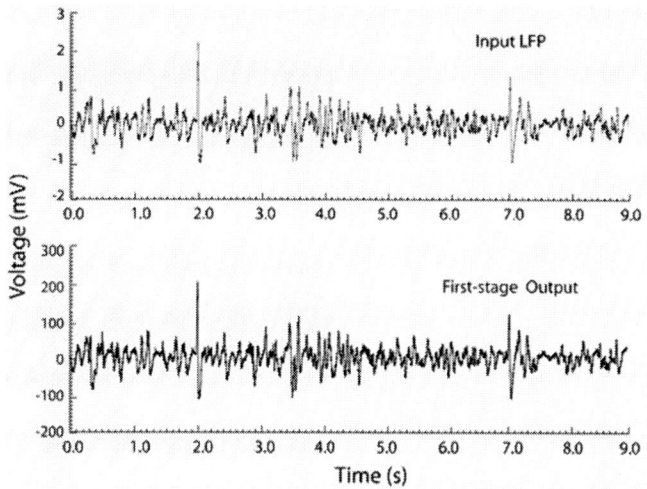

Fig. 4. Simulated input and preamplified local field potential at the first stage.

To acquire spikes, the high-pass-control voltage is set to 700 mV, thus yielding a high-pass corner frequency of approximately 1 kHz to filter out the local field potentials. The gain-control voltage is set to 200mV to provide a total gain of approximately 90 dB to amplify the filtered spikes. Figure 5 illustrates the output of the amplifier in spike acquisition mode.

ACKNOWLEDGMENT

The authors would like to thank Dr. David W. Parent for helping prepare the circuit for fabrication, and Dr. Anatole Bragin for providing the raw neural recordings.

REFERENCES

[1] S. Farshchi, P. H. Nuyujukian, A. Pesterev, I. Mody, J. W. Judy, "A MICA2-enabled TinyOS-based wireless neural interface," *IEEE Transactions on Biomedical Engineering*, to be published.

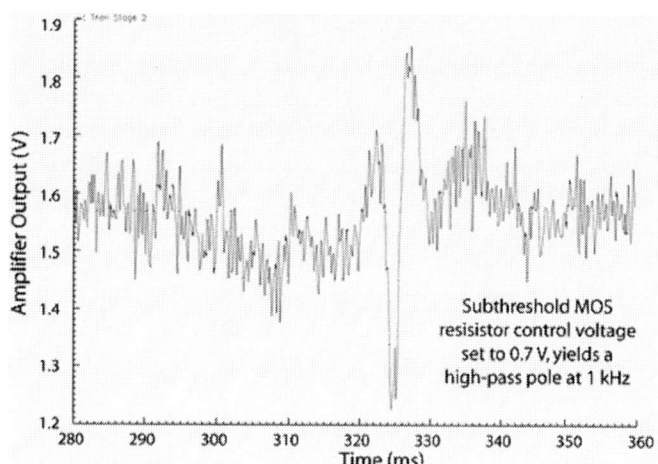

Fig. 5. Spike signal filtered and amplified from LFP input.

[2] H. J. Song, D. R. Allee, and K. T. Speed, "Single chip system for bio-data acquisition, digitization and telemetry," *Proc. of the 1997 IEEE International Symposium on Circuits and Systems*, Hong Kong, June 9-12, 1997, vol.3, pp. 1848-1851.

[3] G. A. DeMichele and P. R. Troyk, "Integrated multichannel wireless biotelemetry system," in *Proc. of the 25th International IEEE-EMBS Conf.*, Cancun, Mexico, September 17-21, 2003, pp. 3372-3375.

[4] J. Parramon, P. Doguet, D. Martin, M. Verleyssen, R. Munoz, L. Leija, and E. Valderrama, "ASIC-based batteryless implantable telemetry microsystem for recording purposes," in *Proc. of the 19th International IEEE-EMBS Conf*, Chicago, IL, Oct. 30 to Nov. 2, 1997, pp. 2225-2228.

[5] P. Irazoqui-Pastor, I. Mody, and J. W. Judy, "Transcutaneous RF-powered neural recording device," in *Proc. of the 24th Annual Conf. and the Annual Fall Meeting of the EMBS/BMES Conf.*, Houston, TX, October 23-26, 2002, vol. 3, pp. 2105-2106.

[6] A. Nieder, "Miniature stereo radio transmitter for simultaneous recording of multiple single-neuron signals from behaving owls," *Journal of Neuroscience Methods*, vol. 101, pp. 157-164, Sep. 2000.

[7] S. Takeuchi and I. Shimoyama, "A radio-telemetry system with a shape memory alloy microelectrode for neural recording of freely moving insects," *IEEE Transactions on Biomedical Engineering*, vol. 51, pp. 133-137, Jan. 2004.

[8] M. Modarreszadeh and R. N. Schmidt, "Wireless, 32-channel, EEG and epilepsy monitoring system," in *Proc. of the 19th International Conference of IEEE/EMBS*, Chicago, IL, Oct 30-Nov 2, 1997, pp. 1157-1160.

[9] I. Obeid, M. A. L. Nicolelis, and P. D. Wolf, "A multichannel telemetry system for single unit neural recordings," *Journal of Neuroscience Methods*, vol. 133, pp. 123–135, February 2004.

[10] J. L. Hill and D. E. Culler, "Mica: a wireless platform for deeply embedded networks," *IEEE Micro*, vol. 22, Issue 6, pp.12-24, Nov/Dec 2002.

[11] S. Farshchi, P. H. Nuyujukian, A. Pesterev, I. Mody, J. W. Judy, "A MICA2-enabled TinyOS-based wireless A TinyOS-Based Wireless Neural Sensing, Archiving and Hosting System," *Proc. of the 2nd Int. Conf. of the IEEE EMBS Conference on Neural Engineering*, March 16-19, 2005, Arlington, VA, USA.

[12] C. T. Charles and R. R. Harrison, "A low-power low-noise CMOS amplifier for neural recording applications," *IEEE Journal of Solid-State Circuits*, Vol 38, Issue 6, pp: 958 - 965. June 2003.

[13] D. A. Johns and K. Martin, *Analog Integrated Circuit Design*. New York: Wiley, 1997.

[14] N. S. Sooch, "MOS Cascode Current Mirror," *US Patent* no. 4550284, October 1985.

[15] C. T. Charles "Electrical components for a fully-implantable neural recording system," M.S. dissertation. Electrical Engineering Program, University of Utah, Salt Lake City, UT, USA, 2003

Si-Based Resonant Interband Tunnel Diode/CMOS Integrated Memory Circuit

Stephen Sudirgo, David J. Pawlik, Karl D. Hirschman, Sean L. Rommel, Santosh K. Kurinec
Department of Microelectronic Engineering
Rochester Institute of Technology
82 Lomb Memorial Dr., Rochester, NY 14623

Phillip E. Thompson
Naval Research Laboratory
Code 6812, Washington, DC 20375-5347

Paul R. Berger
Department of Electrical and Computer Engineering
The Ohio State University
205 Dreese Laboratory, Columbus, OH 43210

Abstract— **The development of Si-based tunneling-based static random access memory (TSRAM) has been described. This multi-institutional research endeavor has successfully demonstrated for the first time an integrated TSRAM that utilizes Si/SiGe resonant interband tunnel diode (RITD) and conventional NMOS. The memory cell exhibits a bistable latching operation at a low power supply voltage below 0.5 V. The key to success in the tunnel diode-based novel memory research at RIT is mutual collaboration between the institutions from the universities, government, and industry, which provides a hotbed for technological innovations and creativity.**

I. HISTORICAL DEVELOPMENT

Rapid developments in information technology, such as the internet, portable computing, and wireless communication, create a huge demand for fast and reliable ways to store and process information. Thus far, this need has been paralleled with the revolution in solid-state memory technologies. Memory devices, such as SRAM, DRAM, and flash, have been widely used in most electronic products. The primary strategy to keep up the trend is miniaturization. CMOS devices have been scaled down beyond sub-45 nm, the size of only a few atomic layers. Scaling, however, will soon reach the physical limitation of the material and cease to yield the desired enhancement in device performance.

A new class of memory devices based on Si/SiGe resonant interband tunnel diode (RITD) has been fabricated. It utilizes the inherently fast band-to-band electron tunneling mechanism, making this device suitable for high-frequency operation. In addition, the non-linear current-voltage characteristics of a tunnel diode can be exploited to create multiple stable latching states in which information can be stored.

The initiative began at the University of Delaware and NRL in 1998 as shown in the timeline given in Fig. 1. The invention of a high performance Si-based tunnel diode that operates at room temperature by Rommel *et al.* [1] brought forth a renewed interest in bringing tunnel diode-based circuitry on to Si platform. Previously, tunnel diode/FET circuits were explored mainly in III-V material system [2]. Seeing the potential in this endeavor and having an established student-run CMOS process, RIT decided to undertake the first ever integration of MBE grown resonant tunnel devices with CMOS. Preliminary studies were conducted to investigate the compatibility of the clean processes and thermal constraints between the two technologies. This effort rapidly evolved into a multi-institutional research and received funding from the NSF in 2001. The Ohio State University focuses on the tunnel diode structure engineering and modeling. NRL has developed a low temperature molecular beam epitaxy (LT-MBE) technique to fabricate tunnel diodes. RIT conducts process integration with its CMOS technology, circuit and layout design, mask fabrication and electrical test and characterizations.

The first Si/SiGe tunnel diodes, fabricated through openings in the field oxide and on top of p^+ implanted regions, were realized in early spring 2002 [3]. This result was a strong indication that integration with CMOS could be possible. One year later, monolithic integration of Si/SiGe RITD with CMOS was achieved [4]. The tunnel diodes exhibit a peak-to-valley current ratio up to 2.8 with a peak current density up to 260 A/cm^2. A simple monostable-bistable logic element (MOBILE), with low operating voltage supply below 0.5 V was also demonstrated [4]. Out of this study, a new type of device structure known as overgrown RITD was conceived.

1-4244-0267-0/06/$25.00 ©2006 IEEE

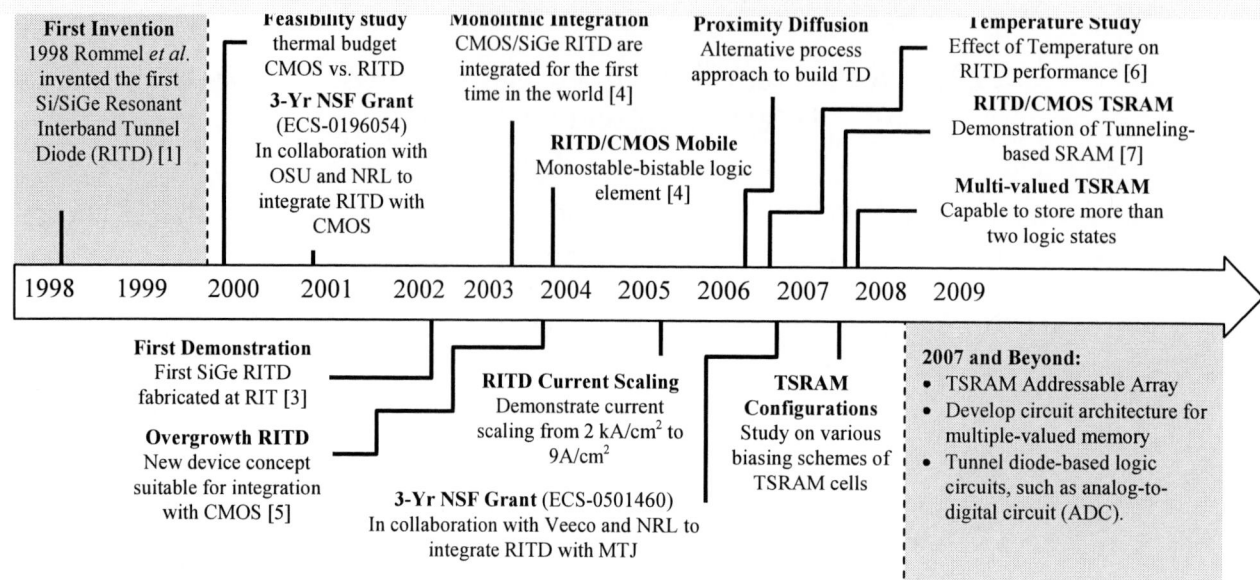

Fig. 1 Timeline of the development of Si-based RITD/CMOS Technology.

Utilizing the inherent nature of epitaxial growth that adopts the crystal structure of underlying film, a polysilicon-isolated device was fabricated [5]. A subsequent experiment shows that the Si/SiGe RITDs are capable to operate with acceptable performance for circuit application at temperature as high as 200°C [6].

II. SI-BASED RESONANT INTERBAND TUNNEL DIODE

Resonant interband tunnel diode (RITD) is a hybrid between Esaki diode and RTD. It incorporates quantum wells on each side of a p-n diode structure using δ-doped planes and the insertion of an *i*-layer to form a double quantum well structure. Fig. 2 illustrates the latest variation of the device structure for the Si/SiGe RITD.

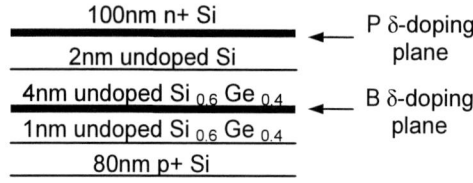

Fig. 2 A typical device structure of a Si/SiGe Resonant Interband Tunnel Diode.

In essence, there are six key points to the SiGe RITD design: (i) an intrinsic tunneling barrier called spacer (ii) δ-doped injectors, (iii) off-set of the δ-doping planes from the heterojunctions interfaces, (iv) SiGe cladding layers to suppress boron diffusion of p$^+$ delta doping plane, (v) low temperature molecular beam epitaxial growth (LT-MBE), and (vi) post-growth rapid thermal annealing (RTA) for dopant activation and point defect reduction. A typical Si/SiGe RITD of the structure shown in Fig. 3 exhibits a PVCR as high as 3.6 and a J_p of 0.3 kA/cm^2.

III. INTEGRATED TUNNELING-BASED SRAM

Recently, a fully-integrated tunneling-based SRAM (TSRAM) has been demonstrated [7]. The cell design consists of two tunnel diodes connected in series, one acting as the drive and the other as the load as shown in Fig. 2. This configuration allows a bistable operation at a particular range of supply voltages (V_{DD}). The information is stored at the sense node, which can be altered by modulating current into the node via a FET. Current injection into the sense node by applying high bias to the bit line and turning on the word line will force the cell to latch into a logic state high. On the other hand, draining the current by grounding the bit line while keeping the gate on will force the sense node to latch to a logic low state. The RITD load-line stores the logic value while the wordline is off.

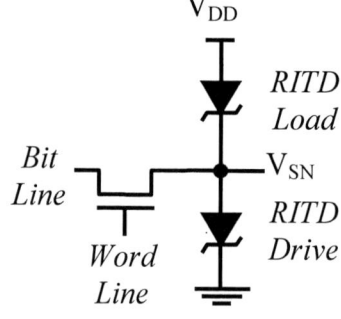

Fig. 3 Circuit diagram of a T-SRAM cell.

The proposed scheme integrates quantum devices, Si/SiGe resonant interband tunnel diodes (RITD), with classical CMOS devices forming a microsystem of disparate devices to achieve higher performance as well as higher density. The CMOS devices used in this study is a second generation CMOS technology developed at RIT, a student-run laboratory [9]. It features a double-well process with LOCOS for electrical isolation between devices. An n^+ polysilicon layer is used as the gate material. The gate dielectric is made of thermally grown SiO_2 with thickness of 31 nm. The source and drain are formed using a self-align technique via ion implantation. Low temperature oxide deposited using PECVD technique is used as the first interlayer dielectric material. The devices are connected through Al metal lines.

A typical CMOS technology has been developed, such that process steps that involve high temperature for a long period of time occurs towards the beginning of the fabrication sequence. The CMOS process used in this study is no exception. The incorporation of Si/SiGe RITD was evaluated based on the thermal budget consideration.

The step with the highest thermal budget in the fabrication of a discrete Si/SiGe RITD is the post LT-MBE rapid thermal annealing process that can vary from 700°C to 825°C for a few minutes depending upon the device structure. The last CMOS process step that involves a higher thermal budget than that of RITD is the source and drain anneal, which is typically done at 1000°C for at least 30 min. Therefore, it is imperative to build the RITD after all CMOS processes up to, and including, the S/D thermal anneal step.

Another major consideration is the material compatibility. CMOS consists of silicon and its derivatives, such as silicon nitride and silicon dioxide. It also has dopant elements, such as boron and phosphorus. The tunnel diode used in this study, fortunately, is also Si-based with a couple of SiGe layers. Therefore, material compatibility should not be an issue. However, special attention has to be given to the cleaning process prior to the epitaxial growth process used to build the RITD layers. The surface has to be relatively defect-free.

In terms of placement with respect to CMOS devices, RITDs are being integrated on the p^+ regions used for the source and drains of the PFET. Therefore, they can share a common well in many cases.

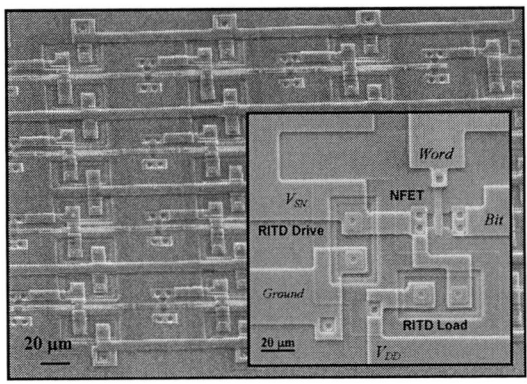

Fig. 4 SEM micrograph of T-SRAM Array. The inset shows a single TSRAM cell.

The device/circuit designs, layouts and masks (involving 12 levels) were fabricated utilizing a process that incorporates nearly a hundred processing steps. Fig. 4 shows the SEM micrograph of the memory array with an inset of a single cell TSRAM. It consists of two tunnel diodes coupled in series

with an NFET connected at the sense node. The I_{DS}-V_{DS} characteristics of the CMOS are illustrated in Fig. 5.

The current-voltage characteristics of the drive and load tunnel diodes is given in Fig. 6. This configuration allows for two stable latching states as indicated by the intersections between the drive and load characteristics. The FET modulates the current going in and out of the sense node, forcing the system to latch high or low, respectively. The ratio of the difference between high and low-state with respect to V_{DD} reaches a maximum of 53.5% at a standby supply voltage of 0.57 V. The cell can either latch to a low-state at 0.13 V or a high-state at 0.43 V.

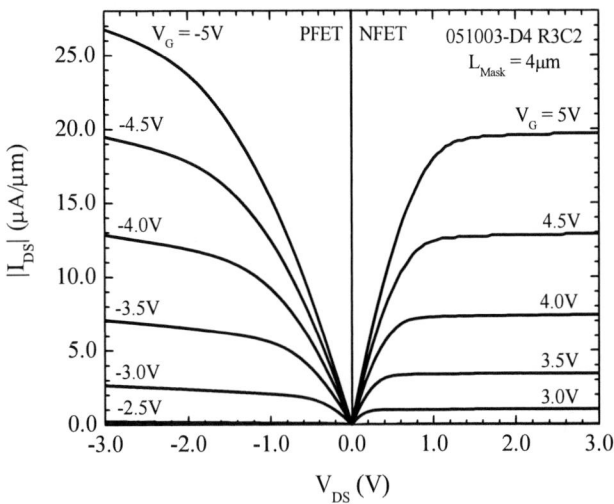

Fig. 5 I_{DS}-V_{DS} characteristics of PFET and NFET integrated with Si/SiGe RITD.

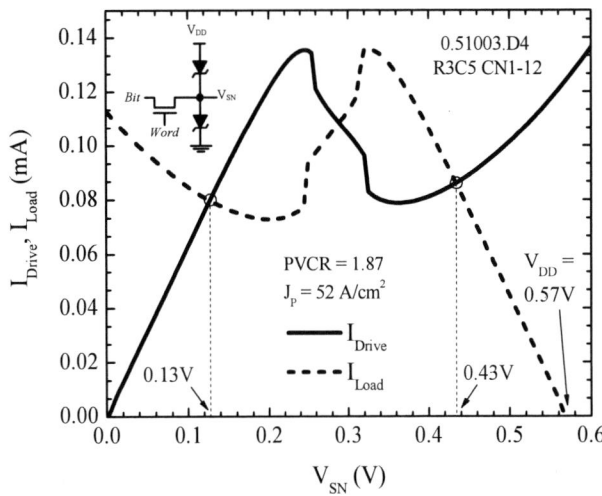

Fig. 6 Load line analysis of back-to-back tunnel diodes, providing two stable latching points [7,8].

Fig. 7 depicts the time diagram during write high, low, and standby operations.

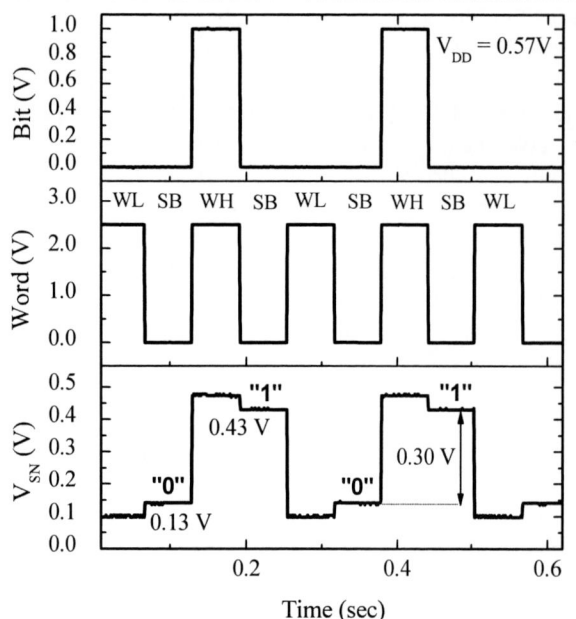

Fig. 7 Time diagram of T-SRAM cell during standby (SB), write high (WH) and low (WL) operations at power supply voltage of 0.57V [7,8]

IV. CONCLUSION

In conclusion, the authors have fabricated and demonstrated the first NMOS/SiGe RITD-based TSRAM cell that successfully operates at power supply voltages below 0.5 V. This work opens up new possibilities for the realization of ultra-low power TSRAM utilizing low current density Si/SiGe RITDs. The key to success in the tunnel diode-based novel memory research at RIT is mutual collaboration between the institutions from the universities, government, and industry, which provides a hotbed for technological innovations and creativity.

ACKNOWLEDGMENT

This work is funded by the National Science Foundation under grant ECS-0196054 and ECS-0501460.

REFERENCES

[1] S.L. Rommel, *et al.*, "Room Temperature Operation of Epitaxially Grown $Si/Si_{0.5}Ge_{0.5}$ Resonant Interband Tunneling Diodes," *Appl. Phys. Lett.*, 73, pp. 2191-2193, 1998.

[2] J.P.A. Van der Wagt, "Tunneling-based SRAM," *Proc. of IEEE*, vol. 87, pp. 571-595, April 1999.

[3] S. Sudirgo, *et al.*, "Challenges in Integration of Resonant Interband Tunnel Devices with CMOS," *Proceedings of the 15th Biennial Uni./Gov./Ind. Microe. Symp.*, pp. 275-278, 2003.

[4] S. Sudirgo, *et al.*, "Monolithically Integrated Si/SiGe Resonant Interband Tunnel Diode/CMOS Demonstrating Low Voltage MOBILE Operation," *Solid-State Electronics*, vol. 48, pp. 1907-1910, Oct.-Nov., 2004.

[5] S. Sudirgo, *et al.*, "Overgrown Si/SiGe Resonant Interband Tunnel Diode for Integration with CMOS," *Proceedings of 62nd Device Research Conference*, pp. 109-110, June 2004.

[6] D.J. Pawlik, *et al.*, "High Temperature Study of Si/SiGe Resonant Interband Tunneling Diodes," *to be presented at the 2005 International Semiconductor Device Research Conference.*

[7] S. Sudirgo, *et al.*, "NMOS/SiGe Resonant Interband Tunneling Diode Static Random Access Memory," *Submitted to 2006 Device Research Conference.*

[8] *S. Sudirgo, Quantum and Spin-based Tunneling Devices for Memory Systems,* Ph.D. Dissertation, Microsystem Engineering, Rochester Institute of Technology, 2006.

[9] K.D. Hirschman, J. Hebding, R. Saxer, K. Tabakman, "Semiconductor Process and Device Modeling: A Graduate Course/Undergraduate Elective in Microelectronic Engineering at RIT," *IEEE 15th Biennial University/ Government/ Industry Microelectronics Symposium,* Boise, ID, June 30-July 2, pp.138-146, 2003

Numerical and Analytical Results for the Polysilicon Gate Depletion Effect on MOS Gate Capacitance

H. Abebe[*], E. Cumberbatch[**], H. Morris[***] and V. Tyree[*], *Member, IEEE*

Abstract—**Analytical and numerical gate capacitance models with polysilicon (poly) depletion effect are studied by directly solving the coupled Poisson equations on the poly and silicon sides (see Fig. 1 and (1)). The poly depletion effect is known to significantly reduce surface potential, channel current and gate capacitance values, [2-4]. Different oxide thicknesses and doping levels of the MOS device are studied, and the final analytical gate capacitance model exhibits an excellent fit with numerical data (see Figures 6 and 7). The analytical model is determined using asymptotic methods, [1]. The models presented here give accurate results for the poly depletion effect and this new information may be used to improve SPICE circuit simulations in advanced VLSI since the gate depletion effect is significant in current nanoscale MOSFET devices.**

Index Terms—**Device modeling, gate capacitance, MOSFETs, polysilicon depletion effect.**

I. INTRODUCTION

One of the assumptions in earlier MOS device modeling approaches was to treat the polysilicon (poly) gate as a perfect conductor which would not deplete because the poly was heavily doped compared to the transistor channel. However the assumption of a perfect conductor is no longer valid for current deep sub-micron processes. Advanced CMOS technology in a deep sub-micron process requires ultra-thin gate dielectrics and higher levels of channel doping in order to maximize the drive current of the transistor. As the oxide thickness gets smaller, the poly depletion problem gets worse, [5]. Moreover, an increase in channel doping and a decrease in the oxide thickness for a given poly doping have a direct effect on poly depletion, [3]. Consequent MOSFET performance degradation of reduced channel current and gate capacitance, [3], is of major concern. This performance degradation is due to the voltage drop across the poly gate as a result of the formation of a depletion layer near the poly/silicon-oxide interface.

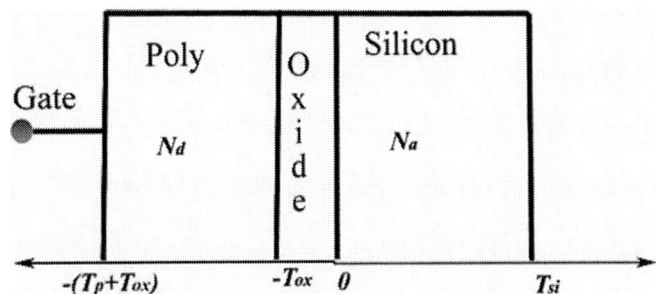

Fig. 1. Poly/Silicon-oxide and Silicon/Silicon-oxide interfaces with oxide tickness T_{ox}, poly thickness T_p, silicon thickness T_{si}, doping levels N_a and N_d.

In [2-8] different MOS device modeling approaches of the poly depletion effect are discussed. In this paper, analytical and numerical gate capacitance models with poly depletion effect are studied by directly solving the coupled Poisson equations on the poly and silicon sides (see Figure 1 and (1)). All computational work is done using MATLAB and our analytical gate capacitance model is determined using asymptotic methods, [1]. A good approximation for the Lambert function is given in the Appendix.

II. GATE CAPACITANCE MODEL WITH POLY DEPLETION EFFECT

For the common parallel plate capacitor there is a linear relationship between the charge Q and the voltage V, and the static capacitance $C = Q/V$ and differential capacitance $C = |dQ/dV|$ will be the same. Since the gate capacitance of a MOS device varies nonlinearly with the gate voltage, the static and differential capacitances are different. Because of this reason in our discussion here only the differential capacitance is applicable.

Using the gradual channel approximation the coupled one-dimensional Poisson equations for the two regions in Fig. 1 are written as

---

This work was supported in part by the USC/ISI MOSIS service.
*USC Viterbi School of Engineering, Information Sciences Institute, MOSIS service, Marina del Rey, CA 90292, USA. Tel: (310) 448-8740, Fax: (310) 823-5624, (e-mail: abebeh@mosis.org and tyree@mosis.org).
**Claremont Graduate University, School of Mathematical Sciences, 710 N College Ave, Claremont, CA 91711, USA. Tel: (909) 607-3369, Fax: (909) 621-8390, (e-mail: ellis.cumberbatch@cgu.edu).
***Department of Mathematics, San Jose State University, San Jose, CA 95192, USA, (e-mail: morris@math.sjsu.edu).

1-4244-0267-0/06/$25.00 ©2006 IEEE

$$V'' = \frac{q}{\varepsilon_{si}} \begin{cases} n - p - N_d & X \le -T_{ox} \\ n - p + N_a & X \ge 0 \end{cases} \qquad (1)$$

where

$$n = n_i e^{(V - \phi_n)/V_{th}}$$

$$p = n_i e^{-(V - \phi_p)/V_{th}}$$

$V(X)$ is the electrostatic potential; X is the perpendicular distance from the gate to silicon substrate. N_d and N_a are the donor and accepter doping densities respectively, q represents electron charge, ε_{si} semiconductor permittivity, n electron density, p hole density, n_i intrinsic density, ϕ quasi-Fermi potential, $V_{th} = k_b T / q$, k_b Boltzmann constant and T temperature. Both the poly and the silicon doping are considered to be uniform and separated by a thin oxide layer. The assumption here is that the doping density $N_d \gg N_a$ in the poly and $N_a \gg N_d$ in the silicon.

The boundary conditions consist of the continuity of electric potential V and electric displacement, $\varepsilon dV/dX$, at the oxide interfaces $X=0$, $-T_{ox}$ (see Fig. 1). Moreover the electric displacements at the two interfaces are equal. The electric potential and electric field are also considered to be zero in the bulk.

In asymptotic methods, scaling the parameters is useful and it is commonly used to simplify and analyse non-linear problems. Here, dimensional voltages and lengths are denoted by capital letters, and lower-case letters denote the same quantities non-dimensionalised. The voltage $V_{th} \ln \lambda$ is used as reference for voltages and potentials, and the length value $L_d \sqrt{\ln \lambda / \lambda}$ is used as reference length, where $\lambda = N_a / n_i$ ranges from 10^6 to 10^9. The size of the parameter λ (or more accurately its logarithm) allows asymptotic approximations to be effective. $L_d = \sqrt{\varepsilon_{si} V_{th} / q n_i}$ is the intrinsic Debye length. The gate oxide capacitance is scaled as $C_{ox} = c_{ox} \varepsilon_{si} / L_d = \varepsilon_{ox} / T_{ox}$.

In scaled variables equation (1) becomes

$$v'' = \begin{cases} \frac{1}{\beta} e^{(v - v_g^*) \ln \lambda} - \beta e^{-(v + 2 - v_g^*) \ln \lambda} - 1/\beta & x \le -t_{ox} \\ e^{(v-2) \ln \lambda} - e^{-v \ln \lambda} + 1 & x \ge 0 \end{cases}$$

$$\qquad (2)$$

where $\beta = N_a / N_d$, $v_g^* = v_{gs} - v_{fb}$ (gate voltage minus flat band) and at thermal equilibrium, $\varphi_n = \varphi_p = \varphi$.

The applied gate voltage generates a difference between the levels of the quasi-Fermi potentials at the extremities of the device. The quasi-Fermi potential is taken to be unity in the silicon substrate ($\varphi = 1$, so that $v \to 0$ at large x), and it is $1 + v_g^* - v_{bi}$ at the poly gate where $v_{bi} = 2 - (\ln(\beta))/\ln \lambda$ is the built-in potential.

The boundary condition on the electrostatic potential v away from the interface at the edge of depleted poly gate is then the applied gate voltage minus the flat band voltage. The flat band voltage represents the built-in potential or work function differences across the oxide interfaces, [9].

It is possible to integrate (2) once: multiplying it by v', and integrate both sides of the equation with respect to x. Applying the boundary conditions at the bulk and poly gate give

$$(v')^2 =$$

$$\begin{cases} \frac{2}{\beta \ln \lambda} (e^{(v - v_g^*) \ln \lambda} - 1) + \frac{2\beta}{\ln \lambda} (e^{-(v + 2 - v_g^*) \ln \lambda} - \frac{1}{\lambda^2}) \\ - \frac{2}{\beta} (v - v_g^*) \qquad x \le -t_{ox} \\ \frac{2}{\ln \lambda} (e^{(v-2) \ln \lambda} + e^{-v \ln \lambda} - \frac{1}{\lambda^2} - 1) + 2v \qquad x \ge 0 \end{cases}$$

$$\qquad (3)$$

Boundary conditions at the oxide interfaces yield

$$v'\big|_{x=-t_{ox}} = v'\big|_{x=0} = c_{ox} \sqrt{\frac{\ln \lambda}{\lambda}} (v_s - v_t) \qquad (4)$$

where $v_s = v(0)$ and $v_t = v(-t_{ox})$

There are three cases for the NMOS device corresponding to negative, positive, or near flat band applied voltages. (See Fig. 2). In the accumulation case holes are at the surface to provide a net positive charge, at depletion the bulk charge dominates, and electrons provide a net negative charge in the inversion case.

Numerical solutions for NMOS device are found from the

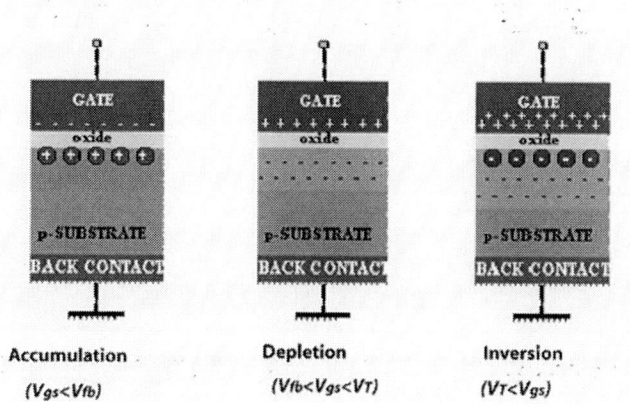

Fig. 2. NMOS device under accumulation, depletion and inversion conditions, where V_T is the device threshold voltage.

transcendental equations of (3) and (4) for given gate voltages. Replacing v_t by v_g^* in (4) and solving (3) for $x \geq 0$ give solutions omitting the poly depletion effect. The poly depletion effect is compared with results without poly depletion effects in Fig. 3-5. The surface electric field in Fig. 4 is determined from the boundary condition in (4). Fig. 5 shows gate capacitance reduction due to poly depletion and it is determined from the derivative of Fig. 4 with respect to the relative gate voltage.

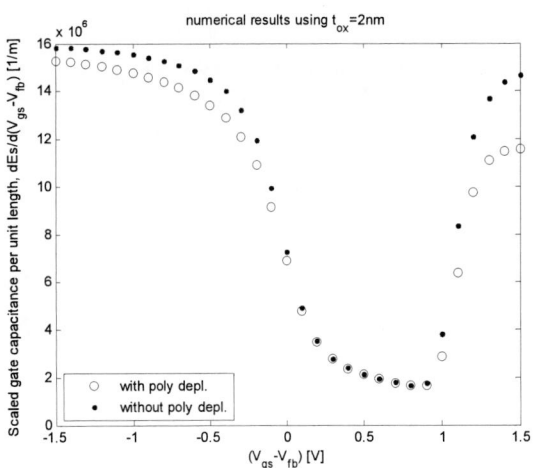

Fig. 5. Scaled gate capacitance versus relative gate voltage ($\lambda = 10^7$ and $\beta = 10^{-2}$).

order approximation lead to (5).

$$f = \frac{2}{\beta \ln \lambda}(e^{(v_t - v_g^*)\ln \lambda} - 1) + \frac{2\beta}{\ln \lambda}(e^{-(v_t + 2 - v_g^*)\ln \lambda} - \frac{1}{\lambda^2})$$

$$- \frac{2}{\beta}(v_t - v_g^*)$$

where   (5)

$$f = \frac{2}{\ln \lambda}(e^{(v_s - 2)\ln \lambda} + e^{-v_s \ln \lambda} - \frac{1}{\lambda^2} - 1) + 2v_s$$

We can split (5) in to two equations: one that is a good approximation for the inversion and depletion regions, another for the accumulation region. An approximate solution for v_t in inversion and depletion, then can be determined from

$$f = \frac{2\beta}{\ln \lambda}(e^{-(v_t + 2 - v_g^*)\ln \lambda} - \frac{1}{\lambda^2}) - \frac{2}{\beta}(v_t - v_g^*)$$

with a solution   (6)

$$v_t = \frac{1}{\ln \lambda}W[\frac{\beta^2}{\lambda^2}e^{(\frac{f\beta \ln \lambda}{2} + \frac{\beta^2}{\lambda^2})}] - \frac{f\beta}{2} - \frac{\beta^2}{\lambda^2 \ln \lambda} + v_g^*$$

where W is the Lambert function that is the principal-branch solution to the equation $W(x)e^{W(x)} = x$, [11, 12]. The remaining first two terms of (5) are used to approximate v_t at accumulation as

$$f = \frac{2}{\beta \ln \lambda}(e^{(v_t - v_g^*)\ln \lambda} - 1) \text{ with solution}$$

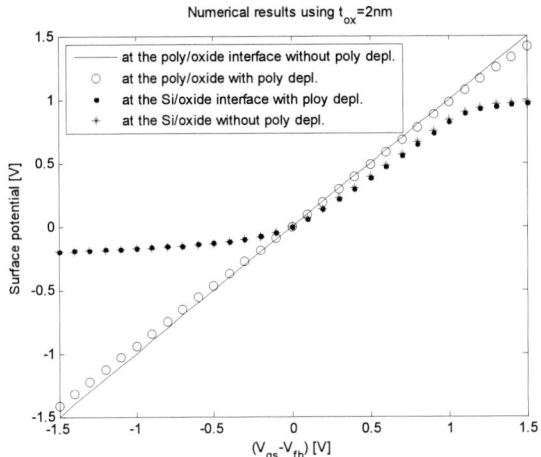

Fig. 3. Surface potential versus relative gate voltage ($\lambda = 10^7$ and $\beta = 10^{-2}$).

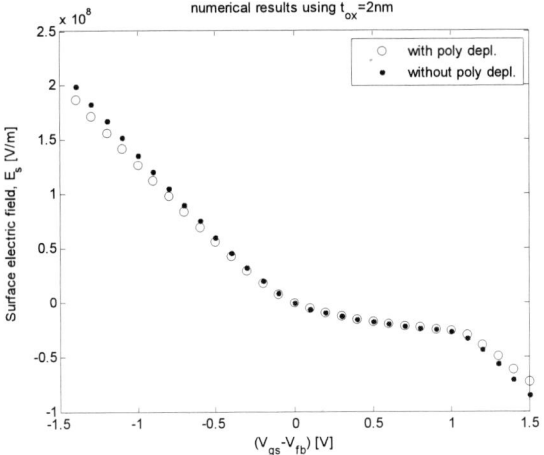

Fig. 4. Surface electric field versus relative gate voltage ($\lambda = 10^7$ and $\beta = 10^{-2}$).

The numerical solutions in Fig. 3 indicate that the poly depletion mainly affects the electrostatic potential v_t rather than v_s. Methods of solving v_s without the poly depletion effect are well established (see [1, 10]). Using equations (3), (4) and neglecting the poly depletion effect on v_s as a first

$$v_t = v_g^* + \frac{1}{\ln \lambda}\ln(\frac{f\beta \ln \lambda}{2} + 1)$$ (7)

1-4244-0267-0/06/$25.00 ©2006 IEEE   115

The surface electric field and gate capacitance are calculated analytically from equations (4), (6) and (7). The results and

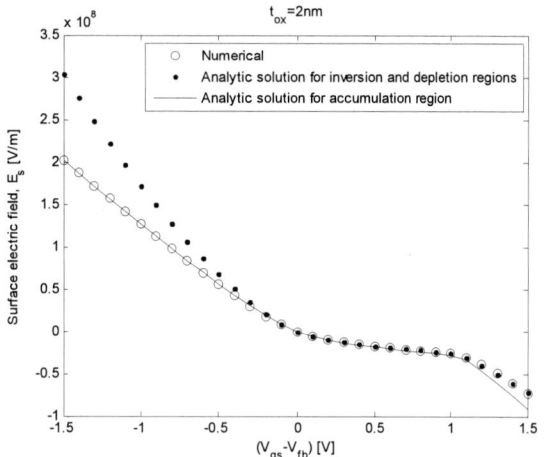

Fig. 6. Surface electric field versus relative gate voltage ($\lambda = 10^7$ and $\beta = 10^{-2}$).

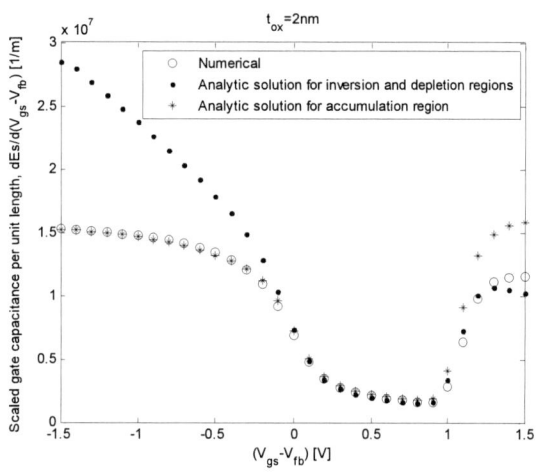

Fig. 7. Scaled gate capacitance versus relative gate voltage ($\lambda = 10^7$ and $\beta = 10^{-2}$).

comparison of the surface electric field with the exact numerical results are shown in Fig. 6. The gate capacitance results are shown in Fig. 7.

III. CONCLUSION

The results shown in Figures 6 and 7 indicate that our analytical formulae give excellent results in most of the device operational regions. The largest error (11%) of our analytical gate capacitance model compared with numerical occurs at strong inversion. Further examination of different oxide thicknesses and doping levels also gives similar excellent results in most of the regions. The largest error results are shown in Fig. 8 (at strong inversion).

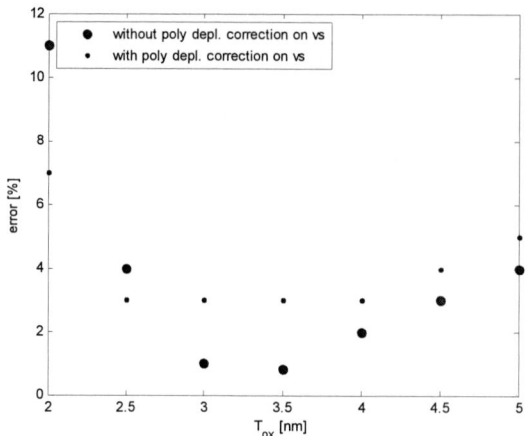

Fig. 8. The largest percentage error between analytical and numerical results versus oxide thickness ($\lambda = 10^7$ and $\beta = 10^{-2}$).

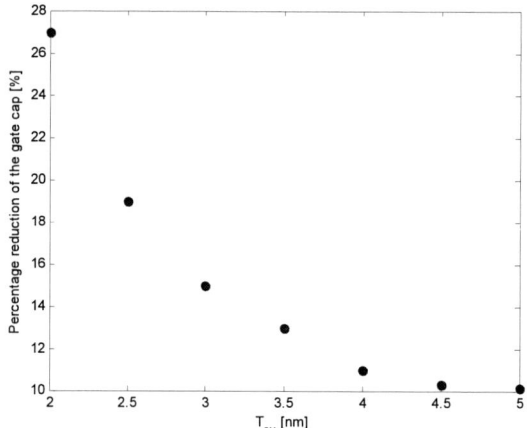

Fig. 9. The largest percentage reduction of the gate capacitance due to poly depletion versus oxide thickness (numerical simulation for $\lambda = 10^7$ and $\beta = 10^{-2}$).

We tried to find out the source of this relatively large error at strong inversion. Initially we considered the error to be due to our assumption of negligible poly depletion effect on v_s at strong inversion. It is shown in [13] that poly depletion has some effect on v_s at strong inversion. A first order approximation using asymptotic methods for the decrease of v_s due to poly depletion is found to be

$$\Delta v_s = -\frac{2}{\ln \lambda} \ln[1 - \frac{\beta c_{ox}^2}{2\lambda}(v_g^* - 1 - \varphi - 2\frac{\ln(\ln \lambda)}{\ln \lambda})\ln \lambda].$$

(8)

Including this correction for the poly depletion effect on v_s at strong inversion does not consistently improve the C-V results and this area needs further investigation (see Fig. 8).

Our analysis (see Fig. 9) also confirms that the main source for the poly depletion effect is a significant reduction of the oxide thickness. MOSFET performance degradation occurs when the oxide thickness is reduced to low nanometer values.

APPENDIX

The rational function

$$
y = z - \frac{z(z-1)}{1+z} + \frac{1}{2}\frac{z(z-1)^2}{(1+z)^3} - \frac{1}{6}\frac{(z-1)^3(z-2z^2)}{(1+z)^5} +
$$

$$
\frac{1}{24}\frac{z(6z^2-8z+1)(z-1)^4}{(1+z)^7} - \frac{1}{120}\frac{z(24z^3-58z^2+22z-1)(z-1)^5}{(1+z)^9} +
$$

$$
\frac{1}{720}\frac{z(120z^4-444z^3+328z^2-52z+1)}{(1+z)^{11}}
$$

$$
- \frac{1}{5040}\frac{z(720z^5-3708z^4+4400z^3-1452z^2+114z-1)}{(1+z)^{13}}
$$

with $z = x/e$, provides a good approximation to the Lambert function $W(x)$ for $x < 8$ and the function

$$
y = L_1 - L_2 + \frac{L_2}{L_1} + \frac{1}{2}\frac{L_2(-2+L_2)}{L_1^2} + \frac{1}{6}\frac{L_2(6-9L_2+2L_2^2)}{L_1^3}
$$

$$
+ \frac{1}{12}\frac{L_2(-12+36L_2-22L_2^2+3L_2^3)}{L_1^4} +
$$

$$
\frac{1}{60}\frac{L_2(60-300L_2+350L_2^2-125L_2^3+12L_2^4)}{L_1^5}
$$

with $L_1 = \ln(x)$ and $L_2 = \ln\ln(x)$ is a good approximation to $W(x)$ for $x \geq 8$. These two formulae together provide a good compact formula for $W(x)$ over the entire positive domain, [12].

REFERENCES

[1] E. Cumberbatch, H. Abebe, and H. Morris, "Current-voltage characteristics from an asymptotic analysis of the MOSFET equations," *J. of Engineering Mathematics*, vol. 39, pp. 25-46, (2001).

[2] N. Arora and C. Huang, "Modeling the polysilicon depletion effect and its impact on submicrometer CMOS circuit performance," *IEEE Transactions on Electron Devices*, Vol. 42, No. 5, May (1995).

[3] R. Rios, N. D. Arora and C. L. Huang, "An Analytic polysilicon depletion effect model for MOSFET's," *IEEE Electron Device Letters*, Vol. 15, No. 4, April (1994).

[4] G. Gildenblat, T. L Chen and P. Bendix, "Analytical application for perturbation of MOSFET surface potential by polysilicon depletion layer," *IEE Electronic Letters*, Vol.. 35, No. 22, 28th October (1999).

[5] C.-H, Choi *et al..*, "Gate length dependent polysilicon depletion effects," *IEEE Electron Device Letters*, Vol. 23, No. 4, April (2002).

[6] S. Lo, D. Buchanan, and Y. Taur, "Modeling and characterization of quantization, polysilicon depletion, and direct tunneling effects in MOSFETs with ultrathin oxides." *IBM J. Res. Develop.* Vol. 43, No. 3, May (1999).

[7] A. Gupta *et al.*, "Accurate determination of ultrathin gate oxide thickness and effective polysilicon doping of CMOS devices," *IEEE Electron Device Letters*, Vol. 18, No. 12, December (1997).

[8] F. Gamiz *et al.*, "Effect of polysilicon depletion charge on electron mobility in ultrathin oxide MOSFETs," *Semiconductor Science and Technology*, Vol. 18, N0. 11, November (2003).

[9] R. F. Pierret, *Field Effect Devices*, Addison-Wesley, second edition, Vol. IV, (1990).

[10] T. L. Chen and G. Gildenblat, "Analytical approximation for the MOSFET surface potential," Solid State Electronics, 45, pp.335-339, (2001).

[11] A. Ortiz-Conde, F. J. Garcia Sanchez, M.Guzman, "Exact Analytical Solution of Channel Surface Potential as an Explicit Function of Gate Voltage in Undoped-body MOSFETs Using the Lambert *W* function and a Threshold Voltage Definition Therefrom," Solid-State Electronics 47 pp. 2067-2074 (2003).

[12] R. Corless, G. Gonnet, D. Hare, D. Jeffrey, and D. Knuth, "On the Lambert W function," Advances in Computational Mathematics 5(4): 329-359 (1996).

[13] E. Cumberbatch, H. Abebe, H. Morris and V. Tyree, "Analytical Surface Potential Model with Polysilicon Gate Depletion Effect for NMOS." *Proceedings 2005 Nanotechnology Conference*, Vol. 3, pp. 57-60, Anaheim, CA, May 8-12, (2005).

1-4244-0267-0/06/$25.00 ©2006 IEEE

Compact Models for the I-V Characteristics of Double Gate and Surround Gate MOSFETs

H. Morris, E. Cumberbatch, H. Abebe, and V. Tyree, Member, IEEE

Abstract— **The models presented by Lu and Taur, [1], for lightly doped double gate and surround gate MOSFETs each require numerical solution of a transcendental equation. In this paper we present compact solutions for the equations based on the Lambert function, [2]. These solutions are shown to be accurate compared with exact numerical solutions.**

Index Terms— **Analytic solutions, compact model, double gate MOSFETs, surround gate MOSFETs**

I. INTRODUCTION

The usual single-gate MOSFET is approaching its minimum channel length due to short channel effects (SCE) and the limit imposed by gate oxide tunneling. As CMOS scaling continues to be beyond the 50nm node several non standard MOSFET designs are being investigated. In particular, the double gate (DG) and surround gate (SGT), and are aimed to minimize the short-channel effects (SCE). Taur and Lu [1] have derived expressions for the I-V characteristics for these devices which have the geometries shown in figure 1.

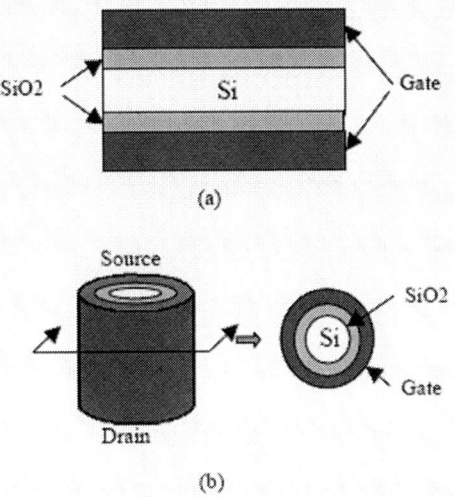

(a)

(b)

This work was supported by MOSIS service. The authors would like to thank Cesar Pina for funding the work.

Henok Abebe and Vance Tyree are with the University of Southern California Information Sciences Institute, MOSIS service, 4676 Admiralty Way, Marina del Rey, CA 90292, U.S.A. (e-mails: abebeh@mosis.org, tyree@ISI.edu).

Ellis Cumberbatch is with Claremont Graduate University, Claremont, CA 91711, U.S.A (e-mail: ellis.cumberbatch@cgu.edu)

Hedley Morris is with San Jose State University, San Jose, CA 95192, U.S.A (e-mail: morris@math.sjsu.edu)

Figure 1. (a) The structure of a double gate (DG) MOSFET (b) The structure of a surround gate (SGT) MOSFET

For each configuration, the formulae obtained in [1] require the solution of a transcendental equation for an intermediate function β. These transcendental equations must be solved numerically. As a result the formulae in [1] are not useful in practice as they cannot be used in device simulators such as SPICE. The objective of this paper is to remedy this situation by providing analytical approximations for these functions expressed in terms of the LambertW function [2] together with fast algorithms for their evaluation. In section II we summarize the results of [1] for the DG and SGT devices. We refer to the original papers for their derivation. This section serves to establish our notation. In section III we construct analytic solutions for both the DG and SGT case in terms of the LambertW function [2]. These approximations are not a set of local approximations blended together, but single, high accuracy, expressions. In section IV we present fast algorithms for all functions required by the formulae in section III. In section V we compute the $I_{ds} - V_g$, $I_{ds} - V_{ds}$ characteristics for DG and SGT devices of various sizes. We close, in section VI, with a brief summary of our results.

II. EQUATIONS AND SOLUTION

A. The Double Gate

For the DG device Taur [1][3-4] has derived the current expression

$$I_{ds} = I_{ds0}[\beta \tan \beta - \frac{1}{2}\beta^2 + r\beta^2 \tan^2 \beta]_{\beta_D}^{\beta_S}$$

$$I_{ds0} = \mu \frac{W}{L} \frac{4\varepsilon_{si}}{t_{si}} (\frac{4kT}{q})^2 \qquad (1)$$

where β is a solution of the equation

$$\ln \beta - \ln(\cos \beta) + 2r\beta \tan \beta = v$$

$$v = \frac{q(V_g - \Delta\phi - V)}{2kT} - \ln(\frac{2}{t_{si}}\sqrt{\frac{2\varepsilon_{si}kT}{q^2 n_i}}) \qquad (2)$$

where t_{si} is the channel thickness, L and W are the device length and width, μ is the effective mobility, $\Delta\phi$ is the work function difference at the gates V_g is the gate voltage and V is the applied source drain voltage. Note that β lies in the range $0 \le \beta \le \pi/2$.

B. The SurroundGate

For the SGT device [1] [5] the current expressions are

$$I_{ds} = I_{ds0}\left[\frac{\eta}{4\beta^2} + \frac{1-\eta/2}{\beta} + \frac{1}{2}\ln\beta\right]_{\beta_D}^{\beta_S}$$

$$I_{ds0} = \mu\frac{4\pi\varepsilon_{si}}{L}\left(\frac{2kT}{q}\right)^2$$

(3)

where β is a solution of the equation

$$\ln(1-\beta) - \ln\beta^2 + \eta\left(\frac{1-\beta}{\beta}\right) = v$$

(4)

$$v = \frac{q(V_g - \Delta\phi - V)}{kT} - \ln\left(\frac{8kT\varepsilon_{si}}{q^2 n_i R^2}\right)$$

In this expression R is the radial thickness and plays a similar role to t_{si} in the DG device. All other variables are the same is described for the DG device. Note that β lies in the range $0 \le \beta \le 1$.

III. ANALYTIC APPROXIMATIONS

A. The Double Gate

Figure 2 shows a schematic of a DG MOSFET. In this case [1][3][4] the defining equation for β is

$$\ln(\beta \sec \beta) + 2r\beta \tan \beta = v$$

(5)

where v is defined in (2).

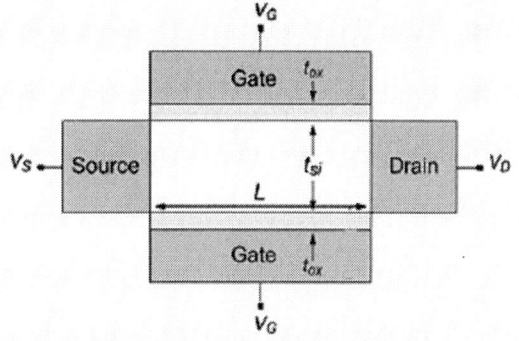

Figure 2. The structure of a double gate (DG) MOSFET

This equation can be recast into the form

$$\chi e^{2r\chi} = e^v \sin\beta$$

$$\chi = \beta \tan\beta$$

(6)

From this we can write

$$\beta = \Phi(\frac{1}{2r}\text{LambertW}(2r\sin\beta))$$

(7)

where $\Phi(z)$ is the solution of $z = \Phi\tan\Phi$. As $0 \le \beta \le \pi/2$ the argument of the LambertW function always remains positive. Accurate approximation for LambertW and the function $\Phi(z)$ are given in the next section. An iterative solution can easily be found to equation (7). For an initial solution we use the low voltage approximate solution

$$\beta = \frac{e^{-v}}{2(1+4r)}\left(\sqrt{1+2(1+4r)e^{2v}} - 1\right)$$

(8)

Using only four iterations one the obtains excellent results

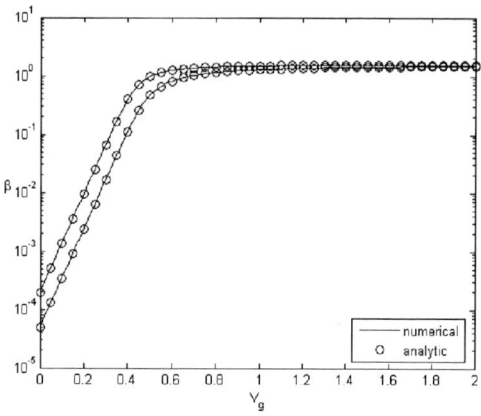

Figure 3. Exact and numerical solution to (5) for $t_{si} = 5\text{nm}$ and 20nm.

Shown in figure 3 which compares an exact numerical solution with out approximate formulae.

B. The Surround Gate

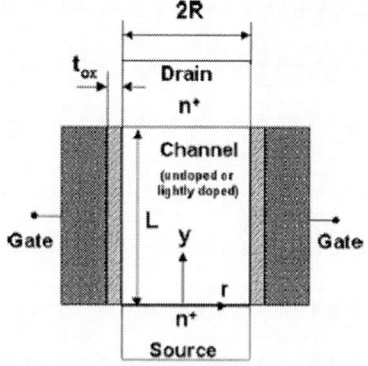

Figure 4. The structure of a surround gate (SGT) MOSFET

In this case the defining equation for β is [1][5]

1-4244-0267-0/06/$25.00 ©2006 IEEE

$$\ln(\frac{1-\beta}{\beta^2}) + \eta(\frac{1-\beta}{\beta}) = \nu \qquad (9)$$

where ν is defined in (4) and $\eta = 4\ln(1+\frac{t_{ox}}{R})$. If we

introduce $\varsigma = \frac{1-\beta}{\beta}$ this equation can be recast into the form

$$e^{\eta\varsigma/2}\varsigma = \frac{1}{\sqrt{(1+\varsigma^{-1})}}e^{\nu/2} \qquad (10)$$

This can further be written in the form

$$\varsigma = \frac{2}{\eta}\text{LambertW}(\frac{1}{2\sqrt{1-\varsigma^{-1}}}\eta e^{\nu/2}) \qquad (11)$$

From this we obtain the equation

$$\beta = \frac{1}{1+\frac{2}{\eta}\text{LambertW}(\frac{1}{2}\sqrt{1-\beta}\eta e^{\nu/2})} \qquad (12)$$

As $0 \le \beta \le 1$ the argument of the LambertW function always remains positive. This equation can be accurately solved by iterative techniques. Figure 4 shows a comparison of an exact numerical solution of (9) with an approximate solution based on (12) with four iterations starting from $\beta = 0$. We follow [1] in using the parameter values $t_{ox} = 1.5\,\text{nm}$ and R=5nm and 20nm. It is necessary to use a log plot in order to see the different curves.

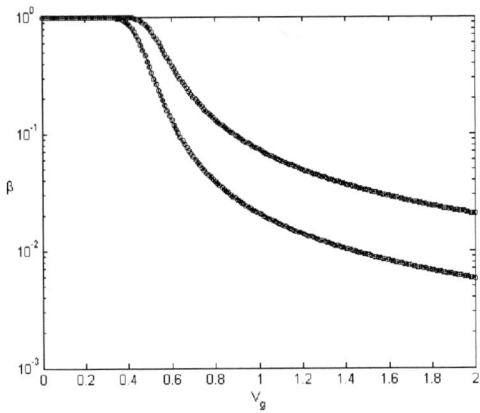

Figure 5. A comparison of the exact and numerical solution to (5) for $R = 5\,\text{nm}$ and $20\,\text{nm}$.

IV. FUNCTION APPROXIMATION

In the previous sections we have introduces the LambertW and function $\Phi(z)$ as the basis for our compact analytic formulae. For our results to be of practical use we have to provide fast computational algorithms for each of those functions.

A. The function $\Phi(z)$

If we define $z = \Phi \tan \Phi$ an accurate approximate solution is given by

$$\Phi(z) = \varsigma - (\varsigma \tan(\varsigma) - z)/(\tan(\varsigma) + \varsigma \sec^2(\varsigma))$$

$$\varsigma = \sqrt{\frac{z\gamma + 1 - \sqrt{(z\gamma - 1)^2 - 4z(z\delta - 1/3)}}{2(z\delta + \gamma - 1/3)}} \qquad (13)$$

$$\gamma = \frac{40}{9\pi^2}, \delta = \frac{16}{9\pi^4}$$

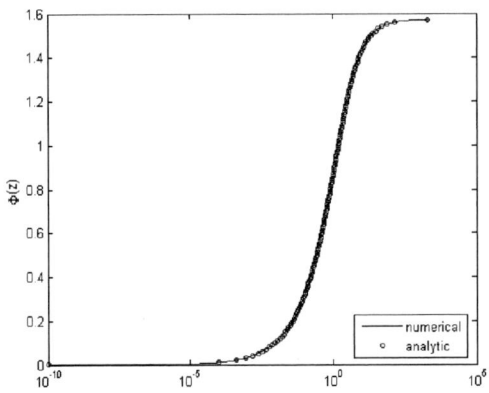

Figure 6. A comparison of the exact and analytic solution to $z = \Phi \tan \Phi$

Figure 6 shows a log-plot comparison between the exact numerical solution for $\Phi(z)$ and the approximate solution (13). The maximum relative error using this two-line function is 0.5344e-5.

B. Approximating the LambertW function

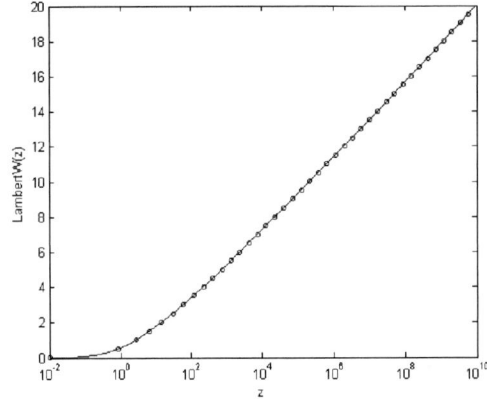

Figure 7. A comparison of the exact and analytic solution to $z = W \exp(W)$.

Figure 7. Shows a log-plot comparison between the exact numerical solution for LambertW(z) and the approximate solution presented in equations (14) and (15). The maximum relative error is 5.4e-3.

The rational function [2]

$$y = z - \frac{z(z-1)}{1+z} + \frac{z(z-1)^2}{2(1+z)^3} - \frac{(z-1)^3(z-2z^2)}{6(1+z)^5} +$$

$$\frac{z(6z^2-8z+1)(z-1)^4}{24(1+z)^7} - \frac{z(24z^3-58z^2+22z-1)(z-1)^5}{120(1+z)^9}$$

$$+\frac{z(120z^4-444z^3+328z^2-52z+1)}{720(1+z)^{11}}$$

$$-\frac{z(720z^5-3708z^4+4400z^3-1452z^2+114z-1)}{5040(1+z)^{13}}$$

(14)

with $z = x/e$, provides a good approximation to LambertW(x) for $x < 8$ and the function

$$y = L_1 - L_2 + \frac{L_2}{L_1} + \frac{L_2(-2+L_2)}{2L_1^2} + \frac{L_2(6-9L_2+2L_2^2)}{6L_1^3}$$

$$+\frac{L_2(-12+36L_2-22L_2^2+3L_2^3)}{12L_1^4} +$$

(15)

$$\frac{L_2(60-300L_2+350L_2^2-125L_2^3+12L_2^4)}{60L_1^5}$$

with $L_1 = \ln(x)$ and $L_2 = \ln\ln(x)$, is a good approximation to LambertW(x) for $x \geq 8$. These two formulae together provide a good compact formula for LambertW(x) over the entire positive domain.

V. RESULTS

A. The Double Gate

Figures 8 and 9 compare the $I_{ds} - V_g$ and $I_{ds} - V_{ds}$ characteristics for a typical DG device.

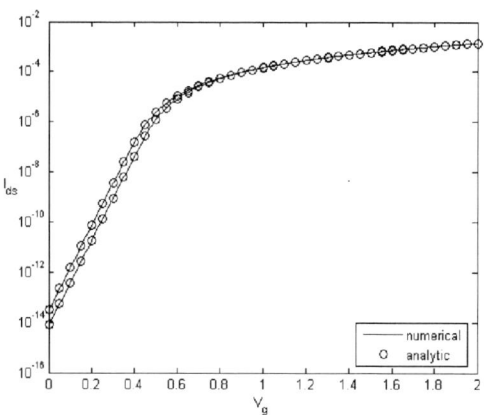

Figure 8. $I_{ds} - V_g$ characteristics obtained from the numerical and analytic solution with $V_{ds} = 2$, and $\Delta\phi = 0$, for t_{si} =5nm (bottom) and 20nm (top) and $\Delta\phi = 0$.

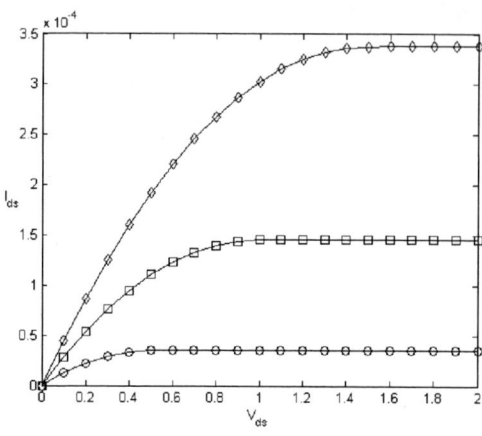

Figure 9. $I_{ds} - V_{ds}$ characteristics with $t_{si} = 5$nm and $\Delta\phi = 0$ and $V_g = 2$(top), 1.5(middle), 1.0(bottom).

B. The Surround Gate

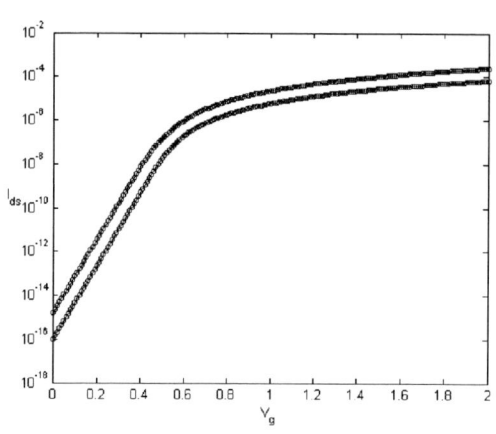

Figure 10. $I_{ds} - V_g$ characteristics with $V_{ds} = 2$, for R =5nm (bottom) and 20nm(top) and $\Delta\phi = 0$.

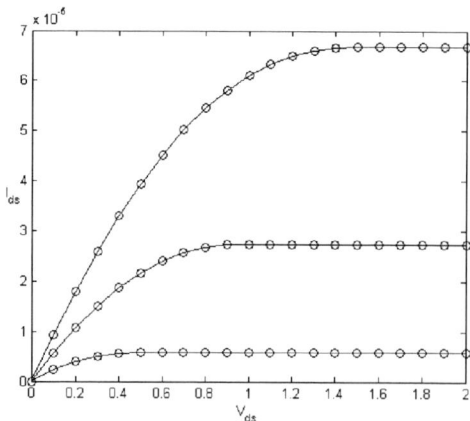

Figure 11. $I_{ds} - V_{ds}$ characteristics with R=5nm, $\Delta\phi = 0$ and $V_g = 2$(top), 1.5(middle), 1.0(bottom).

1-4244-0267-0/06/$25.00 ©2006 IEEE

Figures 10 and 11 compare the $I_{ds} - V_g$ and $I_{ds} - V_{ds}$ characteristics for a typical SGT device. The parameter vales used were those used in [1].

VI. CONCLUSIONS

We have presented compact models for both regular double gate (DG) and surround gate mosfet (SGT) based on the theoretical results of [1]. As a result we have provided practical formulae that can be used in simulators such as SPICE.

REFERENCES

[1] H. Lu and Y. Taur, "Physics-Based, Non-Charge-Sheet Compact Modeling of Double Gate MOSFETs," Nanotech, Anaheim, Ca., 2005.

[2] R. Corless, G. Gonnet, D. Hare, D. Jeffrey, and D. Knuth, "On the Lambert W function", Advances in Computational Mathematics 5(4): 329-359 (1996).

[3] Y. Taur, X. Liang, W. Wang and Hu. Lu, A continuous, analytic drain-current model for DG MOSFETs, IEEE Electron Device Letters, Volume 25, Issue 2, 2004, pp:107 - 109.

[4] Y. Taur, "An analytical Solution to a Double-Gate MOSFET with Undoped Body," IEEE Electron Device Lett , vol 21.,no.5, pp. 245-247.

[5] D. Jimenez, D., B. Iniguez, B., J. Sune, L.F. Marsal, J. Pallares, J..Roig and D. Flores, "Continuous analytic I-V model for surrounding-gate MOSFETs", IEEE Electron Device Lett., Volume 25, Issue 8, Aug. 2004, pp: 571 - 573.

1-4244-0267-0/06/$25.00 ©2006 IEEE

Evaluation of a Double Implanted Diffused MOSFET for Analog Operation

Eric J. Basham*, David W. Parent
Electrical Engineering
San Jose State University
One Washington Square, San Jose, CA 95192-0084
* correspondence to ebasham@email.sjsu.edu

A methodology is presented to evaluate the low power operation of a laterally diffused implanted MOS transistor. The transistor was fabricated in a standard 1.5um CMOS process by shifting the alignment between the p-well and the gate, resulting in an asymmetric channel doping and a lowly doped drift region. Fabrication was necessary to have detailed knowledge of the processing conditions and the resultant junction locations. Following the fabrication, devices were tested and shown to have improved performance as compared to standard CMOS transistors as a function of the heritage of the device.

I. INTRODUCTION

ANALOG design is a required part of heavily integrated systems, and this trend is likely to accelerate. It is clear that scaling for digital performance can have severe impacts on the analog performance characteristics of the transistors and this is poorly predicted by traditional modeling methods [1]. Falling power supplies, part of constant field scaling, necessitate the need for designs which operate with low voltage headroom, but traditional design methodologies fail to facilitate this type of design. A recently developed approach, the "gm/id" or "level of inversion" design methodology [2] provides several convergent solutions. These tools are useful for aiding engineering insight into transistor operation and developing a more intuitive understanding of circuit operation. Due to the closed form nature and the accurate prediction of transistor operation through all levels of channel inversion, this approach allows the comparison of transistor operation across process technologies [3]. This in turn can be applied to allow transistor designers direct insight into the analog performance space the circuits built with these transistors will encompass.

Integration of specialized transistors optimized for analog design is an expensive option. However, it becomes necessary for integration of high voltage applications, such as motor control and I/O functions. Indeed, higher breakdown source-drain voltages are increasingly required as a result of scaling efforts and integrating more diverse systems in a single chip. One common approach to inclusion of these devices is an additional well or a shifted well which is diffused to create an asymmetric device. The class of these power devices is commonly referred to as double implanted or double diffused lateral MOS, first introduced in [4].

Originally conceived as a method of exceeding photolithography limits to achieve smaller channel lengths, only limited exploration of LDMOS analog operation was performed [5], in part due to the difficulty of correctly predicting the junction positions and threshold voltages. Since modern submicron transistors have very narrow thermal budgets and very tight process control is required for nm scale devices, these sensitivities have been largely mitigated.

Thus, system-on-a-chip technology has driven a convergence. Cost and reliability drivers require the inclusion of analog in mixed signal systems. Mixed signal systems in submicron, thin oxide, low Vdd processes require I/O and power devices. Higher speeds and more processing power drive the focus towards low power design. It seems reasonable to explore the low-power performance of devices already included in many processes.

II. DESIGN APPROACH

A. Level of Inversion Design Methodology

A brief introduction to the design methodology follows. The methodology was first elucidated in [6] and further outlined in [2] and [7]. At the core of the approach is the Transconductance Efficiency Curve Figure of Merit [8] (Fig. 1), where the ratio of gain to bias current is plotted versus the inversion coefficient. The function is describing this behavior from weak to strong inversion is equation (1).

$$\frac{g_m}{I_D} = \frac{1}{nU_T} \cdot \frac{(1 - e^{-\sqrt{IC}})}{\sqrt{IC}} \tag{1}$$

Where I_D is the drain bias current, IC is the inversion coefficient defined in (3), U_T is the thermal voltage, n is the subthreshold slope factor or body effect coefficient defined in (4), and g_m is the transistor transconductance. The inversion coefficient (IC) is defined as the bias current normalized by the technology current (Io, defined in (2)) and the width and the length of the transistor, as defined in (3).

$$Io = 2n\mu_n C_{ox} U_T^2 \tag{2}$$

1-4244-0267-0/06/$25.00 ©2006 IEEE

$$IC = \frac{I_d}{Io\frac{w}{l}} = \frac{I_d}{2n\mu_n C_{ox}\frac{w}{l}U_T^{\,2}} \qquad (3)$$

Where Cox is the oxide capacitance, w and l the transistor dimensions and μ_n the mobility. This FOM gives direct insight into the amount of charge in the channel as a function of the bias current applied to the drain. As shown in dark grey, the strong inversion, saturated region, far to the right of moderate inversion is the typical design space that evolves from using the transistor bias current as the main parameter in design and analysis. In bulk this is because the common design methodologies employ a square law modeling of the transistor and these overpredicts the saturation current, and drain voltage V_{dsat} at which this occurs. The portion of the slope between moderate inversion and the design space typifies the headroom associated with conventional design methodologies.

In contrast, the level of inversion is identified and treated as a design variable. The bias current falls out as a result of the inversion coefficient (IC) selection, which is in turn determined by the gain and frequency requirements of the circuit. Length is selected as a function of the desired gain, mismatch, noise, operation frequency and other design considerations. This is nicely presented as a design figure of merit in [9]. A detailed example employing this design approach for a common source amplifier design is included in appendix A.

III. PERFORMANCE ESTIMATION

As can be seen from the Transconductance Efficiency Curve Figure of Merit low frequency gain in weak inversion is a function of the technology factor or substrate factor, n, where n can be approximated by (4) [10]. Interface state charge, C_{it}, is usually negligible in comparison and thus ignored. C_{depl} is the depletion width capacitance in the substrate.

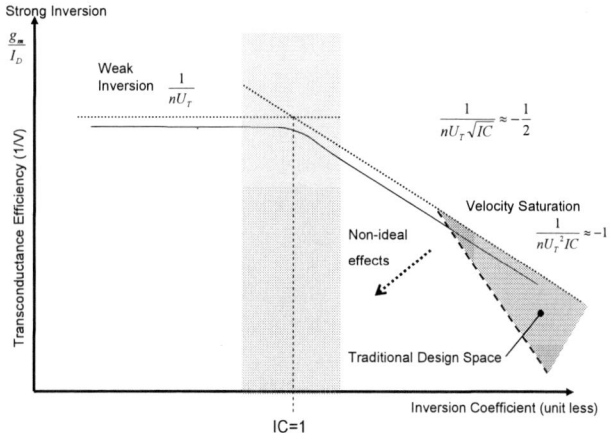

Fig. 1. Transconductance efficiency curve figure of merit. Non-ideal effects bend the curve depicting the gm/Id ratio down, increasing the slope and thus reducing the total achievable gain.

$$n = 1 + \frac{C_{depl} + C_{it}}{C_{ox}} \qquad (4)$$

An increase of the technology current has the effect of reducing the width at which a particular value of IC is achieved. This would in turn reduce the gate capacitance and increase the intrinsic frequency of the transistor.

Referring to Figure 1, two important facts are noted. The maximum intrinsic gain achieved per unit of bias current falls as the transistor enters strong inversion. The second point is that the maximum gain of a unity sized transistor is a function of the process dependant parameter n. Thus, in considering analog transistor design, three variables become of primary concern. The substrate factor n, and the technology leakage current, Io, and the output conductance which affects achievable gain and switching time.

While the gate overlap capacitance produced as a result of the non-self aligned fabrication of power MOS devices limits the upper frequency of these transistors, the reduction in output conductance vastly improves low frequency single stage gain. This improvement in the channel length modulation effect is as a direct result of the asymmetric nature of the channel doping. Near the drain, there is a lowly doped drift region which abuts a higher doped channel region. This has the effect of modulating the depletion region in the drift region, rather than in the channel. Since the drift region is conductive at all levels of inversion, there is little channel length modulation. In addition, because the doping of the channel is higher near the source, and lower near the drain/drift region, this effect is also predicted to have positive effects on the output conductance, according to [11]

$$V_{ea} = V_{ea[HD]} \cdot \left(\frac{g_m}{g_{ds}}\right)_{[LD]} \qquad (5)$$

where V_{ea} describes the channel length modulation effect, and [HD] and [LD] refer to the parameters of interest in the highly doped and lowly doped regions. Unfortunately, there are also rather important consequences to the drain current as this is a function of the total device length.

The substrate factor is improved by reducing the total capacitance in the bulk by reducing the channel doping. In effect, very low doping can be used in the channel region without incurring severe channel length modulation effects. This should also have positive benefits upon the non-surface mobility due to reduced substrate doping, increasing the device current. The net effect is to reduce the area transistors operating in weak inversion occupy, as this can be considerable.

Using an asymmetric device with diffused junctions should also improve the technology current. While it is a nuisance for power devices to have a measurable off state leakage current, operation in the subthreshold region is actually improved by higher leakage current. Below threshold the MOS device conducts mainly due to diffusion rather than drift mechanisms. Because of this leakage across the junction is more appropriately described by diffusion currents describing BJTs and thus an asymmetrically doped device, following a junction gradient from S-C-D similar to E-B-C, should also increase the diffusion currents.

One important consideration is that the substrate current must occur near the surface of the device, rather than deep in the bulk or it does not come under gate control. There is an optimum design region where too low a channel doping results in a punch through like effect and too high a doping results in a drastically reduced depletion width [12]. Both of these cases are well described by the parameter n, where the first case is a virtual increase in C_{ox} as a result of the depth of the depletion region and the second case as an increase of C_{depl}. Both cases increase n.

IV. SIMULATION

Simulation was performed to determine junction positions using Silvaco's Athena process simulation. Due to the inherent variability as part of a processing in a research facility, the simulation was iterated each time a process step was performed to incorporate measured variability with the intention of increasing final yield. Correct meshing around the diffusion region was found to be critical. Process simulation files are available on line at: http://www.engr.sjsu.edu/~dparent/Silvaco/index.htm.

A reduced impact of well alignment on threshold voltage is expected as the threshold voltage is dependant upon the highest doped portion of the channel, which reaches a maximum at the source side of the channel. Simulation showed that the estimated channel length was approximately 2.9uM when the gate and P-well were aligned.

V. FABRICATION PROCESS

N-type (P= 1E14, <100>) wafers were oxidized in the furnace to create an implant protection oxide, and then implanted (Do=1E11, 140KeV, SP=P, Tilt=7) to standardize the surface doping concentration. Mask 1 defined the P-wells

of the CMOS device by protecting the areas that remained N-well (since it is was initially an n-wafer). Wafers were implanted (Do=3E13, 100KeV, Sp=B11, Tilt=7) and then etched in BOE to remove the protective oxide. Wafers were wet oxidized in the furnace to form the field oxide and then a drive-in step was performed defining the channel width in the asymmetric devices and well depth for the self aligned MCOS devices. Mask 2 defined the active area for the transistors, and subsequently a 630 Å dry oxide was grown for the gate oxide. 5000A of polysilicon was sputter deposited and then implanted (Do=1E15, 130Kev, SP=P, Tilt=7) to make the poly conductive. Mask 3 defined the poly lines. Mask 4 defined the P-Select source and drain regions of the PMOS which were opened over the common n-wells. The window included the poly gate, but since the boron implant (Do=1.5E13, 60KeV, Sp=Bf2, Tilt=7) did not penetrate the gate, only the areas to the left and right of the gate were doped. This formed the standard self-aligned PMOS transistors. Mask 5 defined the N-Select regions of the NMOS source and drain. Implant dose was (Do=5E15, 50KeV, Sp=As, Tilt=7). Contact holes were etched by defining Mask 6 and etching through the FOX with Reactive Ion Etching (RIE). Metal was deposited by evaporating Al or sputter deposition of an Al (1%Si) layer onto the wafers. Mask 7 defined the metal traces the wafers were low temperature annealed in forming gas at (450 C). Once removed from anneal, the fabrication processing was completed and the devices were ready for testing.

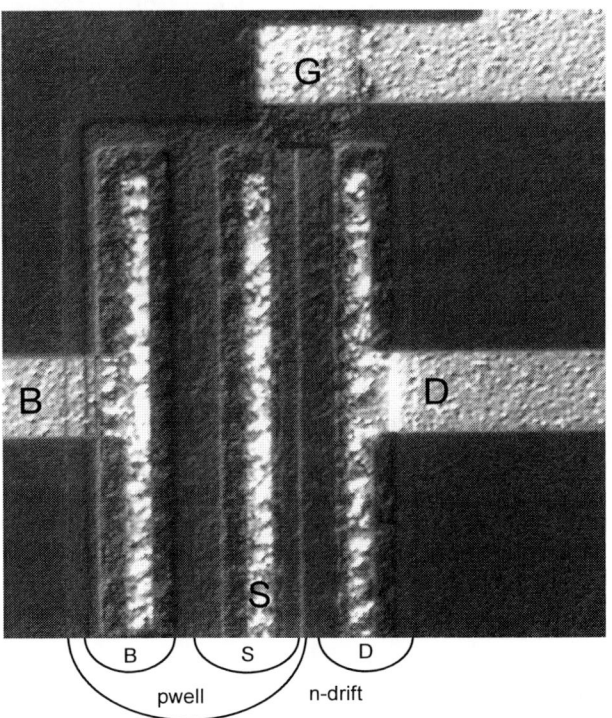

Fig 3. Nomarsky DIC Micrograph of Fabricated LDMOS Device. Overlapping of p-well and source and drain regions is clearly seen. The masking location, not final diffused location can easily be determined visually, but are diagrammed for additional clarity. This device was aligned with no offset, so the edge of the p-well and the left edge of the polysilicon gate are aligned.

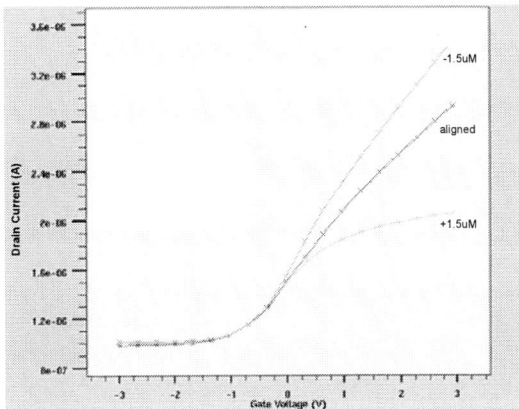

Fig. 2. Estimating the impact of voltage shifts to the threshold voltage.

1-4244-0267-0/06/$25.00 ©2006 IEEE

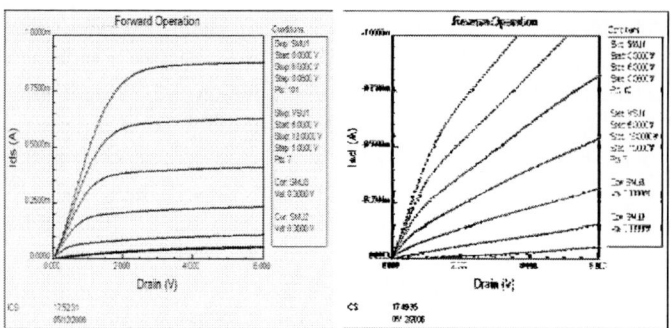

Fig. 4. Forward (drift region on the drain side) versus reverse (drift region on the source side) operation of the LDMOS device.

Some notes on the fabrication of the asymmetric (LDMOS) devices are in order. To form the asymmetric device the Pwell was aligned to the edge of the poly gate. Thus these devices were not self aligned. The Pselect was used to form the body contact in the pwell, next to the source. The source and body contacts were junction isolated. Fig 3. shows the implant windows with a diagram which aligns to the features in the micrograph.

VI. Results and Discussion

The most direct way to demonstrate the asymmetric nature of the LDMOS devices is to observe the results of the Id-Vds and Is-Vsd curves, as shown in Fig 4. As is clearly seen the initial channel current is significantly reduced, and the channel length modulation effect increases significantly when the devices are operated source to drain. This is consistent with the predictions in section III because the doping is much higher on the source side and there is no drift region next to the channel.

The clearer way to illustrate these effects are by referring to the TEC FOM curves. As is seen in Fig. 5 the transconductance efficiency of this device can clearly be measured without detailed information about the processing details. Moreover, even in the case of this device, there is a clear definition of the movement from weak, to moderate, to strong to velocity saturation during operation.

Finally, as predicted the leakage current for forward versus reverse operation are not equal. As suggested in section III, the forward operation of the device actually has a higher diffusion current, and thus the technology current is higher. Technology current is extracted according to [13] and shown in Figure 6.

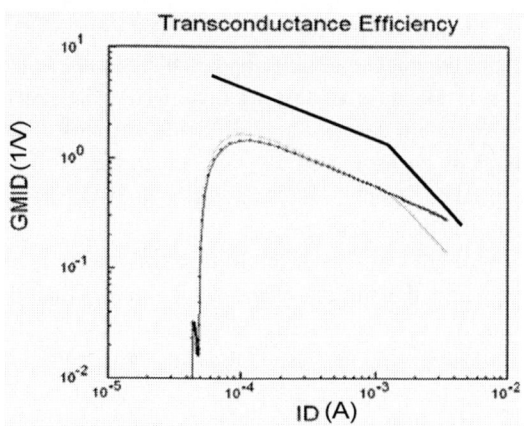

Fig. 5. The bold black line shows the slopes of the strong inversion (slope = -1/2) and velocity saturation (slope = -1) portions of the curve. The light grey curve was collected under conditions which would result in velocity saturation. The weak inversion operation of this device was compromised by high substrate doping. This would have the effect of increasing mobility field reduction effects and increasing the bulk capacitance, which was indeed recorded.

Metric (⇕ goal)	benchmark	LDMOS
Is[A/u,l] ↑	50 – 200	73-75
Vea [v/um,l] ↑	4-8	100 - 150
n [.] ↓	1.3 – 1.5	3.5 @Is

Table 1. Final results for LDMOS device tested according to the method outlined. Arrows indicate desired parameter direction.

Table 1 summarizes the design goals evaluated against a standard CMOS process for 2uM devices as used in the design example in Appendix A. Significant gains were made with respect to channel length modulation effects, but the expected increase in technology current was not observed. Measurement was hampered by a significant increase in the technology factor n.

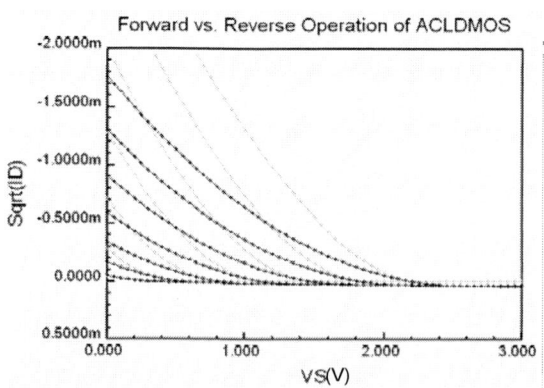

Fig. 6. Specific Current extraction for forward (light grey) and reverse (black) operation. Note the forward operation technology current (Io) is much higher in the forward direction than the reverse. Technology current is calculated as Is=[2U$_t$(slope)]^2 from [13].

VII. Conclusion

A methodology was presented to evaluate the low power operation of a laterally diffused implanted MOS transistor. The transistor was fabricated in a standard CMOS process by shifting the alignment between the p-well and the gate, resulting in an asymmetric channel doping and a lowly doped drift region. This drift region had the effect of drastically increasing the output resistance; however, processing variations resulted in a negative impact to the technology factor and less than expected improvements in the technology current. Both of these impacts were due to the increased threshold voltage as a result of the increased well doping.

Appendix

VIII. An Illustrated gm/Id Design Example

To illustrate this approach an example of gm/id based design for a common source amplifier is shown below, extracted from the general approach presented in [7]. The pedantic approach is used to provide a specific result and show a specific design case to illustrate the methodology.

Technology parameters for an example process are given as:

$$V_{early} = 8 \frac{V}{\mu m}$$

$$U_T = .0259V$$

$$n = 1.35$$

$$\mu_n = 550 \cdot 10^{-4} \frac{V^2}{cm \cdot s}$$

$$\varepsilon_{ox} = 0.345 \cdot 10^{-10} \frac{F}{m}$$

$$t_{ox} = 30 \cdot 10^{-9} m$$

$$C_{ox} = \frac{\varepsilon_{ox}}{t_{ox}} = 0.00115 \frac{F}{m}$$

$$K_n = \mu_n C_{ox}$$

$$Io = 2n\mu_n C_{ox} U_T^2$$

These equations allow the determination of a specific current, denoted (Io), characteristic to the process. Basic specifications are then selected for design:

$$f_t = 10Mhz$$

$$|A_v| = 50dB$$

$$C_{Load} = 10pF$$

Length is chosen as a function of the available technology and which operational region is to be optimized (see figure 3):

$$L_{min} = L = 2\mu m$$

Now the general expression to relate (gm/Id) and the bias current (Ido), inversion coefficient (IC) and bias current scaled as a function of transistor size ratio, denoted (I') are introduced. These equations would hold for all transistors in a circuit.

$$\frac{g_m}{I_D} = \frac{1}{nU_T} \cdot \frac{(1 - e^{-\sqrt{IC}})}{\sqrt{IC}}$$

$$I_{do} = \frac{g_m}{\frac{g_m}{I_D}}$$

$$I' = \frac{I_{do}}{\frac{w}{l}}$$

$$IC = \frac{I_d}{2n\mu_n C_{ox} \frac{w}{l} U_T^2} = \frac{Id}{Io \frac{w}{l}} = I' \frac{1}{2n\mu_n C_{ox} U_T^2}$$

With this, we can revisit the basic design equation for a common source NMOS amplifier, rewritten to emphasize the transconductance efficiency expression (gm/Id).

$$f_t = \frac{g_m}{I_D} \cdot \frac{I_{do}}{2\pi C_{Load}}$$

$$|A_v| = \frac{g_m}{I_D} \cdot V_{early} \cdot L$$

This allows us to determine the transconductance efficiency required from the given specifications:

$$\frac{g_m}{I_d} = \frac{10^{\frac{|Av|}{20}}}{V_{early} \cdot L} = 19.76 \frac{1}{V}$$

$$I_{do} = \frac{2\pi f_t \cdot C_{Load}}{\frac{g_m}{I_d}} = 3.177 \cdot 10^{-5} \approx 32uA$$

This allows determination of the inversion coefficient. However, the form

$$\frac{g_m}{I_D} = \frac{1}{nU_T} \cdot \frac{(1 - e^{-\sqrt{IC}})}{\sqrt{IC}}$$

is difficult to use to find a closed form solution for IC. It has been shown that a good approximation is given by [14]:

$$\frac{g_m}{I_D} = \frac{1}{nU_T} \cdot \frac{1}{\frac{1}{2} + \sqrt{\frac{1}{4} + IC}}$$

This is rearranged to give;

$$IC = \frac{nU_T \frac{g_m}{I_D} - 1}{-\left(nU_T \frac{g_m}{I_D}\right)^2}$$

And in this case,

$$IC = 0.6475$$

Which shows the transistor is in moderate inversion, which is generally described as bounded by weak inversion IC<0.1 and strong inversion IC>10. Now, since

$$I' = I_o \cdot IC$$

$$I' = 7.42 \cdot 10^{-8} \text{ A}$$

Where I' is the current the unit sized transistor is required pass. Since L was already specified, the last parameter to determine is (W). Given the bias current to bias in moderate inversion, the gain and thus the needed transconductance efficiency this is a closed form result;

$$W = L \cdot \frac{I_{do}}{I'}$$

$$W = 856uM$$

ACKNOWLEDGMENTS

Cadence Design Systems for donation of the Cadence Design Laboratory. Silvaco for supply of education copies of Athena. Intel for major upgrade funding for the microprocessing engineering lab. SNF CIS Grant supporting teaching lab processing. Orbit Semiconductor for donation of test and measurement equipment. Dan Hicks and Shao Ng for their tireless processing assistance.

REFERENCES

[1] A. J. Annema, "Analog circuit performance and process scaling," *Ieee Transactions on Circuits and Systems Ii-Analog and Digital Signal Processing*, vol. 46, pp. 711-725, 1999.

[2] D. B. D. Foty, and M. Bucher, "Starting Over: gm/ID-Based MOSFET Modeling as a Basis for Modernized Analog Design Methodologies," presented at Technical Proceedings of the 2002 International Conference on Modeling and Simulation of Microsystems, Nanotech 2002, 2002.

[3] D. Flandre, "The gm/ID synthesis methodology: the missing link between symbolic analysis and design automation for MOS integrated analog circuits," vol. PhD: UCL, 1999.

[4] S. Colak, B. Singer, and E. Stupp, "Lateral Dmos Power Transistor Design," *Electron Device Letters*, vol. 1, pp. 51-53, 1980.

[5] M. Hung, "Double Diffused (DMOS) FETs for Analog Applicaitons," in *Electrical Engineering*, vol. PhD: MIT, 1991.

[6] J. Franca and Y. Tsividis, *Design of analog-digital VLSI circuits for telecommunications and signal processing*, 2nd ed. Englewood Cliffs, N.J.: Prentice Hall, 1994.

[7] D. Flandre, "flandre thesis."

[8] D. M. Binkley, M. Bucher, and D. Foty, "Design-oriented characterization of CMOS over the continuum of inversion level and channel length," 2000.

[9] D. M. Binkley, B. J. Blalock, and J. M. Rochelle, "Optimizing drain current, inversion level, and channel length in analog CMOS design," *Analog Integrated Circuits and Signal Processing*, vol. 47, pp. 137-163, 2006.

[10] K. Rajendran and G. S. Samudra, "Modelling of transconductance-to-current ratio (gm/id) analysis on double-gate SOI MOSFETs," *Semiconductor Science and Technology*, vol. 15, pp. 139-144, 2000.

[11] A. Kranti, T. M. Chung, D. Flandre, and J. P. Raskin, "Laterally asymmetric channel engineering in fully depleted double gate SOI MOSFETs for high performance analog applications," *Solid-State Electronics*, vol. 48, pp. 947-959, 2004.

[12] M. Stockinger, "Optimization of Ultra-Low-Power CMOS Transistors," Institut für Mikroelektronik, , 2000.

[13] M. Bucher, C. Lallement, and C. C. Enz, "An efficient parameter extraction methodology for the EKV MOST model," presented at Proceedings of International Conference on Microelectronic Test Structures. Trento, Italy. IEEE Electron Devices Soc. 25-28 March 1996., 1996.

[14] A. I. A. Cunha, M. C. Schneider, and C. Galupmontoro, "An Explicit Physical Model for the Long-Channel Mos-Transistor Including Small-Signal Parameters," *Solid-State Electronics*, vol. 38, pp. 1945-1952, 1995.

Large Stroke Actuators for Adaptive Optics

Bautista Fernández and Joel.A. Kubby
Department of Electrical Engineering
University of California at Santa Cruz
1156 High Street, MS:SOE2
Santa Cruz, CA 95064

ABSTRACT

In this paper we review the use of a 3-dimensional MEMS fabrication process to prototype long stroke (>10 μm) actuators as are required for use in future adaptive optics systems in astronomy and vision science. The Electrochemical Fabrication (EFAB™) process that was used creates metal micro-structures by electroplating multiple, independently patterned layers. The process has the design freedom of rapid prototyping where multiple patterned layers are stacked to build structures with virtually any desired geometry, but in contrast has much greater precision, the capability for batch fabrication and provides parts in engineering materials such as nickel. The design freedom enabled by this process has been used to make both parallel plate and comb drive actuator deformable mirror designs that can have large vertical heights of up to 1 mm. As the thickness of the sacrificial layers used to release the actuator is specified by the designer, rather than by constraints of the fabrication process, the design of large-stroke actuators is straightforward and does not require any new process development. Since the number of material layers in the EFAB™ process is also specified by the designer it has been possible to gang multiple parallel plate actuators together to decrease the voltage required for long-stroke actuators.

Keywords: Adaptive Optics, long-stroke, electro-deposition, EFAB

1. Introduction

Adaptive optic (AO) systems are used to enhance optical signal quality by compensating for aberrations caused by fabrication errors, thermal effects, and atmospheric turbulence.[1] The latter is a particularly important distortion when using astronomical telescopes. To correct for this in both astronomy and vision science, deformable mirrors that can rapidly change their surface geometry are used. These consist of thousands of actuators that deform the mirror by pulling on the surface, and the component actuators must have a deflection > 10μm.

Other research efforts in such 'large-stroke' actuators have used the Sandia ultra-planar, multi-level MEMS technology (SUMMiT) to fabricate segmented AO mirrors and the MEMSCAP polysilicon surface micromachining process (MEMSCAP is a company based in France and the US) to fabricate continuous-face-sheet AO mirrors.[2] These methods are basic modifications of the two-dimensional surface micromachining process developed by Howe and Muller[3] and consist of 1-2μm thick layers of structural polysilicon and sacrificial oxide. The modifications needed to fabricate large stroke actuators, e.g. thicker sacrificial oxides and additional polysilicon layers, have required a large amount of process development, costing time and money. Nonetheless the goal of 10μm of stroke for continuous-face-sheet mirrors has yet to be achieved by this route.

2. Fabrication

An electrochemical fabrication process[4], EFAB™, —which can have structural heights of up to 1mm— has been used to fabricate comb-drive and vertical plate actuators with calculated deflections of 24 and 4.67μm, respectively.[5] This is a true three-dimensional fabrication process that can be used for rapid prototyping, since the actuators can be designed with computer-aided-design software such as SolidWorks™. Thus, what you see is what you get.

The EFAB™ process consists of multi-layer electrodeposition and planarization of metal layers.[5] The fabrication process starts with a thick alumina substrate (1mm), followed by the deposition of a patterned sacrificial metal, copper, and a blanket deposition of a structural nickel-cobalt layer. These layers are then planarized using chemical-mechanical polishing. The patterned deposition, blanket deposition, and planarization of layers are repeated multiple times with different layer thicknesses to fabricate structures of the required height. This process is showed on Figure 1.

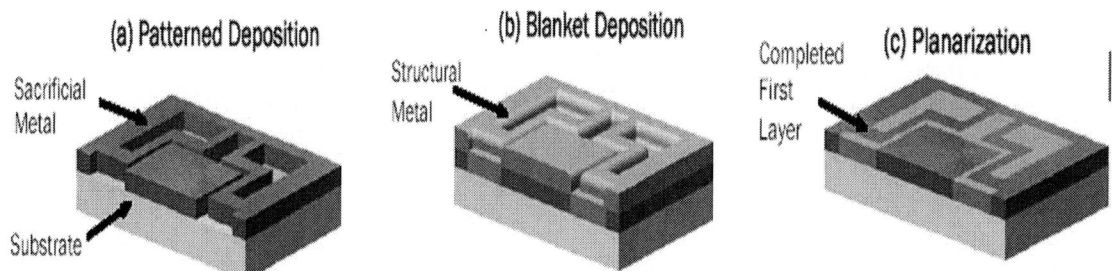

Figure 1: EFAB process overview. Pictures obtained from, *EFAB Access Design Guide*

3. Design

To address the needs of both astronomical and vision science, we designed two types of large-stroke actuator: a vertical ganged parallel-plate actuator and a comb-drive actuator (as shown in Figures 2 and 3, respectively).

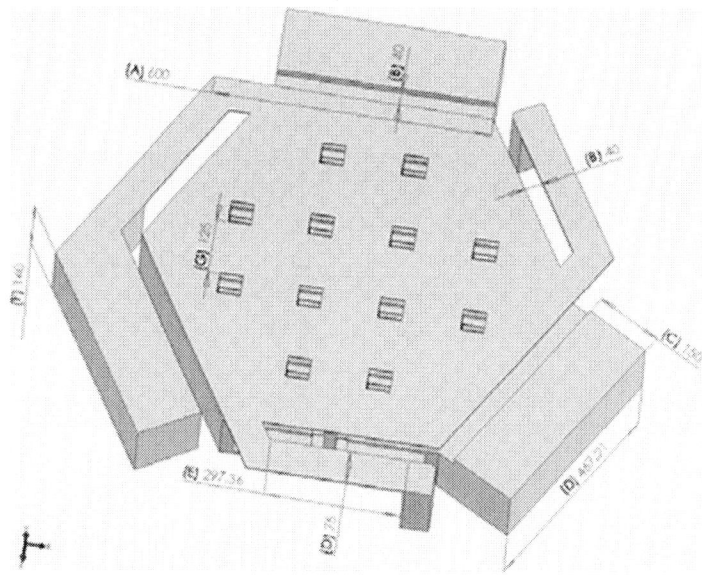

Figure 2. A vertical parallel-plate actuator can be used in deformable mirrors for adaptive optical systems.

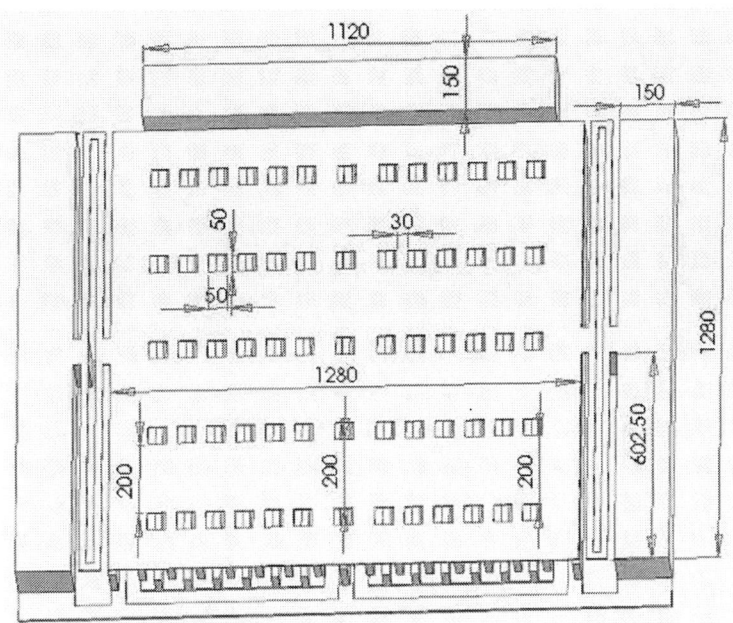

Figure 3. Comb-drive actuators have the advantage of providing larger deflections.

Both types of actuator have a low pull-in voltage (less than 200V) and provide both a large stroke (μ 10μm) and a high bandwidth (μ 10kHz). Created using SolidWorks™ software and fabricated via the EFAB™ process, the results are shown in Figures 4 and 5.

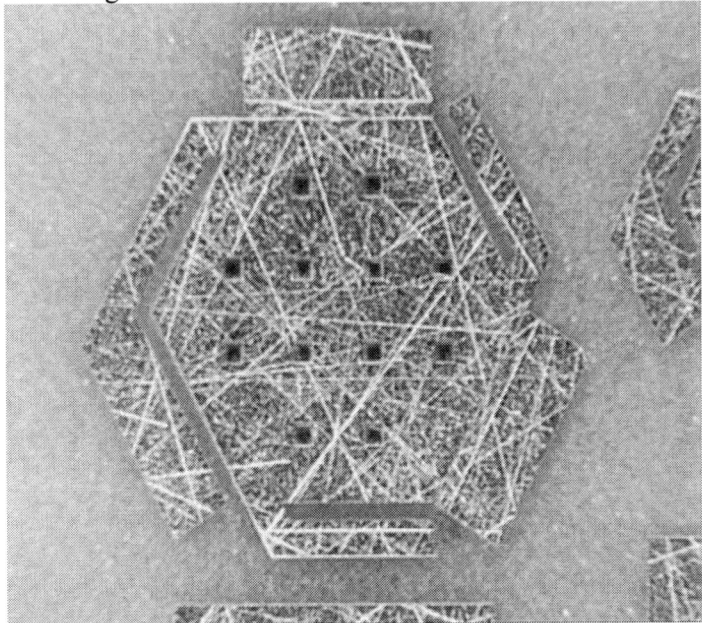

Figure 4. A vertical parallel plate actuator[5] fabricated with EFAB™.

Figure 5. An EFAB™-fabricated comb-drive actuator.[5]

Parallel-plate actuators of 600μm, 500μm, and 400μm were designed to have a 4.67μm stroke before reaching pull-in instability under voltage control at one third of the initial 14μm gap. [5] Table 1 shows the calculated pull-in voltage values for the three scaled parallel actuators. In addition, four scaled models of comb-drive actuators with lengths of 800μm, 900μm, 1120μm, and 1280μm were designed to have a maximum stroke of 24μm. Table 2 contains their calculated pull-in voltage values using a displacement of 20μm. With a laser vibrometer we measured that the largest comb-drive actuator was deflected by 28μm with an applied voltage of 300V.

	600μm	500μm	400μm
Vpi (V)	96	277	1006

Table 1. Calculated pull-in voltages for vertical parallel actuators

	800μm	960μm	1120μm	1280μm
Vpi (V)	203	121	79	55

Table 2. Calculated pull-in voltages for comb-drive actuators

Conclusion

Vertical parallel-plate and comb-drive actuators designs can be used in applications such as astronomy and vision science, where large stoke is required. They can be applied to deformable mirrors that can help enhance an optical system's signal quality by compensating for atmospheric turbulence.[1] Other research programs have used surface micromachining for fabrication of AO mirrors that consists of thin layers were the layer thicknesses are set by the fabrication design rules. EFAB™ allows for rapid prototyping and the structural and sacrificial layer thicknesses are left to the designer rather than being set by design rule restrictions. With this manufacturing process, vertical parallel-plate and comb-drive actuators designs were fabricated with deflections of 4.67μm and 24μm, [5] respectively.

This work has been supported in part by the National Science Foundation Science and Technology Center for Adaptive Optics, managed by the University of California at Santa Cruz under Cooperative Agreement No. AST-9876783. This research was also supported in part by a Special Research Grant from the University of California, Santa Cruz. The authors would also like to acknowledge the help of Dmitry Kozak, Oscar Azucena, and Stacy Barbadillo of the UCSC MEMS group.

References:

1. R. K. Tyson, *Adaptive Optics Engineering Handbook*, 2000.

2. P. Krulevitch, *Moems spatial light modulator development at the center for adaptive optics moems and miniaturized systems* III, Proc. SPIE, Vol: 4983, pp. 227-233, 2003.

3. R. T. Howe, R. S. Muller, *Polycrystalline silicon micromechanical beams*,

   Proc. Electrochemical Society Spring Meeting, pp. 184-185, 1982.

4. A. L. Cohen, *Electrochemical fabrication (EFAB™), The MEMS Handbook*,

   pp. 19-1, 2002.

5. B. Fernández, J. A. Kubby, *Large stroke actuators for adaptive optics*,

   Proc. SPIE, Vol: 6113, 2006.

6. Howe, R. S. Muller, *Surface micromachining for microelectromechanical systems*, Proc. IEEE, Vol: 86, no. 8, pp. 1552-1574, 1998.

Experimental Studies on the Effects of Geometric Parameters in a Planar Pneumatic Microvalve

K. J. Maung, J. Chan, S. J. Lee [*]
Microelectromechanical Systems Laboratory
San José State University
One Washington Square, San José, California 95192

Abstract—The lateral deflection of a microscale elastomer membrane patterned in polydimethylsiloxane (PDMS) has been experimentally characterized in terms of how geometric parameters affect the compliance of the membrane, and consequently the magnitude of deflection for a given pressure. The concept of a laterally-deformable membrane had been previously demonstrated as viable for applications including sealing, mixing, and sorting. The present work specifically applies fractional factorial experiments to examine the effects of thickness, width, and height of the membrane. It was observed that deflection magnitudes in excess of 10 microns could routinely be achieved with actuation pressure less than 200 kPa. In some cases deflection magnitude as high as 20 microns was also demonstrated. Functional membrane fabrication with dimensions as aggressive as 8 microns thick, 200 microns wide, and 45 microns tall has been successfully demonstrated. Thinner membranes and wider span exhibited similar magnitude of effects in terms of improving compliance. Height of the membrane exhibited a notably larger magnitude of effect, attributed to boundary conditions where the foot of the membrane is bonded to the base substrate. Seal reliability at the foot of the membrane was a significant limitation to sample yield, and represents the primary need for future development.

I. INTRODUCTION

MICROFLUIDIC devices take advantage of fluid phenomena at the nanoliter scale to enable rapid, high-sensitivity analysis of biological fluid samples [1]. There is a wide variety of approaches that are used to manipulate the small fluid volumes [2]. As one particular approach, laterally deformable elastomer membrane made of polydimethylsiloxane (PDMS) has been demonstrated in functional microfluidic devices, including pumps, sorters, and mixers [3], [4]. In particular, in-plane configuration enables a diverse set of microfluidic operations with ease of fabrication since multiple masks, layers, and process steps are eliminated. One specific implementation, a Planar Pneumatic Microvalve (PPM), is configured such that a deformable membrane constricts the passage of fluid through an adjacent sample channel. Fig. 1 shows the basic layout for which downward actuation of an input diaphragm reduces the confined volume of air and thereby applies pressure to deform the membrane into an adjacent sample channel.

Combinations of multiple membranes can achieve functions such as sealing, metering, pumping, and mixing. The single-mask approach allows multiple devices to be integrated rapidly and cost-effectively. Fig. 2 shows a photograph of a microfluidic chip with four devices on a glass slide.

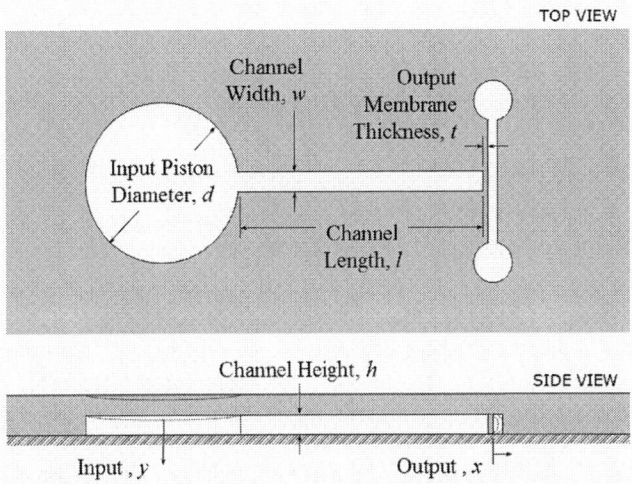

Fig. 1. Top and side layout views of a planar pneumatic microvalve.

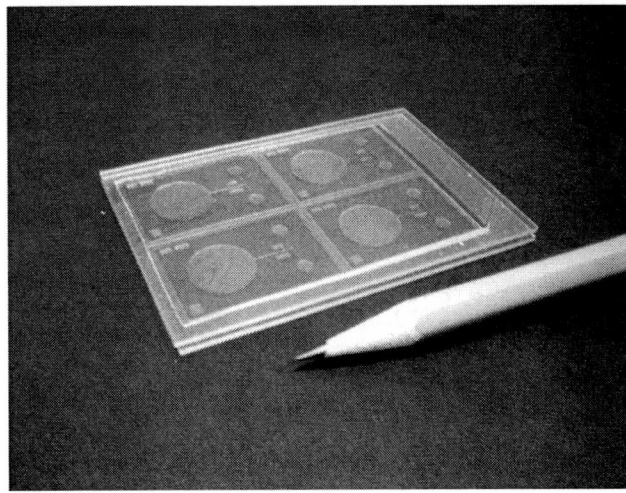

Fig. 2. Example of microfluidic chip device with four planar pneumatic microvalves.

A wide variety of actuation methods may be used to apply pressure, and some alternatives include electromagnetic,

---

[*] Corresponding author: +1-408-924-7167; sjlee@sjsu.edu

1-4244-0267-0/06/$25.00 ©2006 IEEE

piezoelectric, and thermomechanical approaches. Fig. 3 shows a top view of a membrane in a partially actuated position.

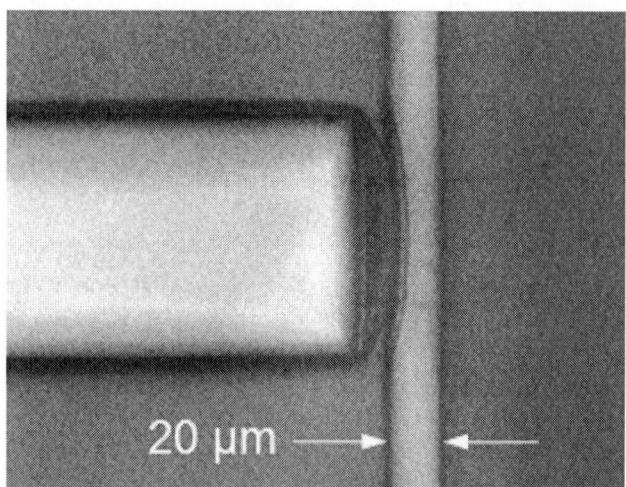

Fig. 3. Applied pressure in the horizontal cavity deforms the elastomer membrane such that it constricts the vertical channel. In this configuration, the membrane may act as a valve seat to seal or open the (vertical) fluid channel.

Although the basic concept has been successfully demonstrated [3], the present work focuses on how geometric parameters affect membrane mechanics. The experimental characterization can serve as design information for new devices that meet target performance.

A 2^{3-1} factorial experiment has been designed to study the effects of varying geometric parameters on the membrane performance. The geometric input variables are thickness, width, and height of the membrane, according to Fig. 1. The response variable is expressed as "compliance", to quantify the ease by which the membrane deforms with respect to applied pressure. Membrane compliance has units of μm/MPa, and is analogous to the inverse stiffness of a spring, expressed in terms of pressure instead of force.

II. EXPERIMENTAL

The fabrication sequence is illustrated in Fig. 4. The main functional layer is polydimethylsiloxane (PDMS, Dow-Corning Sylgard 184), patterned by soft lithography techniques already well-developed in the field of microfluidics [5], [6]. A master is first created by patterning the high-aspect-ratio SU-8 2035 ultrathick photoresist on a 100-mm silicon wafer substrate. The surface is treated with a silane-based vapor to facilitate ease of mold removal [7], and a mild nitrogen stream is blown over the SU-8 master to remove dust particles before casting. The PDMS is poured over the SU-8 master and is subjected to pressure cycles in the vacuum chamber to release trapped air bubbles. The PDMS is cured at 95 °C for 1 hour on a hotplate, then the PDMS layer is released from SU-8 master. Before creating the hermetic bond, the surfaces of PDMS layer and glass microscope slide (75 mm x 50 mm x 1 mm) are cleaned with isopropyl alcohol

rinse. After cleaning, the mating faces are exposed to air plasma running 200 W for 30 seconds with 50 sccm flow, then the PDMS layer is bonded to the glass substrate.

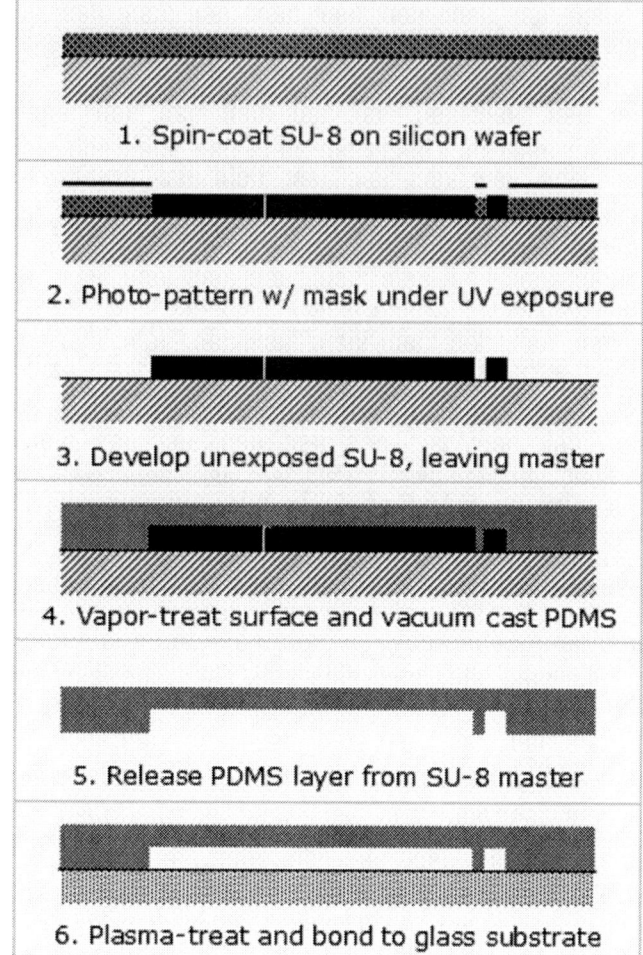

Fig. 4. Fabrication sequence for planar pneumatic microvalve. The device requires only a single photolithography step.

Table I shows the set of combinations that were designed into the photolithography mask. The membrane thickness and width were pre-determined by the mask layout, but channel height (same as membrane height) could be altered by spinning a different thickness of SU-8 photoresist on the master wafers.

TABLE I. FACTORIAL DESIGN FOR THICKNESS, WIDTH, AND HEIGHT

Run	Thickness (μm)	Width (μm)	Height (μm)
1	10	100	30
2	10	100	45
3	10	200	30
4	10	200	45
5	12	100	30
6	12	100	45
7	12	200	30
8	12	200	45

The mask actually provided differences in channel length as

1-4244-0267-0/06/$25.00 ©2006 IEEE

well (30 mm vs. 60 mm), but preliminary measurements confirmed expectations that length of the channel had negligible effect on membrane deflection. Pressure was applied quasi-statically and monitored by digital pressure gage. During experimental runs, the pressurized channel was allowed to reach equilibrium before measurements were recorded.

Channel height was measured by first taking contact profilometer (Tencor P1) scans on the SU-8 masters, followed by post-inspection of sliced cross-sections of the as-cast PDMS, as shown in Fig. 5.

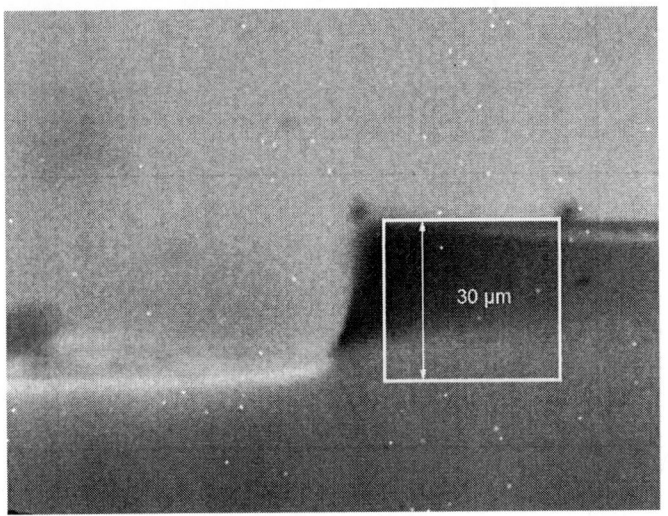

Fig. 5. Cross-sectional slice across PDMS diaphragm step for channel height verification, imaged via optical microscope.

Membrane thickness was the most challenging geometric parameter to verify as well as the most difficult to control. Fig. 6 shows an SEM image of the bottom foot of a well-formed membrane. Successful fabrication requires that the membrane have uniform thickness and a consistently flat surface for bonding.

Fig. 6. Bottom view of a well-formed membrane, exhibiting uniform thickness and a consistently flat surface.

III. RESULTS AND DISCUSSION

Table II shows the relevant subset of raw data that was used for factor effects analysis. Additional measurements were recorded with other sets of input variables, but because of yield loss only the relevant half-fraction is shown. For each run, compliance was quantified by a least-squares linear fit through six data points that measured maximum deflection versus applied pressure, ranging from zero to 207 kPa.(30 psi) in increments of 34 kPa (5 psi). Deflection magnitude for a given pressure can be calculated by multiplying the compliance by the applied pressure.

TABLE II. EXPERIMENTAL RESULTS IN HALF FRACTIONAL FACTORIAL

Run	Thickness (µm)	Width (µm)	Height (µm)	Compliance (µm/MPa)
1	10	100	45	65.7
2	10	200	30	44.9
3	12	100	30	22.3
4	12	200	45	66.5

A graphical representation of factor main effects is shown in Fig. 7. Each plotted point represents average of measurements when each respective variable is at its low or high level. The membrane behavior agrees with physical argument that greater compliance is benefited by membrane geometry that is thinner, wider, and taller. No conclusions are made on interaction effects because in this fractional factorial design they are confounded with main effects.

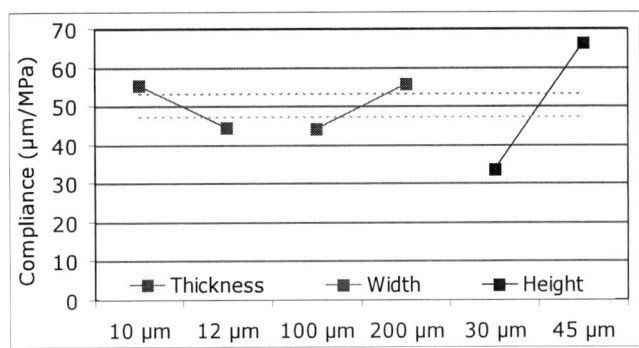

Fig. 7. Factor effects for membrane compliance as a function of thickness, width, and height.

In contrast to elastic thin-plate theory [8], it is not the case that thickness has a greater effect than width or height. Equation (1) below suggests that for thin plates, thickness t of the membrane has greater influence on deflection y than with w or height h. In this equation, q is the applied load (in units of pressure), E is Young's modulus, and α is a coefficient that depends on aspect ratio, mechanical properties, and edge constraints.

$$y_{max} = \alpha \frac{q}{E} \frac{w^2 h^2}{t^3} \qquad (1)$$

However, in the case of the membranes for this study, they are clearly not thin plates because in the extreme case the thickness-to-height aspect ratio may be as bulky as 1/3. This explains why height has such a dramatic effect on the compliance.

Although limited yield prevented an adequate set of replicates for analysis of variance, available replicate samples available did show good consistency in behavior. Fig. 8 shows that two separate samples fabricated with the same geometric parameters behave almost identically when subjected to applied pressure.

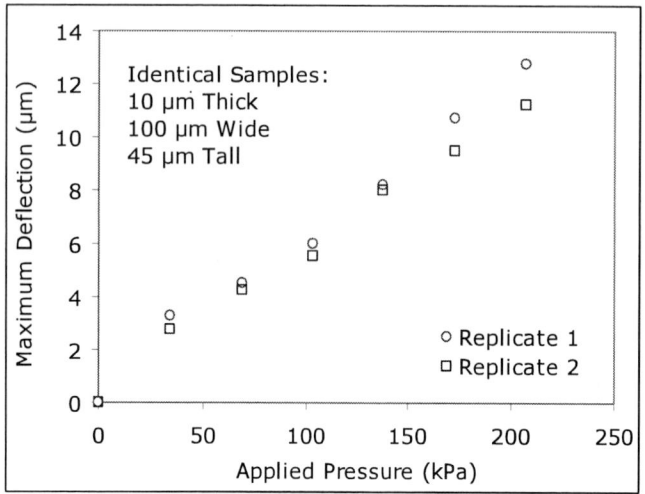

Fig. 8. Comparison of deflection measurements for two distinct samples having identical geometric parameters shows very consistent behavior.

Thinner membranes yield even better results, as demonstrated by a sample with 8-μm thickness in Fig. 9. Notable is that 20 μm coincidentally equals the width of the transverse fluid channel (Fig. 3), so a membrane that is 8 μm thick, 200 μm wide, and 45 μm tall can reach completely across the transverse channel.

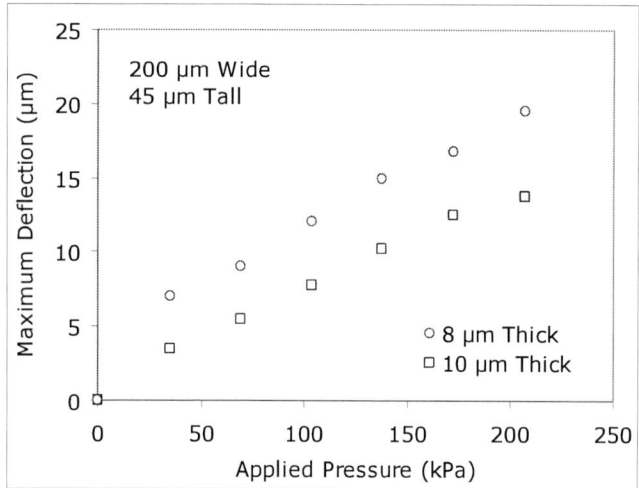

Fig. 9. Very thin membranes (8 μm thick) have even greater compliance and exhibit substantial deflection.

The most critical feature and most difficult challenge in terms of fabrication is the membrane foot. This region, the pressurized actuation channel, must hermetically seal against the base substrate, and high fidelity is needed for accurate, repeatable, and reliable operation. Several of the experimental samples failed to actuate or failed during operation in a "blow-out" mode, because of imperfect sealing with the glass substrate.

Fig. 10. Example of a faulty membrane foot. Imperfections in either the photolithography or the vacuum casting step may result in incompletely formed membrane foot, compromising the seal against the glass substrate.

A shortcoming of the available data in this half-fractional factorial is that interaction effects are confounded with main effects. This hinders the ability to form predictive models, especially considering that the geometric parameters of thickness, width, and height are indeed inter-related in terms of aspect ratio. Improved process yield, in particular at the membrane foot, would allow more comprehensive experimental data and subsequent analysis.

IV. CONCLUSIONS

Experimental study of deformable elastomer membranes has provided a quantitative understanding of how performance depends on membrane geometry. Thinner, wider, and taller membranes contribute to better compliance, and the results are entirely consistent with physical reasoning. The membrane height (same as channel height) has been observed to play a particularly significant role, and boundary constraints are the suspected to be the fundamental reason why it may be more dominant than other geometric parameters.

The characterization has been based only on a two-level half-fractional factorial, so interaction effects and non-linear dependency on the factors are not revealed. A more detailed understanding would demand more than three levels per factor and/or a factorial design with higher resolution. At the present state of development, the most vital aspect is process engineering for more reliable bonding at the membrane foot. High aspect ratio with very thin membranes presents the greatest challenge for high-quality masters in the photolithography step, but the experimental results show good promise that thicker membranes can achieve large deflections if height and width are correspondingly larger.

A parallel effort to be reported elsewhere by the authors is being conducted simultaneously to compare the observed performance against numerical predictions using finite element analysis (FEA). These studies explore a variety of assumptions on boundary conditions, including incorporation of hyperelastic material testing data to capture the unique mechanical behavior of elastomers.

V. ACKNOWLEDGMENTS

The authors thank Dr. Narayan Sundararajan at Intel Corporation for sharing the initial concept of single-layer lateral membranes in PDMS, and for technical advice throughout the study. Key steps in fabrication were made possible by the facilities, equipment, and staff support in the Microelectronics Process Engineering Laboratory at San José State University (SJSU). SEM imaging was provided by the Materials Characterization and Metrology Center at SJSU.

REFERENCES

[1] T. M. Squires and S. R. Quake, "Microfluidics: Fluid physics at the nanoliter scale", *Reviews of Modern Physics*, vol. 77, pp. 977-1026, July 2005.

[2] P. Gravesen, J. Branebjerg, and O. S. Jensen, "Microfluidics-a review", *Journal of Micromechanics and Microengineering*, vol. 3, pp. 168-182, 1993.

[3] N. Sundararajan, D. Kim, A. A. Berlin, "Microfluidic operations using deformable polymer membranes fabricated by single layer soft lithography", *Lab-on-a-Chip*, 2005 (3) 350-354.

[4] S. J. Lee, C. Y. Chan, N. Sundararajan, "Characterization of In-Plane Pneumatically-Actuated PDMS Membranes for Microfluidics", presented at *MEMS & BioMEMS 2005 - 7th International Conference*, November 2005, San Francisco, CA.

[5] Y. Xia and G. M. Whitesides, "Soft Lithography", *Annual Review of Material Science*, vol. 28, pp. 153-184, 1998.

[6] M. A. Unger, H..P. Chou, T. Thorsen, A. Scherer, S. R. Quake, "Monolithic Microfabricated Valves and Pumps by Multilayer Soft Lithography", *Science*, vol. 288, pp. 113-116, 2000.

[7] E. E. Parker, W. R. Ashurst, C. Carraro, and R. Maboudian, "Adhesion Characteristics of MEMS in Microfluidic Environments", *Journal of Microelectromechanical Systems*, vol. 14 No. 5, pp. 947-953, Oct. 2005.

[8] W. C. Young and R. Budynas, Roark's Formulas for Stress and Strain, 7th ed., McGraw-Hill, 2002.

1-4244-0267-0/06/$25.00 ©2006 IEEE

Advanced Studies into the Hermeticity
of
Micro-Electronic & Micro-Ordnance Devices

Karl K. Rink, Ph.D
University of Idaho

George R. Neff & Jimmie K. Neff
IsoVac Engineering, Inc.

ABSTRACT

This paper presents the preliminary findings of research studies initiated due to hermeticity problems associated with micro-cavity hermetic devices, some of which have resulted in field failures. The technical shortcomings and even misapplication of hermeticity test procedures when used for small or zero-cavity device testing has been allowed to prevail with the re sultant escape of non-hermetic devices into critical systems applications. The authors have decades of experience in the fields of hermetic device evaluation, and they have assembled this paper to couple some of IsoVac Engineering's seal-defect findings with fundamental research at th e University of Idaho involving defectiv e hermetic seals and gas flow studies.

Both groups have initiated studies to re-evaluate herm eticity test procedures and practices, and associate them with the scientific theory upon which they were assumed to have been originally developed. This paper attempts to show a need for an approach that can perhaps bridge the gap between industry and academia in a very critical field. There needs to be a strong effort to understand and use proper laboratory failure analysis; isolate and/or derive and correct the theory that applies; and initiate experimental studies to verify that theory. These studies need to then be incorporated into upgraded and credible specifications to replace those being misapplied.

This paper presents some of the failure analyses that have isolated and defined problems in non-hermetic failed devices. The devices included microelectronic devices, as well as micro-cavity ordnance devices used in both military and automobile restraint system applications. These studies have investigated the state of the art hermeticity or leak testing procedures that were used to qualify the devices later found to be non-hermetic. The laboratory data and basic theory are showing that the commonly applied testing methodologies are flawed, and considerable effort must be put into the sophistication of the methods used in commercial hermeticity testing. These investigations are confirming that industry is relying on historical practices which are decades old, rather than adva ncing those practices to accommodate the changes in device designs, and their critical applications. Some of the devices evaluated in these studies have internal cavities in the ranges of 10^{-3} to 10^{-6} cc. Such micro-cavities are virtually impossible to leak test using conventional methods such as Helium Mass-Spectrometry (M/S).

The two primary leak detection methods being studied, (Helium M/S, and Radioisotope Kr85), are being analyzed for their applicability to small cavity devices. The theory and limits of tracer-gas leak detection for the entire leak rate range is being studied for small cavity devices. Special techniques are being studied to enhance the use of these gases for this type of leak testing. The theoretical relationship between measured leakage using the conventional tracer gases, and the permeation of other undesirable gaseous media into a non-hermetic device is under study at the university. These studies are requiring some of the fundamental theories of gas flow to be reexamined, and in many cases equations reported on the literature are being re-derived, since most reviews were conducted decades ago.

Presentation Preference: Oral
Author contact Information:

Karl K. Rink, Ph.D.
Assistant Professor
Department of Mechanical Engineering
University of Idaho
Room 324K EPB
Moscow, Idaho 83844-0902
208-885-9447
karlrink@uidaho.edu

George R. Neff
Jimmie K. Neff
IsoVac Engineering, Inc.
Glendale, CA, 91201
818-552-6200
gneff@isovac.com

Low Cost, Tailored Polymer-Metal Nanocomposites for Advanced Electronic Applications

Abhijit Biswas and Pramod C. Karulkar

Office of Electronic Miniaturization, University of Alaska Fairbanks, AK 99775-8090

Abstract-The primary purpose of this paper is to introduce a nanomaterials fabrication facility that has recently been set up at the University of Alaska Fairbanks' Office of Electronic Miniaturization (OEM). The secondary purpose is to outline our research programs on nanoengineered materials for advanced electronic applications. We present a versatile, simple, and single-step high-vacuum e-beam codeposition technique for the controlled fabrication of nanostructured materials. The method allows evaporation of up to four different polymers, metals, semiconductors, insulators, and ceramics simultaneously or sequentially. Film properties can be controlled by controlling a number of deposition conditions. The system can be switched over rapidly to use a different set of materials. The process has a great potential to produce materials for many different practical applications for both R&D and scaled-up pilot production. This is the first nanomaterial fabrication facility of its kind in the State of Alaska and provides exciting opportunities for university-government-industry-collaboration that will be highlighted.

1. Introduction

Nanotechnology has the promise to provide us the breakthroughs that are needed to realize the next generation advanced electronics. Particularly, materials designed and engineered at the nanoscale have a great potential for practical application. In this regard, nanostructured composites of metal particles within a dielectric polymer medium are of interest. Depending on the composition and structure, they exhibit a unique combination of desirable properties that are otherwise unattainable. Polymers are attractive matrix materials to embed metal nanoparticles due to the ease of processing, inertness and stability over long term, low cost, and wide acceptance for use in state-of-the-art electronics.

Ability to precisely tailor and optimize several desirable material properties in one material simultaneously creates opportunities for addressing a wide range of electronic applications. For example, one class of nanostructured materials is capable of absorbing the whole solar spectrum and may prove valuable in converting it into electrical energy more efficiently and cost effectively. This is unlike the existing solar cell materials which absorb only a part of the solar spectrum. Enhancing insulating and magnetic properties does not go hand-in-hand in commonly used magnetic materials. But nanostructured materials would allow coexistence of soft magnetic and electrical insulating properties that valuable in making the next generation magnetic cores and high frequency devices. Further, extremely high interfacial area in specially structured nanocomposites just near the percolation threshold can be leveraged to form high value capacitors for embedded applications.

On the practical side, like the development of complex alloys during the last century, the development of nanostructured materials is very arduous due to the numerous parameters and constituents. This is particularly encountered in the case of combination of highly dissimilar materials such as a polymer and a metal. Existing methods to fabricate such nanostructured materials involve mostly multiple or tedious fabrication steps. Hence, there is an intense interest in developing a straightforward, highly controllable process for the rapid fabrication of many variations of nanoscale materials, investigating their properties, and adapting the processes and materials for practical applications.

We have taken advantage of the versatility offered by the established technique of e-beam evaporation to develop and optimize nanocomposites comprising metal nanoparticles dispersed in a polymer matrix. Here, we present a versatile, single-step electron beam (e-beam) vacuum codeposition technique that we have set up at the University of Alaska Fairbanks' Office of Electronic Miniaturization (OEM). While the technique is capable of evaporating different materials including metal, polymer, ceramics, and semiconductors, the present paper describes the work done on polymer-metal composites. The following sections describe OEM'S nanomaterials fabrication facility, and some of its ongoing research programs based on nanostructured materials for advanced electronic applications.

2. Experimental

Our deposition system (Fig. 1) consists of a four-pocket e-beam source, where up to four different materials can be simultaneously or sequentially evaporated at different rates by controlling the electron current used to heat each individual pocket [1]. An effusion cell evaporation process is used to evaporate the polymer where a metal crucible is bombarded with electrons thus preventing the decomposition of the polymer by electrons. Earlier processes used multiple, resistively heated sources or multiple, conventional electron beam or sputtering sources [2]. The codeposition process used in this work is simpler, highly controllable, and allows:

- Good control of the deposition parameters

- Variation of structure and morphologies of the deposited films
- A wide range of nanoparticle to polymer ratios
- A broad range of deposition parameters
- Multilalyer films
- Rapid tryouts using different materials

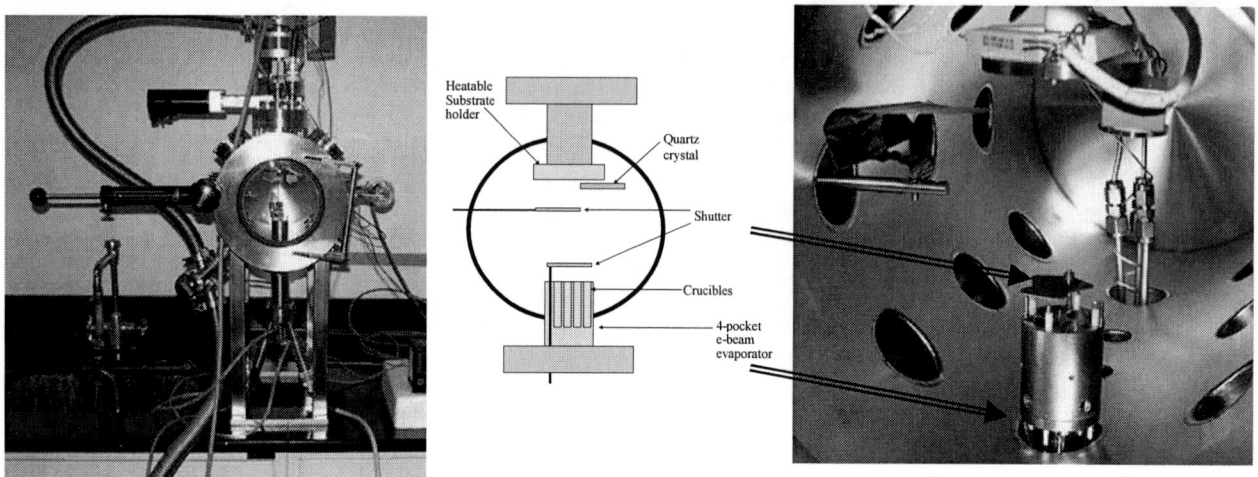

Fig. 1. A photograph and a schematic of nanomaterials fabrication chamber and a close-up view of the crucible assembly with the shutter. This system is now operational at the OEM. We have started experiments on nanostructured materials for electronic miniaturization.

Vacuum deposited polymer thin films are formed by polymerization of vaporized organic components at the substrate surface. The vaporized components can either be monomers vaporized from a monomer that is used as the source or be monomer-like radicals formed by thermal breaking of the covalent bonds in a more complex polymer used as an evaporation source. [2]. The key feature of the process used in our work is the preservation of the polymer structure without total decomposition or disintegration after repolymerization. Vapor-phase deposition of polymers has a number of advantageous features:

- Conformal coating of complex topographies
- Good film uniformity over large-diameter wafers
- Environmentally safe processing due to the absence of solvents [2].

It is possible to make large quantities of nanoscale colloidal solutions using wet chemical processes. They can be inexpensive, and in principle, can be used to make nanocomposites. However, wet processes are complex and have a discrete ("sweet") operating point at which they produce acceptable results. A particular wet process generally fails if the process parameters or compositions are varied away from the "sweet" operating point. The risk of contamination from unwanted reaction products or unused reactants imposes purification process before using the desirable reaction products. Vacuum e-beam coevaporation process allows a much broader variation of process parameters and compositions while maintaining the nanoscale structure of deposited films as was verified in our work. The process has a great potential for both R&D and scaled-up pilot production.

3. Applications

Figure 2 shows schematics of different polymer-metal nanostructures [2]. These materials one day may not only usher technology breakthroughs, but they offer an opportunity to carry out fundamental studies of nanoscale phenomena involving two very dissimilar constituents, a metal and a polymer dielectric.

Dilute metal-polymer nanocomposites containing nanopartciels widely separated by the polymer matrix are insulators. As the fraction of the metal in a nanocomposite increases, the particle separation decreases. At a certain composition, known as the percolation threshold, the material becomes highly conductive. Depending on the nature of the nanocomposite, the percolation threshold is characterized by other interesting properties that are beyond the scope the discussion here. Nanoparticles are tightly packed and may be connected to each other very close to percolation threshold. The surface area (area per gram) of the embedded material (e.g. metal) is very large near the percolation threshold. The particle shape also affects the surface area of the particles which may depend on their density in the nanocomposite and fabrication method. Under certain conditions the columnar structure and under some other conditions the fractal-like structure (both can be seen in Fig. 2) can be attained. The later is characterized by very large surface area of the metal nanoparticles. Specific advantages of each structure can be leveraged for different practical applications.

There have been attempts to form nanocomposites with high dielectric constant for fabricating capacitors by mixing either bulk produced nanoparticles of barium titanate or of a metal with an epoxy [3, 4]. Daiwan Choi et. al. employed nanocrystallites of vanadium nitride as a dielectric to make high value capacitors [5]. Various forms of nanostructured films based on nanoparticles of silver are being pursued for detection of biomolecular materials through enhanced fluorescence [6, 7]. Nanostructured materials comprising semiconductor nanorods in a conjugated polymer matrix are being pursued for high-efficiency solar cells through enhanced solar radiation absorption [8]. Further, nanostructured materials comprising ferromagnetic nanocolumns embedded in a polymer matrix are promising candidates for perpendicular high-density magnetic data storage media [9]. Diverse materials for all such applications can be pursued by e-beam codeposition.

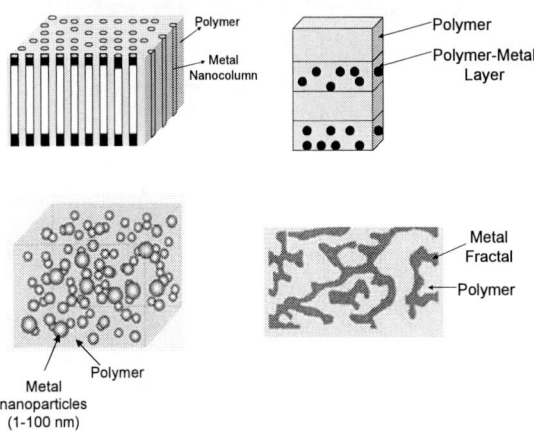

Figure 2. Schematics of polymer-metal nanocomposites with various nanostructures.

4. Nanostructured materials deposited by e-beam coevaporation

Scanning electron micrographs in Fig. 3 and Transmission electron micrograph in Fig. 4, based on one of the authors' (AB) previous work, illustrate different nano composite structures. The bright and dark contrast regions in the images represent Silver (Ag) and the polymethylmethacrylate (PMMA) matrix, respectively. One can see a gradual increase (Fig. 3, top to bottom) in the metal volume fraction from 45% to 60% in the composite and a transition from isolated spherical Ag nanoparticles to an interpenetrating fractal-like network (near percolation threshold concentration) of Ag nanoparticles embedded in a PMMA matrix [2]. The uncertainty in the estimation of the metal volume fraction in each case is ± 5%. A better estimate of the metal volume fraction can be obtained by a combination of physical characterization (thickness, volume, and density), stereo electron microscopy, and chemical analyses.

Highly one-dimensional parallel arrays of

1-4244-0267-0/06/$25.00 ©2006 IEEE

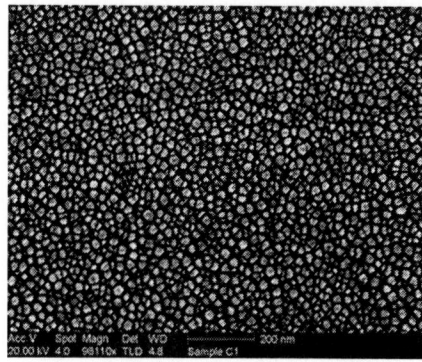

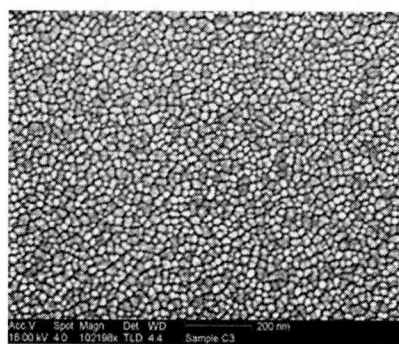

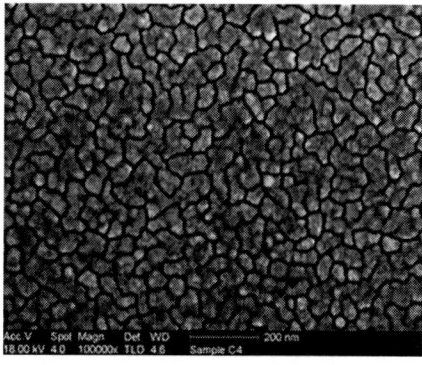

Figure 3. SEM images (~10000X) of PMMA/Ag nanocomposites fabricated by e-beam codeposition with increasing metal fraction

nanofilaments can also be fabricated under certain deposition conditions. Figure 4 shows a cross-sectional TEM image of a uniform distribution of gold (Au) nanocolumns (dark region in the TEM picture) with high-aspect ratio (> 30) and an average diameter of about 10 nm embedded in a Teflon AF matrix [2]. Such parallel arrays of high-density Au nanocolumns are of significant interest in novel applications

such as nanoelectronics and sensors [10]. Similar materials can be fabricated using soft or hard magnetic alloys as one of the constituents leading to high performance magnetic shielding or information storage material.

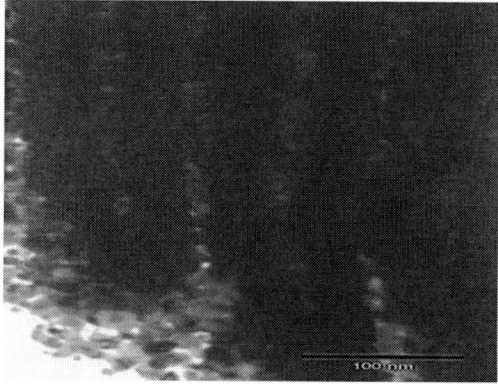

Figure 4. Au nanocolumns in a Teflon AF® matrix fabricated by e-beam codeposition.

4. Summary and Opportunites for Collaboration

The potential of nanostructured materials for advanced applications will have to be verified by applying these materials to specific applications and benchmarking them against the materials that are currently in use. This requires capabilities and expertise far beyond the level needed for fabrication and characterization of test structures. For example, nanostructures containing silver at a certain dilute composition exhibit very broad absorption behavior which can be tested in the laboratory. However, the participation of an expert partner would be required to fabricate and benchmark broadband absorbers or sensors based on this technology. The same applies for columnar nanostructures of hard or soft magnetic materials. The expertise of a partner working on the current technology in this field would be valuable to thoroughly evaluate the promise of such materials.

We have set-up a versatile, simple, and single-step nanofabrication technique based on e-beam vacuum deposition process for controlled fabrication of a variety of nanostructured materials. The process has a great potential for both R&D and can be scaled-up for pilot production. We seek to form collaborative

1-4244-0267-0/06/$25.00 ©2006 IEEE

partnerships with complementary know-how and project needs to transition from research to practical applications. We welcome any enquiries on our advanced electronics research programs to discuss opportunities for collaborations.

5. Acknowledgement

A. Biswas gratefully acknowledges his collaborators/coworkers Hergen Eilers, Fred Hidden, M. Grant Norton, Daniel Skorsky and Chris Davitt for their contributions made to the results presented in this paper. The work carried out at the Washington State University, Spokane was supported by the Office of Naval Research (N00014-03-0247). The work at the University of Alaska Fairbanks was supported by the Defense Microelectronics Activity (DMEA). The United States Government is authorized to reproduce and distribute reprints for Government purposes, notwithstanding and copyright notation thereon. The views and conclusions contained herein are those of the authors and should not be interpreted as necessarily representing the official policies or endorsements, either expressed or implied, of the DMEA.

6. References

[1] A. Biswas, H. Eilers, F. Hidden, O. C. Aktas and C. V. S. Kiran, *"Large broadband visible to infrared plasmonic absorption from Ag nanoparticles with a fractal structure embedded in a Teflon AF matrix"*, Appl. Phys. Lett. 88, 013103 (2006).

[2] A. Biswas, P. C. Karulkar, H. Eilers, M. Grant Norton, D. Skorsky, C. Davitt, H. Greve, U. Schuermann, V. Zaporojtchenko, F. Faupel, *"Vapor phase deposition of nanostructured polymer-metal composites for advanced technology applications"*, invited feature article, Vacuum Technology & Coating, April, 54 (2006), and the references therein.

[3] Hitesh Windlas, P. Maarkondeya Raj, Devarajan Balaraman, Swapan K. Bhattacharya, and Rao R. Tummala, *"Polymer-Ceramic Nanocomposite Capacitors for System-on-Package Applications,"* IEEE Transactions on

Advanced Packaging, Vol. 26, No.1, 10-16, (2003).

[4] Yi Li, Suresh Pothukuchi and C. P. Wong, *"Development of a Novel Polymer-Metal Nanocomposites Obtained through the Route of In Situ Reduction and it's Dielectric Properties,"* Electronic Components and Technology Conference, 507-513, (2004).

[5] D. Choi, G. E. Blomgren, P. N. Kumta, *"Fast and Reversible Surface Redox Reaction in Nanocrystalline Vanadium Nitride Supercapacitors"*, Adv. Mater. 18, 1178 (2006).

[6] Chris D. Geddes, Haishi Cao, Ignacy Gryczynski, Zygmunt Gryczynski, Jiyu Fang, and Joseph R. Lakowicz, *"Metal-enhanced fluorescence (MEF) due to silver colloids on a planar surface: Potential applications of indocyanine green to in vivo imaging"*, J. Phys. Chem. A 107, 3443 (2003).

[7] Chris D. Geddes, Alexander Parfenov, David Roll, Ignacy Gryczynski, Joanna Malicka, and Joseph R. Lakowicz, *"Silver fractal-like structure for metal-enhanced fluorescence intensities and increased probe photostabilities"*, J. of Fluoroscence"*, 13, 267 (2003).

[8] Wendy U. Huynh, Janke J. Dittmer, A. Paul Alivisatos, *"Hybrid Nanorod-Polymer Solar Cells"*, Science 295, 2425 (2002).

[9] T. Thurn-Albrecht, J. Schotter, G. A. Kaestle, N. Emley, T. Shibauchi, L. Krusin-Elbaum, K. Guarini, C. T. Blackk, M. T. Tuominen, and T. P. Russell, *"Ultrahigh-Density Nanowire Arrays Grown in Self-Assembled Diblock Copolymer Templates"*, Science 290, 2126 (2000).

[10] S. Ge, K. Jiang, X. Lu, Y. Chen, R. Wang, and S. Fan, *"Orientation-Controlled Growth of Single-Crystal Silicon-Nanowire Arrays"*, Adv. Mater. 17, 56 (2005).

1-4244-0267-0/06/$25.00 ©2006 IEEE

Wuhan National Laboratory for Optoelectronics and Its Collaboration with Georgia Tech

Zhiping Zhou, *IEEE Senior Member*
School of Electrical and Computer Engineering
Georgia Institute of Technology
Atlanta, Georgia 30332-0250
and
Wuhan National Laboratory for Optoelectronics
Huazhong University of Science and Technology
Wuhan, Hubei 430074, China
E-mail: wnlo8@mail.hust.edu.cn

Abstract- A new National Laboratory for Optoelectronics located in Wuhan, China is introduced and its collaboration model with Georgia Tech is outlined. Their joint effort is establishing an international platform for scholars from the world to do research and education on new frontiers of photonics and electronics.

I. INTRODUCTION

WUHAN National Laboratory for Optoelectronics (WNLO) is one of the five national laboratories initiated and sponsored by the Ministry of Science and Technology, China in 2003. WNLO is an important part of the Wuhan • Optics Valley of China (WOVC). It is also an integrated part of Chinese national scientific innovation system. The mission of the WNLO is to become a top innovation base for the field of optoelectronics in China, to promote and lead the complete commercialization system (R&D&P) for "Wuhan • Optics Valley of China", and to contribute to the growth of optoelectronic industries through technology transfer. So far, eight research thrusts (RTs) are designed for the WNLO, which are related to Microelectronics, Optoelectronics, Photonics, Microelectronics, Microsystems, or Nanotechnology. One of the major research themes in the Laboratory is miniaturization and integration.

Similarly, GT has been part of the NSF's National Nanotechnology Infrastructure Network (NNIN) and is planning a new 160,000 sq. ft. Nanotechnology Research Center, on top of its already well-known micro/nano optoelectronics research facilities. Georgia Tech's campus is becoming one of the nation's most advanced facilities for nanotechnology research and development.

We believe that there is a strong need to support and strengthen fields of microelectronics and optoelectronics and ensure their future vitality through strong, coordinated international collaborations at both research and education levels. This vision has led to contacts between Georgia Institute of Technology (GT) and Huazhong University of Science and Technology (HUST), a major participant of the WNLO, for many years. Particularly, after a few exchanged

visits between faculty and administrative members from HUST and GT in 2004 and 2005, a formal agreement has been signed to start a HUST-GT Center for Photonics and Electronics (HGCPE).

This paper is to introduce the Wuhan National Laboratory for Optoelectronics and to describe its partnership with Georgia Tech.

II. WUHAN NATIONAL LABORATORY FOR OPTOELECTRONICS

Wuhan is the capital city of Hubei Province, China, with a population of over 8.5 millions. It is also one of the famous ancient cities and has been the center of industry, finance, commerce, science, technology, and education in central China since its establishment years ago. Since 1980's, Wuhan has been developed into an important optoelectronics industrial base through the following signatures:

- *The biggest optical fiber and cable production base in China —— YOFC, Fiberhome,*
- *The biggest optoelectronic devices production base in China —— Accelink, WTD, Tikom, Zhengyuan, Teamsun, etc.,*
- *The biggest telecommunication R&D base in China ——Wuhan Research Institute of Posts and Telecommunications (WRI),*
- *35% of laser industry in China is located in Wuhan — including influential enterprises, such as HGL, Chutian Laser, Unity Laser, etc.,*
- *One of the largest optoelectronics education bases in china ----- HUST.*

To recognize the achievement, both Ministry of Science and Technology and National Development and Reform Commission of China certified and approved Wuhan as the National Optoelectronics Industrial Base in 2001. Since then, a $132km^2$ land with ready infrastructure has been developed and allocated specifically for the industrial base and known as "Wuhan • Optics Valley of China (WOVC)".

1-4244-0267-0/06/$25.00 ©2006 IEEE

Will WOVC sustain? How does the WOVC adjust its portfolio as the optoelectronics industry evolves? By answering these questions, the Ministry of Science and Technology of China initiated and approved 5 National Labs to be built by Chinese government in 2003, and the only one in the field of optoelectronics was designated to Wuhan (Figure 1), therefore, the Wuhan National Laboratory for Optoelectronics (WNLO).

Figure 1: Birdview of Wuhan.

WNLO is co-sponsored by China's Ministry of Science and Technology and Ministry of Education, Hubei Provincial Government, and Wuhan Municipal Government. The management team consists of Huazhong University of Science and Technology (Leader), Wuhan Research Institute of Posts and Telecommunications, Wuhan Institute of Physics and Mathematics, and Huazhong Institute of Optoelectronic Technology, Figure 2.

The National Laboratory is located at the heart of HUST and the heart of National Optoelectronic Industry Base – Wuhan • Optics Valley of China. Its 405,000 sqft new building, Figure 4, was completed in June 2005. About 60% of the building is lab space, including 18,000 sqft cleanroom.

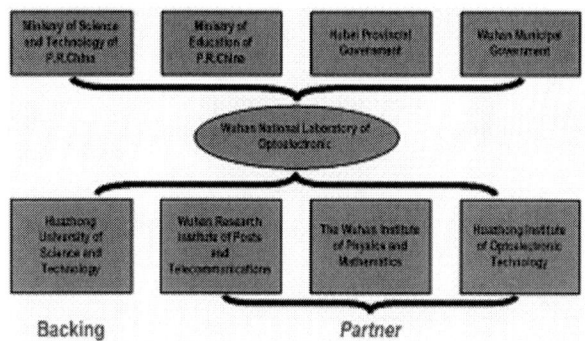

Figure 2: WNLO's Supporting System.

Figure 3: The new building for the Wuhan National Laboratory for Optoelectronics.

WNLO has six Research Divisions and an International Division.

The function of the international division is to look into any forms of collaborations with International communities in a mutually beneficial fashion. These include but not limited to joint centers, joint conferences, joint projects, faculty and student exchanges, etc.

The six research divisions are focused on Laser Science & Technology, Optoelectronic Devices and Integration, Nanophotonics and Microsystems, Data Storage System, Optical Communication Systems and Intelligent Network, and Biomedical Photonics. One of the main goals of the WNLO is to develop the next generation of compactly integrated low cost optoelectronic systems that may be used for real time sensing/detection, high-density data communications, and high-speed control/actuation.

Current research and development at WNLO are summarized below.

A. Optoelectronic Devices and Integration

The main research areas covered by this division are:

1) Micro/nano-optoelectronics devices and integration.

2) Optical communication and all optical processing devices and integration.

3) LED and display. The highlights and achievements from this division are described below.

B. Nanophotonics & Microsystems

The research areas in this division include MEMS Packaging, MEMS Measurement and reliability, Novel Optoelectronics materials, Micro-Optoelectronics devices, SOC, MOEMS, and Bionic systems.

C. Data Storage System

High density optical recording, hybrid recording, magnetic recording, and RAID (Redundant Array of Independent Disks) networking storage are three major research areas in the Data Storage System division. Heavy optoelectronics involvements are described below.

1-4244-0267-0/06/$25.00 ©2006 IEEE

D. Laser Science and Technology

The main research areas in this division include high power laser technology, laser-matter interaction, laser physics, solid state laser technology, and laser advanced manufacture.

E. Biomedical Photonics

Biomedical photonics provides photonics solution for Biotechnology and medicine. Research in this division includes the study of the measurement of human brain function and activity, the diagnostics, prophylaxis and the healing process monitoring of some brain diseases, and the brain mechanism of some psychology activity, the creation of quantitative optical instrumentation for biophysical and biomedical research based on multiphoton fluorescence microscopy which is capable of intrinsic diffraction-limited 3-D resolved imaging of dynamical processes in living cells and thick tissue. Strength of this division is in the development and application of novel optical methods for making real-time, in vivo, high spatial resolution measurements of tissue function and status.

F. Communication and Intelligent Network

The research areas in this division are listed as followings:

1) ***The new-generation optical electronics communication and intelligence network***
2) ***Multimedia polymorphism complexity distortion theory in wireless and optical combined environment***
3) ***Quantum information theory and its combination with broadband wireless communication***
4) ***Media information processing and bionic optoelectronic imaging system***
5) ***Radio over fiber (ROF).***

III. COLLABORATION WITH GEORGIA TECH

In recent years, "nanoscience", "nanotechnology", and "nanofabrication" have been used in all facets of the science and technology community in both academia and industry. This technological trend points to a general interest in ever *smaller and more compact devices and products.* Driven by this desire, U.S., E.U., and other countries have been investing huge amounts of money and resources in the development and integration of micro/nano devices and systems.

Furthermore, as integrated circuit feature sizes continue to shrink and chip sizes to expand, conventional electrical interconnects and switching technology are rapidly becoming critical issues in device and system integration. New devices, new interconnects, and new integration schemes must be developed to meet the demand for high-density data communication and high-speed data processing.

To address these issues, the Chinese government invested over 60 million U.S. Dollars to create its top Photonics and Electronics research laboratory, the "Wuhan National Laboratory for Optoelectronics (WNLO)" managed by HUST in 2003. Basic research on micro/nano scale optoelectronic device and system integration is one of its most important thrust areas. Similarly, GT has been part of the newly established NSF's National Nanotechnology Infrastructure Network (NNIN) and is planning a new 160,000 sq. ft. Nanotechnology Research Center, on top of its already well-known micro/nano optoelectronics research facilities. Georgia Tech's campus is becoming one of the nation's most advanced facilities for nanotechnology research and development.

Although the field of optoelectronics is growing at a rapid rate, its ability to lead efficiently to new devices and new technologies is significantly challenged by the multidisciplinary nature of its underlying fundamental science and engineering. There is a strong need to support and strengthen these fields and ensure their future vitality through strong, coordinated international collaborations at both research and education levels.

After a few exchanged visits between faculty and administrative members from HUST and GT in 2004 and 2005, many expressed their interests in participating in a formal collaboration format between HUST and GT. There are over 20 GT faculty members from School of Chemistry & Biochemistry, School of Electrical & Computer Engineering, School of Materials Science and Engineering, School of Mechanical Engineering, Department of Biomedical Engineering and same amount faculty from WNLO/HUST, which covers research areas of Optoelectronics, Photonics, Microelectronics, Mechanics, Material Science and Engineering, Life Science, Electrical Engineering, etc., are actively involved in this endeavor.

Figure 4: GT delegation visits HUST in 2004.

Based on above facts, both GT and HUST are agreed to work towards a **new HUST-GT Center for Photonics and Electronics (HGCPE)**, nested in the Wuhan National Laboratory for Optoelectronics, with "**New Frontiers in Photonics and Electronics**" as its key research theme.

Figure 5: Agreement signing ceremony between GT and HUST

The goal of this HUST-GT Center is to establish a **unique international platform**, recognized and supported by the international optoelectronics industry community, for research and education on new frontiers of photonics and electronics. **The specific objective** is to develop the next generation of compactly integrated low cost optoelectronic systems that may be used for real time sensing/detection, high-density data communications, and high-speed control/actuation.

The research is planned on two fronts: 1) nanostructures and nanodevices for semiconductor-based optoelectronic systems and 2) nanostructured materials for organic-based optoelectronic systems.

Both fronts share the same research thrust (RT) structures:

RT1: Theoretical Simulation and Device/System Design, focusing on nanostructure diffraction, nonlinear properties, near field properties, and ultra fast phenomena;

RT2: Materials Design, Synthesis, and Processing and Device Fabrication, dealing with novel materials and devices on the micro/nano scale such as photonic crystal structure, NEMS devices, ring resonator, binary blazed grating coupler, wavelength converter, waveguide couplers, electroluminescent diodes, organic solar cells, optical biosensor, quantum-cascade lasers, microlasers, SOA, RAMAN amplifier, 3D optical circuits, switches and routing devices;

RT3: System Integration, to integrate the fabricated devices with electronics so that compact, low cost, integrated optoelectronic systems, which may be used for real time sensing/detection, high density data communications, and high speed control/actuation, may be developed.

Faculty members from both sides have already begun to work together to produce proposals, papers, and to participate in symposia and conferences, such as <u>mini workshop</u> on GT campus in 2004 and <u>2005 Optics Valley of China International Symposium on Optoelectronics</u> in Wuhan, China, November 2 - 3, 2005.

Figure 6: The opening of the 2005 Optics Valley of China International Symposium on Optoelectronics

Figure 7: Scholars from the world discussion photonics and microelectronics related issues at the 2005 Optics Valley of China International Symposium on Optoelectronics.

IV. CONCLUSION

The **broader impacts** resulting from the collaborations between HUST and GT are that the research results on novel optoelectronic systems may be applied in information technology, computer science, nanoscience, engineering, medicine, biology, surveillance, space research, and other forefront science and engineering areas. There may also be no better teaching tool as many of the best young scientists in the optoelectronics area in the U.S.and China were trained to work on novel devices and system integration in a global setting - such as the joint **HUST-GT Center for Photonics and Electronics**. The most important is that the HUST-GT Center is to establish a **unique international platform** for scholars from the world to do research and education on new frontiers of photonics and electronics.

1-4244-0267-0/06/$25.00 ©2006 IEEE

V_T Adjustment by L_{eff} Engineering for LSTP Single Gate Work-function CMOS FinFET Technology

Vidya Varadarajan and Tsu-Jae King Liu

Department of Electrical Engineering and Computer Sciences, University of California, Berkeley, CA, 94720, USA

Email: vidya@eecs.berkeley.edu, Tel: (510) 643-2558

Abstract

Engineering of the electrical channel length (L_{eff}) is investigated as an alternative means for adjusting the threshold voltage (V_T) in double-gate (DG) MOSFETs, to allow a single mid-gap work function gate material to be used with undoped CMOS channels. Device simulations for 18nm gate length (L_G) indicate that ITRS performance specifications for low standby power (LSTP) applications can be met with this approach. Different source/drain (S/D) doping profiles yielding the same L_{eff} can be used to achieve the desired V_T, allowing for flexibility in process design.

Introduction

Non-classical transistor structures such as thin-body MOSFETs and multiple-gate FETs will eventually be necessary to adequately suppress off-state leakage current (I_{OFF}) while meeting on-state drive current (I_{ON}) specifications [1]. The FinFET [2] is a promising structure because it offers the superior scalability of the double-gate structure together with a fabrication process and layout similar to that of the conventional bulk-Si MOSFET. An undoped channel/body is desirable for immunity to V_T variations due to statistical dopant fluctuations (SDF) in the channel, and for the highest possible carrier mobilities to achieve high I_{ON} [3]. However, it necessitates an alternative means for adjusting V_T. Gate work function (Φ_M) engineering, *e.g.* by masked ion implantation, is one such method [4,5]. For compact circuit layouts such as those used in static memory (SRAM) cells, however, it is not possible to separately implant the gate electrodes in the n- and p-channel regions if the FinFET structure is employed. This is because the gate layer fills the entire region in-between the n-channel and p-channel fins (Fig. 1), making it impossible to selectively change the workfunction of one of the FET gates. Thus, a single gate work function must be used, and an alternative means for adjusting V_T is needed.

In this paper, we investigate L_{eff} engineering as a means for adjusting the V_T of a DG MOSFET. The dependences of I_{ON}, I_{OFF} and short-channel effects on L_{eff} are studied via 2D device simulation for devices with $\Phi_M = 4.7eV$ and $L_G = 18nm$ (sub-32nm half-pitch technology node). Through 3D atomistic device simulations, we demonstrate that V_T variation due to SDF in the S/D gradient regions is not expected to be an issue at $L_G = 18nm$.

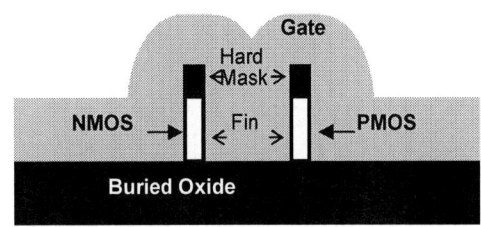

Fig. 1: Schematic cross-section of a CMOS FinFET inverter. If the separation between NMOS and PMOS active areas (Si fins) is less than twice the gate thickness, the gate completely fills the region in-between, so that it is difficult to achieve different gate work functions for NMOS *vs.* PMOS FinFETs.

V_T-tuning Methodology and Simulation Setup

The impact of S/D doping profile on DG-MOSFET performance was studied via 2-D device simulations in Taurus-Device using drift-diffusion transport and 1D-Schrödinger solutions [6]. The simulated structure and design parameters are shown in Fig. 2 and Table I respectively. L_{eff} is defined to be the lateral separation between the locations at which the S/D doping falls to 1×10^{19} cm^{-3}. It is adjusted by varying the S/D doping lateral abruptness (σ_{SD}) and the distance of the starting point of the gradient from the gate-edge (L_{SP}). In practice, σ_{SD} is controlled by the lateral straggle of the S/D ion implantation and L_{SP} is determined by the gate sidewall spacer thickness.

1-4244-0267-0/06/$25.00 ©2006 IEEE

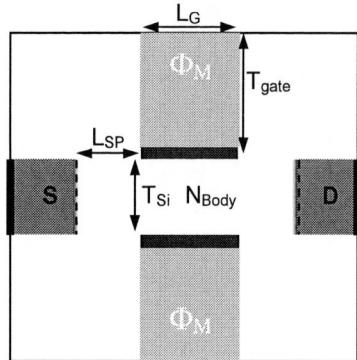

Fig. 2: Schematic cross-section of the simulated double-gate FET structure. L_{SP} denotes the location (referenced to the gate edge) at which the S/D doping begins to fall (from 10^{20} cm^{-3}).

Device Parameters	
Gate length L_g	18 nm
Gate-dielectric EOT	9 Å
Body thickness, T_{Si}	9 or 12 nm
Gate thickness, T_{Gate}	27 nm
Body doping, N_{body}	1E16 cm^{-3}
S/D doping, $N_{S/D}$	1E20 cm^{-3}
Gate ϕ_M	4.7 eV
V_{DD}	0.9 V
Target I_{OFF}	80 pA/μm
Target I_{ON}	880 pA/μm

Table I: Nominal parameter values used for the simulations.

Variations in V_T due to SDF in the S/D gradient region, among other sources of variations like line-edge roughness, are likely to be a concern for highly scaled devices [7]. Impact of SDF was studied via 3-D atomistic device simulations [8]. The atomistic simulations were implemented in Taurus-Device in conjunction with a MATLAB interface to model the atoms accurately. The atoms were randomly generated and placed in the S/D gradient region using a Monte Carlo algorithm applicable to non-uniform doping profiles [9]. In order to avoid charge and potential singularities caused by point dopants, the atoms are defined through their long-range charge densities [10] with proper dose normalization.

Results and Discussion

Fig. 3 summarizes the tradeoff between I_{ON} and I_{OFF} for different combinations of L_{SP} and σ_{SD}. L_{SP} was varied from 6nm to 14nm and the S/D gradient lateral abruptness was varied from σ_{SD} = 1.5nm-5nm.

There is a clear distinction between the short channel effect (SCE)-limited and S/D series resistance (R_{series})-limited regimes and multiple combinations of L_{SP} and σ_{SD} can yield the same I_{ON} and I_{OFF} for given device dimensions. Furthermore, a thinner body is more effective for suppressing leakage due to better gate control, and thus provides higher I_{ON} for a given I_{OFF}. For I_{OFF} = 80pA/μm (the specification for low-standby-power applications [1]), I_{ON} is within 20% of the ITRS target value of 880μA/μm. Since mobility enhancement techniques are projected to provide 30% improvement in I_{ON}, this indicates that I_{ON} and I_{OFF} specifications can be met (for T_{Si} = L_G/2), with multiple combinations of L_{SP} and σ_{SD}. As shown in Fig. 4, all these possibilities correspond to the same L_{eff}.

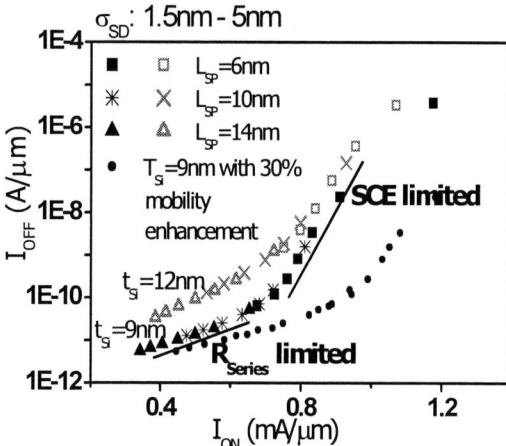

Fig. 3: Trade-off between I_{ON} and I_{OFF} showing SCE- and series-resistance-limited regimes. ITRS specifications can be met through L_{eff} engineering (I_{OFF}) and mobility enhancement techniques (I_{ON}).

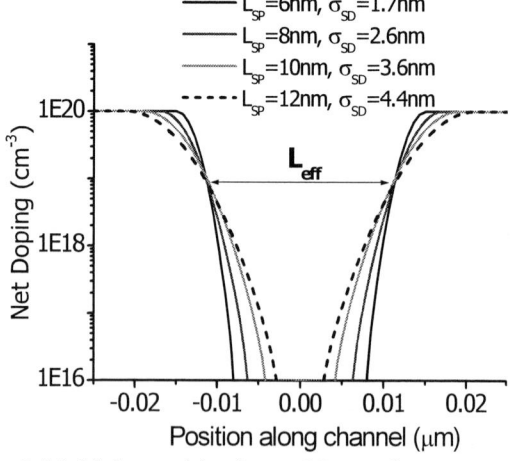

Fig. 4: Multiple combinations of L_{SP} and σ_{SD} are optimal, all with the same L_{eff} (distance between the points where the S/D doping concentrations = 1×10^{19} cm^{-3}).

1-4244-0267-0/06/$25.00 ©2006 IEEE

V_T as a function of L_{eff} is plotted in Fig. 5. In order to avoid sensitivity to variations, it is preferable to tune the V_T in the R_{series}-limited regime and about 0.2V of V_T tuning is possible in this regime. The subthreshold swing (S) and drain-induced barrier lowering (DIBL) are compared for two extreme combinations of L_{SP} and σ_{SD} that yield the same L_{eff}, in Fig. 6. Similar short channel immunity is achieved by both designs and for L_g=18nm the FETs have good short channel control.

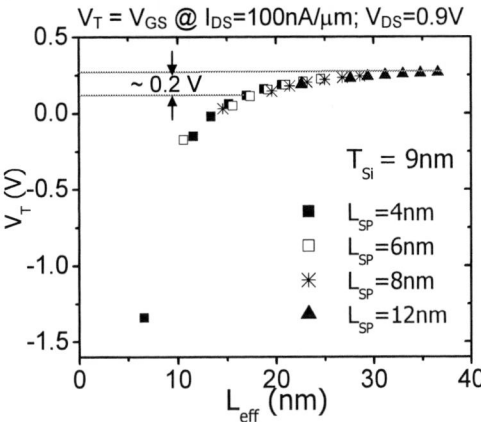

Fig. 5: V_T as a function of electrical channel length (L_{eff}) showing the range of V_T tuning. It is possible to adjust V_T by ~0.2V with reasonable short channel immunity.

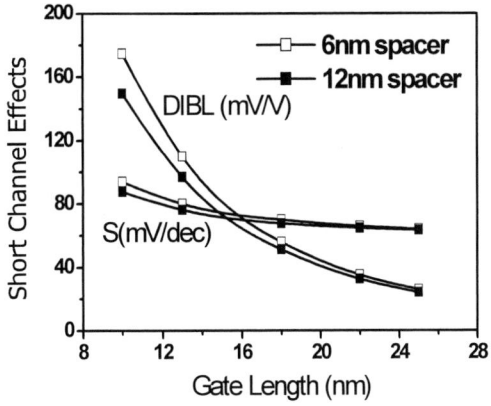

Fig. 6: SCE comparison for two optimal S/D doping profile designs. A thin gate-sidewall spacer (small L_{SP}) with steep doping gradient (small σ_{SD}) yields similar short channel behavior as a thick gate-sidewall spacer with relaxed σ_{SD}.

In a CMOS technology, L_{SP} should be the same for NMOS and PMOS devices and it should be possible to achieve the required V_T for PMOS FinFETs using the same strategy. Fig. 7 shows that PMOS performance specifications can be met (within 20%

of target without any mobility enhancement) using the same L_{SP} as for NMOS.

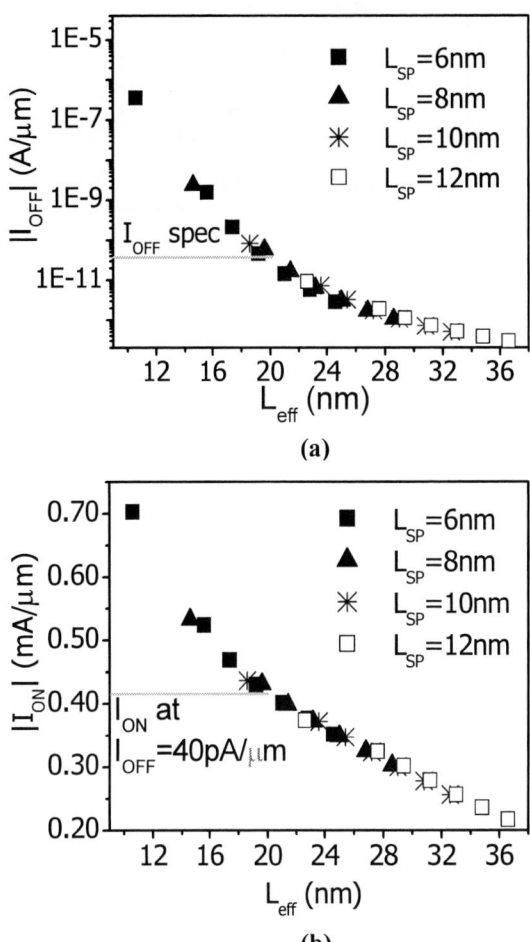

Fig. 7: PMOS performance specifications, I_{OFF}=40pA/μm and I_{ON}=440μA/μm, can be met (within 20% of target) with the same value of L_{SP} as for the NMOS devices.

Since V_T variation is of increasing concern for nanoscale CMOS technologies [7], 3D atomistic simulations were performed to determine the impact of statistical dopant fluctuations (SDF) in the S/D gradient regions, for the two extreme combinations of L_{SP} and σ_{SD}. The statistical V_T data is obtained from an ensemble of ~100 atomistic simulations. As can be seen in Fig. 8, the L_{eff} engineering technique is robust against SDF with σ_{VT} < 5mV, so that statistical V_T variation should not be an issue for this approach at L_G = 18nm. However, as it can be observed, a steeper σ_{SD} offers better immunity to V_T variations from SDF. Therefore, at shorter gate lengths, small-σ_{SD} designs will be necessary to suppress the effects of random dopant fluctuation [8].

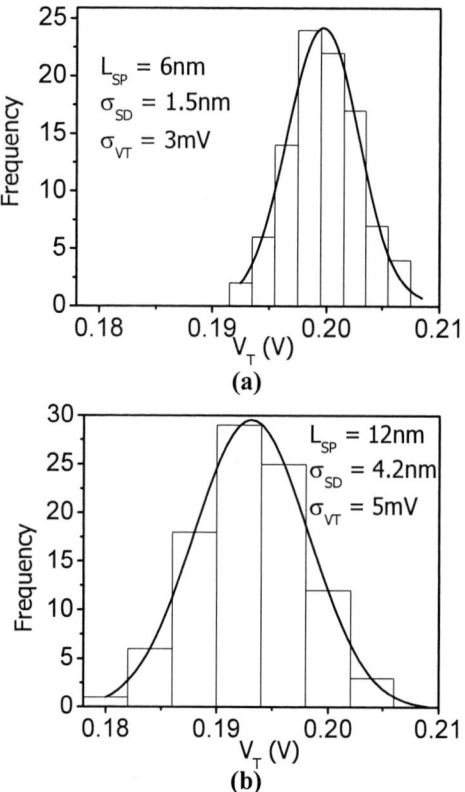

Fig. 8: Statistical V_T data obtained through 100 3-D atomistic device simulations indicate that SDF-induced variations will not be of concern at L_G = 18nm. The Si fin height (*i.e.* the channel width) was assumed to be 50nm.

Conclusions

L_{eff} engineering is a promising approach for adjusting V_T in sub-20nm L_G CMOS FinFETs. It is relatively simple to implement, and allows a single mid-gap-Φ_M gate material (*e.g.* TiN) to be used with undoped channels, to meet future ITRS LSTP performance specifications. There are multiple possible implementations to achieve the same performance, allowing for flexibility in process design. Variations due to random dopant fluctuations is not of concern for L_g = 18nm or higher. But for shorter gate length devices, L_{eff} solutions with steeper S/D gradient and ultra-thin spacers will be necessary to have better inherent immunity to statistical dopant fluctuation effects.

References

[1] International Technology Roadmap for Semiconductors, 2003 Edition.

[2] D. A. Antoniadis, "MOSFET scalability limits and 'New Frontier' devices," *2002 Symposium on VLSI Technology, Technical Digest*, pp. 2-5, 2002.

[3] N. Lindert, L. Chang, Y.-K. Choi, E. H. Anderson, W.-C. Lee, T.-J. King, J. Bokor, and C. Hu, "Sub-60-nm quasi-planar FinFETs fabricated using a simplified process," *IEEE Electron Device Letters*, Vol. 22, No. 10, pp. 487-489, Oct. 2001.

[4] P. Ranade, Y.-K. Choi, D. Ha, A. Agarwal, M. Ameen, and T.-J. King, "Tunable work function molybdenum gate technology for FDSOI CMOS," *International Electron Device Meeting Technical Digest*, pp. 363-366, 2002.

[5] J. Kedzierski *et al.*, "Metal-gate FinFET and fully-depleted SOI devices using total gate silicidation," *International Electron Device Meeting Technical Digest*, pp. 247-250, 2002.

[6] Taurus-Device v.2003 Users Manual (Synopsys, Inc.)

[7] A. R. Brown, A. Asenov, and J. R. Watling, "Intrinsic fluctuations in sub-10 nm double-gate MOSFETs introduced by discreteness of charge and matter," *IEEE Transactions on Nanotechnology*, Vol. 1, No. 4, pp. 195-200, Dec. 2002.

[8] V. Varadarajan, L. Smith, S. Balasubramanian, and T. -J. King Liu, "Multi-gate FET design for tolerance to statistical dopant fluctuations," *Si Nanoelectronics Workshop*, 2006.

[9] D. J. Frank, Y. Taur, M. Ieong, and H.-S. P. Wong, "Monte Carlo modeling of threshold variation due to dopant fluctuations," *1999 Symposium on VLSI Technology, Technical Digest*, pp.169-170, 1999.

[10] N. Sano, K. Matsuzawa, M. Mukai, and N. Nakayama, "Role of long-range and short-range Coulomb potentials in the threshold characteristics under discrete dopants in sub-0.1 μm Si-MOSFETs," *International Electron Device Meeting Technical Digest*, pp. 275-278, 2000.

1-4244-0267-0/06/$25.00 ©2006 IEEE

Impact of Millisecond Anneals on CMOS Scaling – A Device Simulation Study

Sunderraj Thirupapauliyur

Applied Materials Inc.,
Front End Products Division, 3050 Bowers Ave, Santa Clara, CA 95054, USA

Abstract: With the rapid scaling of CMOS devices towards the nanoscale regime as facilitated by lithography and strain engineering, the impact of parasitic series resistance is becoming a bigger issue. The need for shallower junctions to meet the short channel control requirements of the scaled transistors also aggravates the series resistance problem. Advanced technologies like millisecond laser anneal, which produce abrupt, highly activated junctions with negligible diffusion, are being proposed to meet junction requirements of 45nm technology node and beyond. But integrating the millisecond anneals into an existing spike baseline CMOS flow has not yet been fruitful. The increased overlap resistance due to the lack of lateral diffusion in the case of millisecond anneals is considered to offset the benefits of the otherwise abrupt and highly activated junctions. Thus for successful implementation of these advanced anneal technologies it is essential to understand the relative contributions of different components of series resistance and how it is impacted by the different USJ parameters like gate-source/drain overlap, lateral abruptness, junction depth and peak active doping concentration.

Methodology: Analytical doping profiles were generated for the nMOS devices with channel lengths varying from 45-500nm. The device simulations were performed using the drift-diffusion models along with quantum correction on the structure shown in Fig 1. The junction depth, X_j was fixed at 20nm in order to get the minimum gate length devices at Lg=45nm, work with good sub-threshold characteristics. The lateral and vertical doping profiles were described using two independent Gaussian functions. The lateral profile definition used in this study is shown in Fig 2. In order to isolate the effects of overlap and abruptness, a uniform channel doping was used.

Two approaches were used to compare the relative impacts of overlap and lateral abruptness. In the 'Lg' approach, the gate length Lg was used as a free parameter and the improvements in saturation current were compared at a constant leakage of 100nA/um, while the channel doping was kept constant at 4.8e18cm-3. In the 'Nch' approach, the channel doping was varied to maintain a constant leakage current of 100nA/um, while the gate length was kept constant at 50nm.

Results and Discussion: The Ion-Ioff comparison using the Lg approach comparing devices with two overlaps and lateral abruptness varying from 1nm/dec to 12nm/dec is shown in Fig 3. The 'Lg' approach shows marginal improvement in Idsat when compared at an Ioff of 100nA/um. Thus even at the same overlap, a junction with LA=3nm/dec representing a junction obtained by millisecond anneals gives same or marginal improvement compared to a junction with LA=8nm/dec which could be obtained by spike anneal. In Fig 4, the DIBL (Drain Induced Barrier Lowering) characteristics indicate that even at the same overlap, abrupt junctions suffer from severe short channel effects compared to gradual ones and the effect is pronounced at smaller gate lengths. This offsets the series resistance benefits of abrupt junctions and hence no improvement in Ion-Ioff. Fig 5 shows the Idsat obtained by the 'Nch' approach and shows a clear advantage of using an abrupt junction compared to gradual. Fig 5 also shows that abrupt junctions are able to tolerate lower overlap at the same Idsat, which could be used to minimize the overlap capacitance and hence improve CV/I delay. These results indicate that a direct integration of millisecond anneals in a spike baseline would not result in any improvement unless the channel is redesigned carefully to counter the short channel degradation of the abrupt junctions.

In some CMOS integration schemes, millisecond anneals are used in addition to the standard spike anneal to achieve higher peak doping concentration while the lateral abruptness and overlap remain more or less the same compared to a spike annealed junction. Fig 6, shows the trade-off obtained between overlaps and peak concentration at a constant lateral abruptness of 7nm/dec, which indicates that a highly activated source/drain junction could be used to obtain higher drive current at a given overlap or some overlap could be traded at the same Idsat to improve the overall delay.

1-4244-0267-0/06/$25.00 ©2006 IEEE

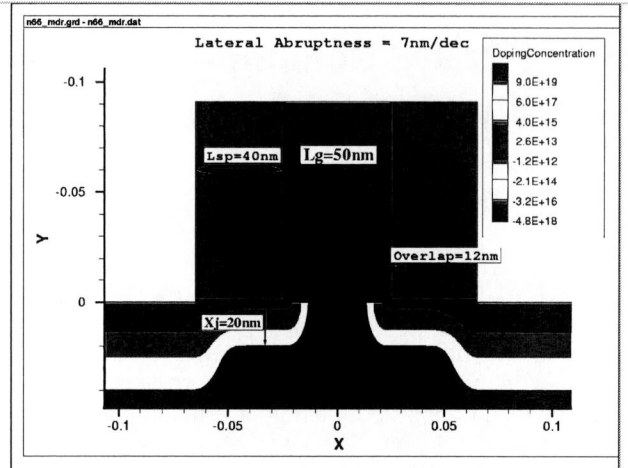

Fig 1: Simulated device structure indicating the gate length Lg=50nm, junction depth Xj=20nm and overlap OL=12nm for a lateral abruptness LA=7nm/dec. The spacer thickness Lsp=40nm. The channel doping was 4.8e18 cm^{-3}.

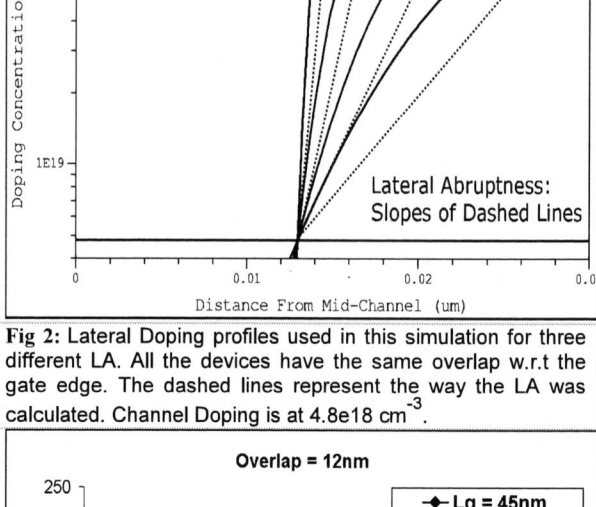

Fig 2: Lateral Doping profiles used in this simulation for three different LA. All the devices have the same overlap w.r.t the gate edge. The dashed lines represent the way the LA was calculated. Channel Doping is at 4.8e18 cm^{-3}.

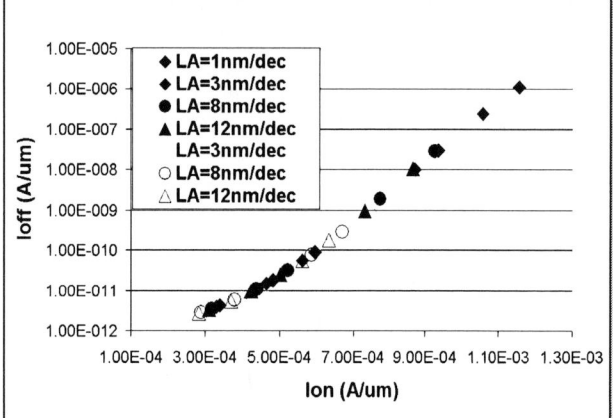

Fig 3: Ion-Ioff plot for devices with gate lengths from 45nm to 500nm using the 'Lg' approach. The solid symbols have an overlap of 12nm and the open symbols have an overlap of 7nm.

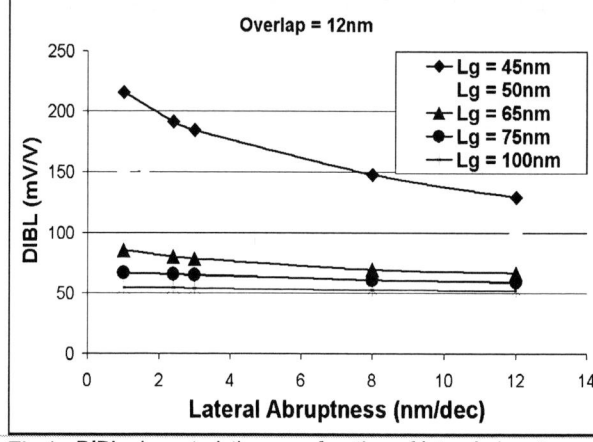

Fig 4: DIBL characteristics as a function of lateral abruptness LA for different gate lengths. Very abrupt junctions seem to have severe short channel effects at very small gate lengths.

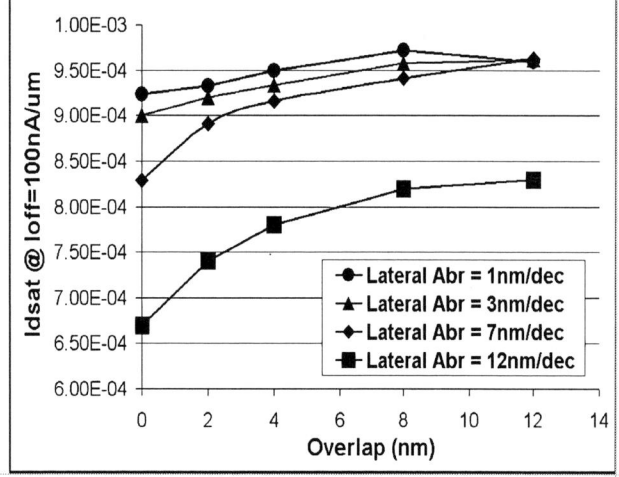

Fig 5: Idsat measured at Vg=Vd=1.0V, at a constant Ioff=100nA/um obtained by varying the channel doping Nch. The gate length was kept constant at Lg=50nm.

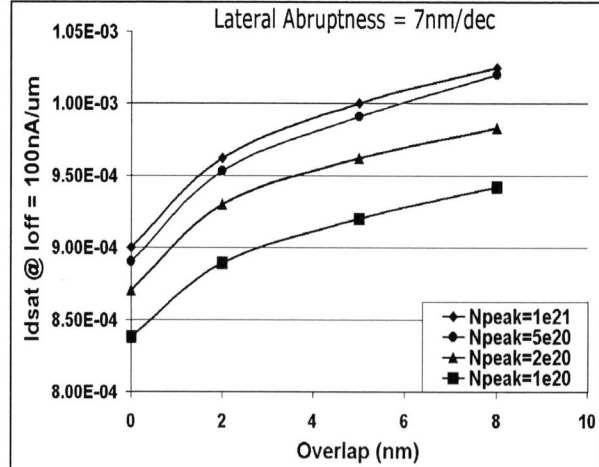

Fig 6: Idsat measured at Vg=Vd=1.0V, at a constant Ioff=100nA/um obtained by varying the channel doping Nch. The gate length was kept constant at Lg=50nm.

1-4244-0267-0/06/$25.00 ©2006 IEEE

Low Temperature Dopant Activation for Integrated Electronics Applications

Eric M. Woodard, Robert G. Manley, Germain Fenger, Robert L. Saxer, and Karl D. Hirschman
Department of Microelectronic Engineering
Rochester Institute of Technology

David Dawson-Elli and J. Greg Couillard
Corning, Incorporated

Abstract— A major area of research for integrated electronic systems is the development of systems on glass or plastic. These alternative substrate materials impose significant constraints on electronic device fabrication, including limitations on chemical and thermal processes. This work presents an investigation on the activation of ion-implanted dopants without using the high temperature processes of conventional CMOS. The annealing temperature applied was 600°C, which could potentially enable integrated microelectronics on high-quality glass. Additional factors studied included the annealing technique (furnace or rapid thermal processing), and the use of pre-amorphization implants. Ion-implant modeling along with SIMS and SRP data was used to develop a comprehensive understanding of the experimental results. The performance of transistors fabricated with low-temperature constraints on both bulk silicon and thin-film SOI will be presented.

Index Terms—low temperature dopant activation, thin film transistors (TFTs), pre-amorphization, solid phase epitaxy (SPE)

I. INTRODUCTION

IN conventional thin-film transistor (TFT) technology, conductive silicon regions are formed via in-situ doping during the deposition of amorphous silicon. In CMOS fabrication on silicon (i.e. bulk Si, SOI), the standard method of adding impurities is ion implantation, after which an anneal must be performed to remove lattice damage and allow the dopants to occupy lattice sites and become electrically active.

While dopant activation at low temperatures is possible, the activation anneal is typically done at 900-1000°C in order to provide full activation and ensure that damage removal is complete. For high-dose implants, solid-phase epitaxy (SPE) initiates at around 500°C, and takes advantage of the lattice rebuilding process to incorporate dopants at lattice sites [1]. However imposing a 600°C temperature constraint provides limited thermal energy for atomic rearrangement, and can result in the formation of secondary defects which are attractive regions for dopants to segregate and limit the level of activation. The activation of low-dose implants is further challenged by the virtual elimination of thermal diffusion, relying on incorporation into the lattice through point defects.

This study investigates the ability to activate boron, phosphorus and arsenic under low temperature conditions. Experiments consisted of a continuum of doses spanning almost four orders of magnitude (approximately $10^{12}cm^{-2}$ to $10^{16}cm^{-2}$). In addition, the use of a high-dose pre-amorphization implant (silicon, argon and fluorine) was explored. While the motivation for this study was process development for low-temperature CMOS, treatment combinations that went outside of a practical "process window" were included in order to better understand the dopant activation process. Experimental results demonstrate that dopant activation under low temperature constraints can be quite effective, and is further established by the realization of low temperature CMOS.

II. DOPANT ACTIVATION THEORY

There is perhaps no other material system as well understood as silicon, and yet there continues to be new areas of exploration and prior studies revisited. While there has been extensive research on dopant activation in silicon, almost all of the recent work has been in the traditional high temperature regime. Modern CMOS technology demands ultra-shallow junctions, for which the strategy has been to minimize the thermal budget and suppress transient-enhanced diffusion (TED) using extremely short-time rapid thermal processing (RTP). More recent studies of dopant activation have been done in this regime [2]. Classical experimental results on low-temperature dopant activation were published during the development of silicon process technology [3,4]. Figures 1 and 2 show isochronal annealing results for boron and phosphorus, respectively, and demonstrate clear differences in the activation process of these substitutional impurities. While both impurities show reasonable low-dose activation at similar levels, there appears to be significant differences in the mechanisms involved with high-dose activation.

A. Primary Crystalline Damage

Under low-dose conditions ($\phi < 10^{13}cm^{-2}$) the activation behavior appears quite straightforward, with the 600°C

1-4244-0267-0/06/$25.00 ©2006 IEEE

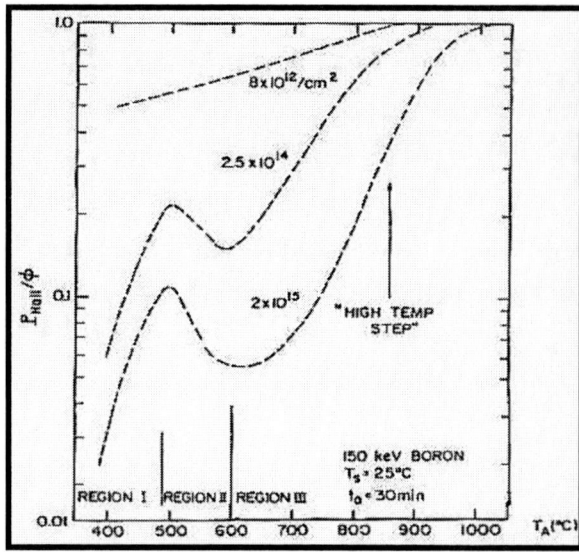

Fig. 1. Isochronal (30min) annealing results of boron (^{11}B). The activation levels are quantified as a ratio of the Hall-dose over the implanted dose. Low, medium, and high-dose characteristics show distinctly different activation behavior. After [3].

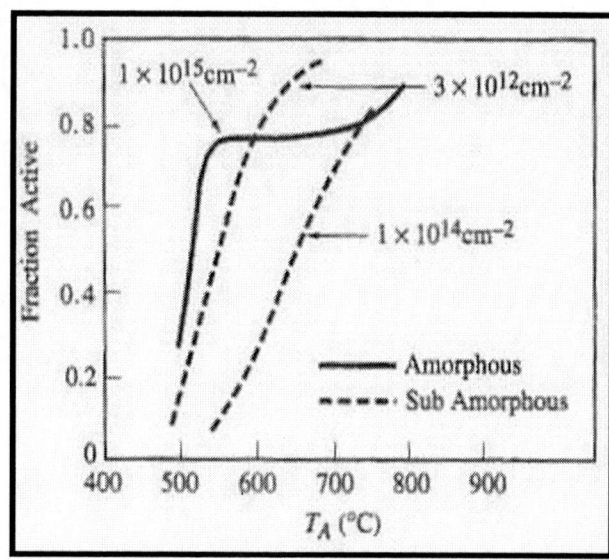

Fig. 2. Isochronal (30min) annealing results of phosphorus (31P). The activation levels are quantified as a ratio of the Hall-dose over the implanted dose. Low, medium, and high-dose characteristics show distinctly different activation behavior. After [1], adapted from [4].

activation levels of boron and phosphorus at approximately 60% (see fig. 1) and 80% (see fig. 2), respectively. In this low-dose range, the implant damage that is created is limited to primary crystalline damage. Low doses of light ions (e.g. boron) create isolated point defects or point defect clusters in essentially crystalline silicon, whereas low doses of heavy ions (e.g. arsenic) create completely amorphous material in an otherwise crystalline layer [5].

In the case of both phosphorus and boron, as the implant dose is increased to the "medium-dose" range ($\sim 10^{14}$cm^{-2}), the 600°C activation levels are decreased significantly; only 25% of the phosphorus dose and 15% of the boron dose becomes electrically active. Phosphorus demonstrates effective self-amorphization at implant doses higher than 5×10^{14}cm^{-2}; doses approaching this level are difficult to activate without the formation of a continuous amorphous region from the surface to the end of range damage. This is due to the formation of crystal dislocations that form during the anneal (referred to as secondary defects) that inhibit dopant activation, as well as the reliance on thermal diffusion for dopant atoms to find available lattice sites [1,2]. Also note the decrease in boron activation in this dose range as the temperature is increased from 500°C to 600°C, shown in fig. 1.

B. Solid Phase Epitaxy

Under high-dose conditions, a completely amorphized surface layer can result in a solid phase epitaxy (SPE) regrowth of the silicon crystal, using the underlying crystal structure as the seed. Dopants become incorporated into the lattice structure during this rebuilding process, requiring significantly less energy than primary defect recombination [1]. The lattice regrowth rate has been shown to be highly dependent on the concentration of substitutional dopants; high concentrations can cause an increase by factors of 6-25 [6].

In the case of phosphorus, a dose of 10^{15}cm^{-2} demonstrates 77% activation at 600°C. However, boron demonstrates a further decrease in the ability to achieve activation (only 5-6%), with the deactivation effect even more pronounced; quite different from the corresponding phosphorus result. The deactivation process occurs because of a competition between silicon interstitials and boron atoms for substitutional lattice sites, or by the formation of inactive complexes due to interstitials pairing with boron [1].

SPE is not effective in activating high-dose boron implants when the boron is introduced as a single ion species (^{11}B^{+}). The success of boron activation can be increased by implanting a molecular species that self-amorphizes the crystal; BF$_2^{+}$ is commonly used for the formation of low thermal-budget shallow p+/n junctions. Another method is the use of a pre-amorphization implant.

C. Pre-amorphization

A continuous amorphous region can be formed without relying on self-amorphization by implanting a non-reactive ion to create a damaged region prior to implantation of the dopant ions. This allows for control on the depth of the amorphous region that is independent from the location of dopant atoms. Pre-amorphization can suppress channeling of the dopant ions, and result in significant levels of activation through SPE. Certain considerations when using this technique are the damage and dopant distributions, and the possibility of secondary defects that may form during the annealing process and remain thermally stable under low temperature constraints.

Germanium (^{74}Ge) and silicon (^{28}Si) are common elements to use for silicon pre-amorphization, which themselves do not modify the electronic properties. In addition to these elements, fluorine (^{19}F) has been investigated as a pre-amorphization implant, preceeding ^{11}B in the formation of shallow p+/n junctions [7]. While fluorine ($^{19}F^+$) may not be an obvious choice for the creation of implant damage due to its low atomic mass, the precedence of using fluorine with boron has been established in the use of BF_2 molecular implants.

The use of ^{19}F & ^{11}B co-implants have been investigated by a number of research groups, mostly in the context of suppressing ion-channeling and TED using low-energy implants along with high temperature /short time RTA [7-9]. This technique appears to successfully reduce TED [8] and enhance boron activation [9]. However in certain cases the junction integrity is inferior to junctions formed using BF_2 molecular implants or pre-amorphization with ^{74}Ge or ^{28}Si [7]. Previous work on the use of ^{19}F & ^{11}B co-implants activated at 600°C has not been reported.

III. EXPERIMENTAL PROCEDURE

This section provides process details on the systematic investigation on low-temperature (600°C) activation of phosphorus, arsenic and boron. All sample preparation and measurement was done at the Semiconductor & Microsystems Fabrication Laboratory (SMFL) at RIT [10]. Implant conditions resulted in primary crystalline damage (low-dose range) and amorphization; created by self-amorphization (^{31}P, ^{75}As and BF_2) or through the use of a pre-amorphization implant. The effectiveness of pre-amorphization using ^{28}Si was studied on the activation of both phosphorus and boron. The use of ^{19}F and ^{40}Ar pre-amorphization implants with ^{11}B were also done in order to compare with ^{28}Si pre-amorphization and BF_2 self-amorphization results.

Argon is not a traditional element used for pre-amorphization due to issues with argon bubble formation during the implant and annealing processes [11], and the inhibition of SPE at high argon concentrations [12]. Regardless, ^{40}Ar was investigated due to the appropriate atomic mass, the attractiveness of an inert gas species, and the immediate availability of argon on the ion implanter.

All phosphorus, boron, fluorine and argon implants were done at RIT on a Varian 350D ion implanter with a fixed 7° tilt. Arsenic and silicon implants were outsourced to Implant Sciences Corporation.

A. N-type Dopant Implant Experiments

The activation of phosphorus was studied by implanting ^{31}P at a constant energy over a wide range of doses. A 1000Å screen oxide was used to prevent ion channeling. The implanted doses range from $5x10^{12}$ to $8x10^{15}cm^{-2}$, in order to capture the extremes of crystal disorder from primary damage to self-amorphization. The implant energy was 92 keV in order to place the peak concentration below but near the oxide-silicon interface; allowing for accurate sheet resistance measurements even under low-dose treatment combinations. A similar dose range was explored for arsenic. The implant energy for arsensic was 120keV, and the screen oxide thickness was scaled to 500Å, providing an equivalent dopant profile.

Co-implantation of ^{28}Si & ^{31}P was also investigated in order to compare with self-amorphization results, and to determine if the activation levels of ^{31}P implant doses below the critical dose to self-amorphize ($\phi_{crit} \sim 5x10^{14}cm^{-2}$) could be increased. The ^{28}Si pre-amorphization implant conditions for these experiments were $1x10^{15}cm^{-2}$ at 120keV. The dose and energy conditions were chosen to provide a disorder distribution similar to successful self-amorphization conditions at $\phi_{phos} > \phi_{crit}$, using SRIM™ simulation software [13] to model the damage density profile.

B. Boron Implant Experiments

Boron activation was studied as a single-element species and with other elements via a molecular implant or co-implant. The initial experiment investigated ^{11}B implants at a constant energy (34keV) over a wide range of doses, through a 1000Å screen oxide. This results in a peak concentration below but near the silicon surface; comparable to the phosphorus and arsenic profiles. High-dose BF_2 molecular implants were done at 155keV without the use of a screen oxide to provide a comparison to ^{11}B and co-implant results.

Co-implantation was investigated over various ^{11}B doses using a variety of pre-amorphization species/dose combinations. The pre-amorphization implants were simulated using SRIM; the dose and energy for each species was chosen to provide a similar damage distribution and density profiles. Fluorine, silicon and argon were implanted using the following dose/energy combinations: ^{19}F $3x10^{15}cm^{-2}$ @ 75keV, ^{28}Si $1x10^{15}cm^{-2}$ @ 120keV, and ^{40}Ar $5x10^{14}cm^{-2}$ @ 170keV. The implants were done through the 1000Å screen oxide prior to ^{11}B implantation.

C. Thermal Activation & Sample Measurement

Both inert-ambient furnace annealing and RTA (AG Associates Heatpulse 610) was used to activate the dopants. Initial RTA experiments varied both temperature (550-650°C) and time (short time increments) in order to investigate the kinetics of dopant activation. Furnace anneals were typically done for 60min at 600°C in N_2. Any screen oxide was removed in HF prior to sample measurement.

Electrical four-point probe sheet resistance (Rs) measurements were taken using a CDE ResMap system, providing a quantitative response for each treatment combination that is ultimately dependent upon the available carriers and the dopant distribution. While not exhaustive, several samples were analyzed via Secondary Ion Mass Spectroscopy (SIMS), performed at Corning, Inc., and Spreading Resistance Analysis (SRA), performed at Solecon Laboratories, Inc. In addition, Transmission Electron Microscopy (TEM) images were obtained from select samples using a JEOL JEM-2000FX, also done at Corning, Inc.

IV. ASSESSMENT OF DOPANT ACTIVATION

The fraction or percentage of dopant activation is usually defined as the ratio of the active dose to the implanted dose. Such a response is shown in figures 1 and 2, where Hall measurements were used. SIMS and SRA can ultimately lead to a similar comprehensive metric. However like Hall effect characterization, the techniques do not come without sources of error and complexity in interpretation under certain sample conditions [14, 15]. In most cases, SIMS and SRA profiles are used for a graphical representation of the data for comparisons between total (chemical) and electrically active dopant profiles, and relative comparisons between treatment combinations. Due to the number of samples involved in this study, and the desire for consistency in the evaluation of dopant activation, an approach referred to as *reverse modeling* [9] has been used to quantify the percentage of dopant activated. As the name suggests, reverse modeling takes a modeled profile that has been adjusted to yield the results of an actual measurement in order to *"back out"* the value of, in this case, the active dopant dose.

There were several assumptions made that are inherent to this technique. Silvaco SUPREM-IV simulation software (Athena™) was used to provide a model for the implanted doping profile. The model profile was dose-adjusted in order to provide a match between the modeled and measured sheet resistance. Fundamental assumptions made include: (1) the active doping profile is a scaled version of the total doping profile; (2) the mobility models used to determine the electrically active carrier concentrations are consistent with crystalline silicon.

Whenever possible, the total and active (dose-scaled) modeled profiles were compared to SIMS and SRA profiles, respectively, to validate the accuracy of the model. In certain cases there was remarkable consistency between the modeled and measured Rs and active dose, while in other cases there was a discrepancy. In cases where the discrepancy was significant, the percent activation was quantified by the SRA/SIMS dose ratio. In cases where there was reasonable correlation, the level of dopant activation as determined by reverse modeling was, at a minimum, useful for relative comparisons. The results of the dopant activation experiments are presented in the next section, with select comparisons between models and measurements provided as appropriate.

V. RESULTS & DISCUSSION

While RTA experiments were performed at different temperature/time combinations in order to investigate the dopant activation kinetics, the analysis is still in progress; results from which will be published elsewhere [16]. This report will be limited to details on the levels of dopant activation achieved via furnace annealing at 600°C in N_2.

A. N-type Dopant Activation

Phosphorus activation results are displayed in fig. 3, over an implanted dose range from $5x10^{12}cm^{-2}$ to $8x10^{15}cm^{-2}$. The x-axis labels provide the modeled dose in silicon; note that this axis is categorical and not an equal-interval scale. An example of the reverse modeling method is shown in fig. 4 and

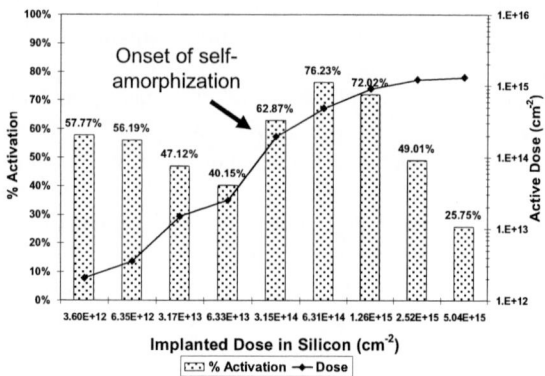

Fig. 3. Activation of phosphorus implants at 600°C. The x-axis labels indicate the modeled implanted dose within the silicon (not to scale). The bars indicate the % activation (Y1-axis) as determined using the measured sheet resistance with a scaled-dose model. The markers indicate the active dose (Y2-axis) used in the % activation calculation.

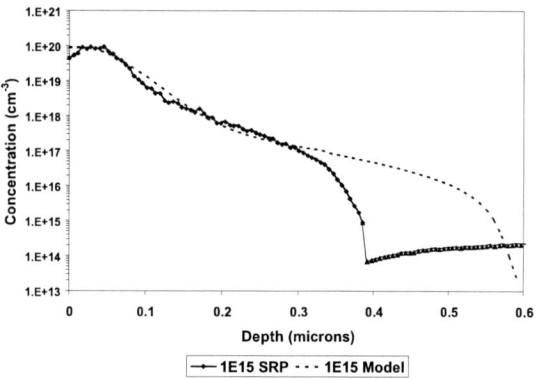

Fig. 4. Comparison of phosphorus profiles from SRA and SUPREM. The SRA active profile was integrated to a dose of $5.8x10^{14}cm^{-2}$. The modeled implanted silicon dose was $1.26x10^{15}cm^{-2}$, and the modeled active dose was scaled to $4.8x10^{14}cm^{-2}$ to match the Rs value of $134\Omega/sq$, yielding an activation of 72% shown in fig. 3.

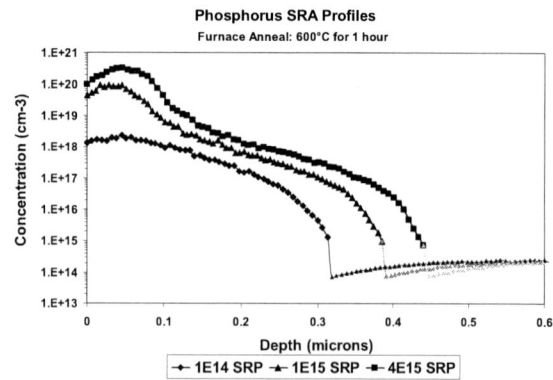

Fig. 5. Comparison of phosphorus SRA profiles. The high dose implants exhibit similarly shaped electrically active profiles, whereas the lower dose implant demonstrates an activation limit of $N \sim 2x10^{18}cm^{-3}$ due to lack of effective self-amorphization.

1-4244-0267-0/06/$25.00 ©2006 IEEE

associated caption. A dual-Pearson implant model was chosen which provides an adequate fit to the shape of the SRA profile. The discrepancy in the tail region of the profile has a minimal impact on the resulting sheet resistance and active dopant assessment. In addition, using a scaled-dose approach seems appropriate when comparing the SRA profiles over different doses, as long as the phosphorus dose exceeds the self-amorphization threshold (see fig. 5).

In cases where the scaled-dose model is not suitable, such as the low-dose phosphorus SRP shown in fig. 5, the reverse modeling approach can contribute up to 10-20% error in the active dose assessment. Although this translates to error in the ultimate % activation metric, due to the lack of a perfect model to match SRA profile information under all dose conditions the use of this method is reasonably well justified.

The explanation of the results displayed in fig. 3 is quite straightforward. Under low dose conditions ($\phi < 1 \times 10^{13} \text{cm}^{-2}$) the annealing of primary damage results in approximately 55-60% activation. As the dose is increased, but remains below the point of self-amorphization ($\phi \sim 1 \times 10^{14} \text{cm}^{-2}$), the formation of secondary defects results in a decreased level of activation to approximately 40%. As the dose is increased above the onset of self-amorphization, the activation level increases to over 75%, consistent with fig. 2. As the dose is further increased, an activation limit causes a decrease in the % activation and the active dose appears to saturate.

The results of arsenic activation are shown along side phosphorus in fig. 6. Under high-temperature annealing conditions, arsenic is well known for having a high solubility limit and a low diffusivity, making it attractive for high concentration shallow n+ source/drain junctions. However, arsenic does deactivate and cluster at supersaturated concentrations if allowed to diffuse [17]. The behavior of arsenic is quite similar to phosphorus, noting higher levels of activation at $\phi < 1 \times 10^{15} \text{cm}^{-2}$. However, at the high-dose range the activation is markedly lower. The comparison of treatment combinations with $\phi = 4 \times 10^{15} \text{cm}^{-2}$ and $8 \times 10^{15} \text{cm}^{-2}$ shows that the amount of electrically active arsenic can actually decrease with a higher implanted dose; evidence of clustering that degrades activation rather than simply reaching an activation limit.

Silicon pre-amorphization was investigated as a method for improving the activation of phosphorus, with unexpected results. The silicon implant was designed to be as effective as self-amorphization, while extending the amorphous region beyond the phosphorous end-of-range. The treatment combinations with the phosphorus dose in silicon at $6.3 \times 10^{14} \text{cm}^{-2}$ and $6.3 \times 10^{13} \text{cm}^{-2}$ demonstrated a decrease in the level of activation from 76.3% to 63.2% and 40.1% to 6.8%, respectively. Figure 7 shows some difference in the implanted phosphorus SIMS profiles; the extended tail appears to be suppressed in the profile with silicon pre-amorphization. Also shown in fig. 7 are SRA profiles for the non-amorphized low-dose sample ($\phi = 6.3 \times 10^{13} \text{cm}^{-2}$), shown previously in fig. 5, and the same dose with silicon pre-amorphization. The reduction in activation can be associated with the formation of secondary defects shown in the TEM image in fig. 8.

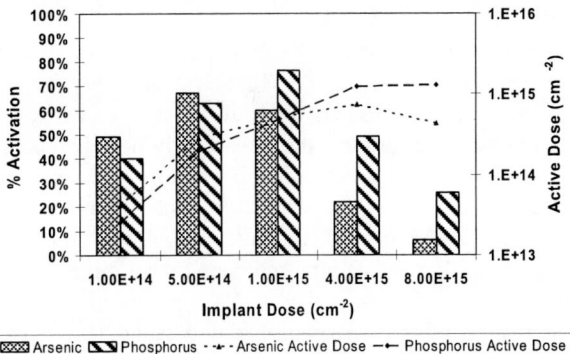

Fig. 6. Comparison of arsenic and phosphorus activation at 600°C. High-dose activation of phosphorus is superior, whereas arsenic shows degradation in activation due to clustering. Note that the x-axis is labeled with the implanted dose rather than the integrated silicon dose, since the treatment combinations did not yield the same integrated dose values.

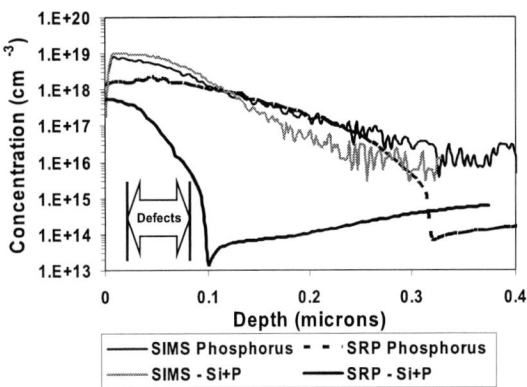

Fig. 7. Comparison of SIMS and SRA profiles on phosphorus implanted samples prepared with and without silicon pre-amorphization. The SIMS profiles show a decreased tail in the pre-amorphized sample, however the SRP shows a dramatic decrease in electrical activity, associated with the formation of secondary defects during the annealing process.

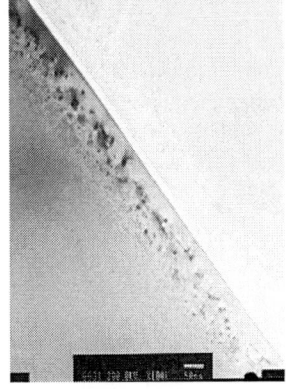

Fig. 8. TEM Image of the silicon pre-amorphized phosphorus implanted sample with greatly reduced electrical activation shown in fig. 7. The highly defective region extends approximately between 25-80nm.

1-4244-0267-0/06/$25.00 ©2006 IEEE

B. Boron Activation

Initial experiments with ^{11}B implants demonstrated a clear relationship between dose and activation capability, as shown in fig. 9. While 50% activation at 600°C can be achieved under low-dose conditions, this was clearly not an acceptable strategy to achieve low-resistance p+ regions. Measurements from high-dose BF_2 molecular implants were also not encouraging, with boron activation levels below 15% as shown in fig. 10. These results motivated the study on pre-amorphization.

The results of the pre-amorphization study are shown in figures 10 and 11. Figure 10 shows that very high levels of boron activation can be achieved at 600°C. Silicon, fluorine and argon pre-amorphization all provided significant enhancement; well above that achieved through self-amorphization using BF_2. For each condition, the amorphous region was designed to extend beyond the ^{11}B end-of-range. The implant model used for % activation calculations on amorphized samples was a Pearson-IV (amorphous model) which provided a reasonable match to SIMS verified profiles. SRA profiles also verified that the reverse modeling approach is reasonable model for a scaled-dose estimate of active dopant.

Further investigation focused on ^{19}F pre-amorphization experiments, due to the common gas source and the initial success shown in fig. 10. Experimental results of boron activation over a wide dose range are shown in fig. 11. There appeared to be a dose-dependent activation effect under low dose conditions ($\phi < 1x10^{14}cm^{-2}$), and the Rs measurements were not reliable. This may coincide with a donor effect associated with the use of fluorine, discovered through SRA [16]. However, as the ^{11}B dose is increased, activation levels at 90% are achievable. As the ^{11}B dose in silicon is increased above $1x10^{15}cm^{-2}$ there appears to be an activation limit similar to phosphorus, shown by a decrease in the % activation and a saturation of the integrated active dose.

VI. LOW TEMPERATURE CMOS

Dopant activation at 600°C has been shown to be highly successful; the levels of activation achievable address the needs of a viable low-temperature (LT) CMOS process. However, it is the demonstration of high-quality n+/p and p+/n junctions that will enable integrated microelectronics under LT constraints. A detailed study is currently under investigation [16], however electrical characteristics of transistors fabricated under 600°C thermal constraints are very encouraging.

Due to LT constraints, isolation schemes become difficult; only NFETs or PFETs can be fabricated on bulk silicon depending on the starting substrate doping. At temperatures 600°C and below, diffusion is insignificant and thus the formation of doped wells is not possible. However, with SOI substrates, CMOS can be easily realized using mesa isolation techniques.

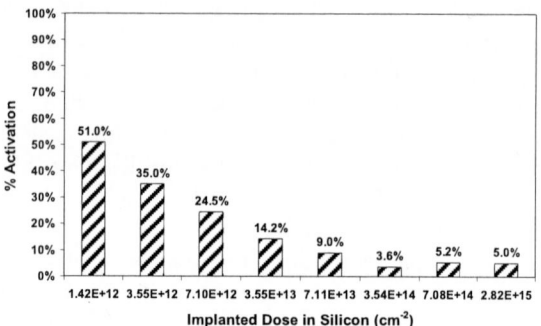

Fig. 9. Activation of boron implants at 600°C. The x-axis labels indicate the modeled implanted dose within the silicon (not to scale). The level of boron activation becomes significantly limited as the dose increased above $\phi \sim 1x10^{13}cm^{-2}$.

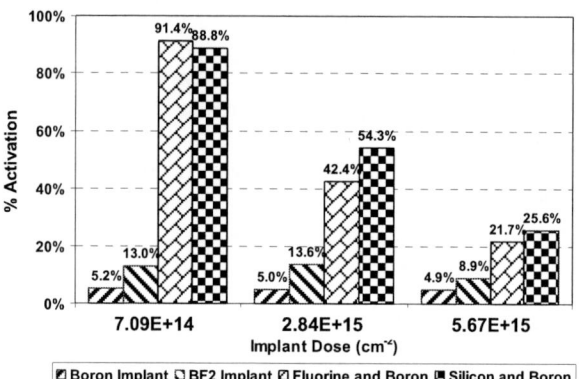

Fig. 10. The influence of pre-amorphization implants on the activation of boron at 600°C. BF_2 self-amorphization results are shown for comparison. ^{28}Si and ^{19}F had very similar activation enhancement effects, achieving $\sim 90\%$ prior to reaching an activation limit. Argon (^{40}Ar) did provide a boron activation peak between 65-70% (results not displayed).

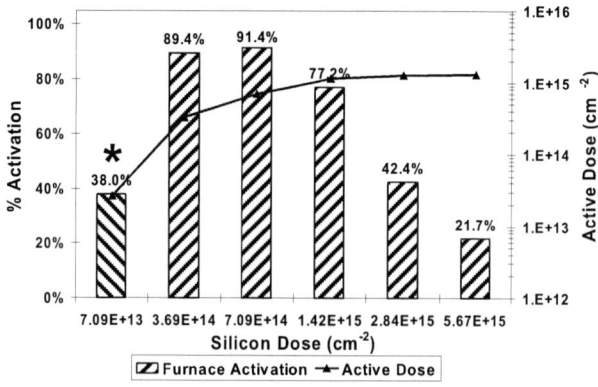

Fig. 11. Activation of ^{19}F pre-amorphized boron implants at 600°C. The bars indicate the % activation (Y1-axis) as determined using the measured sheet resistance with a scaled-dose model. The markers indicate the active dose (Y2-axis) used in the % activation calculation.
* indicates % activation as determined by the SRA/SIMS ratio.

A. Device Fabrication

Device fabrication at low temperatures creates limitations in several areas including: isolation schemes, gate material, gate dielectric, as well dopant activation. Because of these constraints a slightly different approach in fabricating CMOS must be taken.

The low temperature constraint reduces the available options for a high quality gate dielectric. Thermally grown oxides are not possible and thus all dielectrics must be deposited by some method. On bulk silicon, this eliminates such processes as LOCOS. Shallow-trench isolation (STI) can be used, however the advantages of device density and planarization were not required for this study. For device fabricated on bulk silicon, a thick SiO_2 field oxide was deposited at 425°C via low pressure chemical vapor deposition (LPCVD) in a silane-oxygen reaction (low-temperature oxide, or LTO). To prevent field inversion with lightly dope substrates, a low dose blanket boron implant (channel-stop) was performed while the active regions were masked. Isolation for the SOI devices was relatively simple. The SOI film was etched to the buried oxide to form active silicon mesas (thickness = 2000Å) for NFETs and PFETs.

The elimination of the thermally grown oxides also means that the gate dielectric must be deposited. For the fabrication of the bulk and SOI devices an LTO oxide gate dielectric film was deposited via LPCVD. Because the quality is not as good as a thermal oxide, 500Å was used. Interface quality issues also arise but are beyond the discussion of this paper.

To allow for a self-aligned source/drain process the gate material must be able to mask an implant and withstand processing up to 600°C. Doping of polysilicon at low temperatures is difficult. To achieve sufficient levels of active dopant, excessively large implant doses are required as much of the dopants will segregate at grain boundaries. In-situ doped polysilicon can be used however it was not available for this study. Also any annealing done at low temperatures may cause deactivation or dopant segregation at grain boundaries. To avoid these issues, a molybdenum gate metal was used; it is excellent for masking implants and is thermally stable at process temperatures of 600°C and higher.

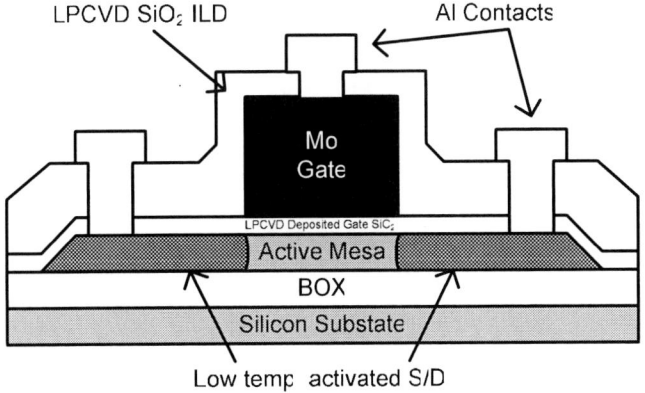

Fig. 12. Cross-section of SOI FET fabricated a low temperature. All dielectric films were dopsited. For CMOS operation, the NFET and PFET were isolated via a Si mesa etch.

For NFET source drain implants, phosphorus was used. As described earlier, at high doses phosphorus will self-amorphize, allowing for reasonable levels of activation at 600°C. To achieve highly doped n+ source/drain regions, a $4E15cm^{-2}$ ^{31}P dose was implanted. To achieve sufficiently doped PFET source/drain regions, a ^{19}F & ^{11}B co-implant was required. A fluorine dose of $3E15cm^{-2}$ was implanted to cause pre-amorphization, followed by a $4E15cm^{-2}$ boron implant.

Following the n+ and p+ implants, an LPCVD SiO_2 inter-level dielectric (ILD) was deposited. An anneal at 600°C for two hours in a N_2 ambient was performed to activated the source drain regions as well as helping to densify the ILD and gate dielectrics. Contacts were then patterned and etch followed by an aluminum deposition and etch. Finally, a sinter at 425°C in forming gas (5% H_2 in N_2) was done for 15 minutes to complete the fabrication process.

It should be noted that on the SOI substrate, the NFET and PFET were fabricated in the same type film. Because of the low temperature constraint an n-type well could not be fabricated because the lack of significant diffusion at 600°C. However, with a thin SOI film and a lightly doped substrate, it was still possible to make an PFET enhancement mode threshold voltage. Details surrounding the operation of this device will be discussed elsewhere [16].

B. Device Results

1) Low Temperature NMOS on Bulk Silicon

As mentioned before, CMOS fabrication on bulk Si substrates was not possible because of limitations related to diffusion and well isolation. Figure 13 show a I_D-V_G characteristic of NFET fabricated at temperatures not exceeding 600°C. The threshold voltage of this device was designed to be 1V, but resulted in being around 1.8V to 2V. The reason for the high shift in this device is due to high levels of interface states at the gate dielectric-silicon interface. Large densities of negatively charged interface traps cause a positive shift in the MOS device threshold characteristics.

The subthreshold swing (SS) of the device in fig. 13 is 157 mV/dec. Typical results of devices fabricated using normal process temperatures are below 100 mV/dec. This degradation in the turn-off response can also be attributed to interface traps. However, the off-state leakage current is quite reasonable; verification that the 600°C annealing process can yield good pn junctions. The field-effect mobility was lower than anticipated. For bulk device process at high temperatures at the same doping level, electron field effect mobilities of 600 to 650 cm^2/Vsec are typical. An electron field effect mobility of 405 cm^2/Vsec was extracted from the LT NFET. This lower value is again attributed to an inferior oxide-silicon interface.

2) Low Temperature CMOS on SOI

LT CMOS has been demonstrated on SIMOX SOI substrates. Both NFETs and PFETs have enhancement mode threshold voltages, made possible due to a 200nm silicon layer. Device isolation is provided by a silicon mesa etch. The NFET is a surface channel device. It exhibits excellent

1-4244-0267-0/06/$25.00 ©2006 IEEE

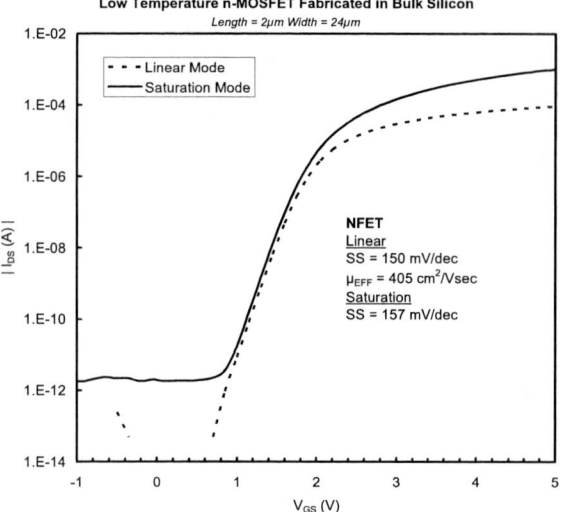

Low Temperature n-MOSFET Fabricated in Bulk Silicon

Length = 2μm Width = 24μm

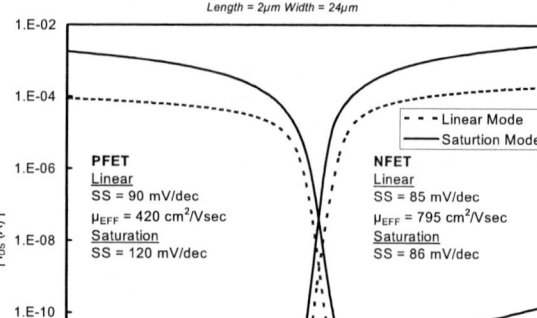

Low Temperature n-MOSFET Fabricated on SOI

Length = 2μm Width = 24μm

Fig 13. I_D-V_G characteristic for an n-MOSFET fabricated a 600°C in a bulk p-type silicon substrate. The electron field effect mobility, as extracted from the maximum transconductance, is 405 cm²/Vsec.

Fig. 14. I_D-V_G characteristic for an n-MOSFET fabricated a 600°C in a SOI p-type silicon substrate. Both the NFET and PFET show enhancement mode threshold voltages, $V_{Tn} = 0.6V$ and $V_{Tp} = -0.56V$

characteristics with a low subthreshold swing of 85 mV/dec and a high electron field effect mobility of 795 cm²/Vsec. In the saturation mode of operation some gate induced leakage is observed, which may be enhanced by the junction integrity.

The PFET is fabricated in a p-type substrate, and thus operates as a buried channel device. When the device is off, the gate must deplete the body region of the device in order for no current to flow. The field-effect hole mobility extracted was 420 cm²/Vsec. The device shows a considerable difference in the subthreshold swing with it being 90 mV/dec in the linear regime and 120 mV/dec in the saturation regime of operation. The reason for the large discrepancy is due to lack of gate control at biases around 0V. The larger bias on the drain changes how the body region of the device depletes, causing it to turn of more gradually. Additional details on this LT CMOS strategy will be presented elsewhere [16].

ACKNOWLEDGMENT

The authors would like to thank the SMFL staff at RIT, and technical staff at Corning that provided analytical services. Financial support has been provided by Corning, Inc. and NYSTAR, through the New York State Center for Advanced Technology (CAT) program.

REFERENCES

[1] J.D. Plummer, M.D. Deal and P.B. Griffin, *Silicon VLSI Technology; Fundamentals, Practice and Modeling.* Upper Saddle River, NJ: Prentice Hall, 2000, ch. 8.

[2] A. Mokhberi, P.B. Griffin, J.D. Plummer, E. Paton, S. McCoy and K. Elliott, "A Comparative Study of Dopant Activation in Boron, BF₂, Arsenic and Phosphorus Implanted Silicon," *IEEE Trans. Electron Devices,* vol. 49, pp. 1183-1191, July 2002.

[3] T.E. Seidel and A.U. MacRae, "The Isothermal Annealing of Boron Implanted Silicon," in *First International Conference on Ion Implantation*, R. Eisen and L. Chadderton, Eds., Gordon and Breach, 1971.

[4] B.L. Crowder and F.F. Morehead, "Annealing Characteristics of N-type Dopants in Ion Implanted Silicon," *Appl. Phys. Lett.,* vol. 14, pp. 313-315, May 1969.

[5] S. Wolf and R.N. Tauber, *Silicon Processing for the VLSI Era, Volume 1 – Process Technology, 2nd ed.* Sunset Beach, California: Lattice Press, 2000, ch. 10.

[6] L. Csepregi, E.F. Kennedy, T.J. Gallagher, J.W. Mayer and T.W. Sigmon, "Reordering of amorphous layers of Si implanted with ³¹P, ⁷⁵As, and ¹¹B ions," *J. Appl. Phys.,* vol. 48, pp. 4234-4240, 1977.

[7] C.P. Wu, J.T. McGinn and L.R. Hewitt, "Silicon Pre-amorphization and Shallow Junction Formation for ULSI Circuits," *J. Electron. Mater.,* vol. 18, pp. 721-730, 1989.

[8] J. Liu, D.F. Downey, K.S. Jones and E. Ishida, "Fluorine Effect on Boron Diffusion: Chemical or Damage?" *Proc. 1998 International Conference on Ion Implantation Technology,* Kyoto, Japan, IEEE, pp. 22-26, June 1998.

[9] E.N. Shauly and S.Lachman-Shalem, "Activation Improvement of Ion Implanted Boron in Silicon Through Fluorine Co-Implantation," *J. Vac. Sci. Technol. B,* vol. 22, pp. 592-596, 2004.

[10] RIT SMFL website http://smfl.microe.rit.edu

[11] A.G. Cullis, T.E. Seidel and R.L. Meek, "Comparative Study of Annealed Neon-, Argon-, and Krypton-Ion Implantation Damage in Silicon," *J. Appl. Phys.,* vol. 49, pp. 5188-5198, Oct 1978.

[12] P. Revesz, M. Wittmer, J. Roth and J.W. Mayer, "Epitaxial Regrowth of Ar-Implanted Amorphous Silicon," *J. Appl. Phys.,* vol. 49, pp. 5199-5206, Oct 1978.

[13] J. Ziegler, Stopping and Range of Ions in Matter (SRIM), www.srim.org

[14] J. Boussey, "Stripping Hall Effect, Sheet and Spreading Resistance Techniques for Electrical Evaluation of Implanted Silicon Layers," *Microel. Eng.* vol. 40, pp. 275-284, 1998.

[15] D.H. Dickey, "Poisson Solver for SRA," *J. Vac. Sci. Technol. B,* vol. 10 pp. 438-441, Jan/Feb 1992. www.solecon.com

[16] K.D. Hirschman *et al.*, to be published.

[17] M. A. Sahiner, S. W. Novak, J. Woicik, J. Liu, and V. Krishnamoorty, "Arsenic Clustering and Precipitation Analysis in Ion-implanted Si Wafers by X-ray Absorption Spectroscopy and SIMS," *Proc. 2000 International Conference on Ion Implantation Technology,* Alpbach, Austria, IEEE, pp. 17-22, Sep 2000.

1-4244-0267-0/06/$25.00 ©2006 IEEE

Modeling and Analysis of the Charging Dynamics in Si-quantum Dots Based Non Volatile Flash Memory Cells

Pavan Singaraju[1], Rama Venkat[1], and Samar Saha[2]

[1]Department of Electrical and Computer Engg, University of Navada Las Vegas
Las Vegas, NV 89154

[2] Silicon Engineering Group
Synopsys, Inc. 700 E. Middlefield Road, Mountain View, CA 94043, USA

Abstract— A model including the presence and effect of discrete quantum energy levels and trap states in nanocrystals is proposed in order to describe the anomalous peaks observed in current-voltage characteristics of emerging Si quantum dot based Floating Gate flash memory cells. The model is employed to investigate the effect of energy levels in quantum dots with a size distribution in the range of 0 to 12nm in explaining the charging dynamics and current versus time characteristics. The simulated results are in close agreement with the experimental results. It is speculated that the additional peaks observed in the experimental current versus voltage characteristics above threshold voltage are because of the filling up of nanocrystals with more than one electron into quantum levels, shifted to higher energy levels due to the increase in charging energy determined by self capacitance.

I. INTRODUCTION

Floating gate nonvolatile memories have a wide array of applications. Continued scaling of these devices into the sub-100nm region poses several technological challenges[1]. One of the main disadvantages of the continuous floating gate based memory devices is the loss of information due to charge leakage through the tunnel oxide in presence of defects in the oxide. New device architectures such as the nanocrystal floating gate (NC)[2] and Silicon Oxide Nitride Oxide Silicon (SONOS)[3] have been proposed to mitigate the gate leakage problem. In these two architectures, devices with reduced gate leakage current and lower programming and erase voltages have been demonstrated and used for various applications. In the NC architecture, the nanocrystals of Si are dispersed in the tunnel oxide act as quantum dots with charge storage capability in the quantum energy levels. These charge nodes loose their charge only if there is a defect in the vicinity of the dot in the tunnel oxide. Thus the discrete nature of the floating gate reduces the net charge loss.

[1]singaraj@egr.unlv.edu

The focus of this paper is to describe the charging dynamics, specifically, the current versus time (I v-s t) characteristics, of the NC based memory device. A recent model proposed by Busseret et al.[4] explains the general I-V and I-t characteristics, but it fails to explain observed peaks in I-V characteristics at gate voltages greater than the threshold voltage. We attribute these peaks to filling up of the nanocrystals with more than one electron into the quantum levels, shifted to higher energy levels due to the increase in charging energy determined by self capacitance. A model including the presence and effect of these energy levels and trap states along with the size distribution of the quantum dots is developed as an extension of the work by Busseret et al.[4] The model is employed for the investigation and analysis of the charging dynamics i.e., I-t, characteristics. Results are compared with the experimental data of Ref. [4].

II. DEVICE ARCHITECTURE

A schematic cross section of the device used for investigation in this paper is shown in Fig. 1. The structure is identical to that used by Busseret et al.[4] The devices are MOS capacitors built on a p-Si substrate. The thickness of the tunnel oxide is (d_{fb}) 2.5nm and Si nanocrystals are obtained using a LPCVD system and are self-assembled. The Si nanocrystals are assumed to be spherical and their average diameter (d_{si}) is 5nm. The nanocrystal size distribution reported in Ref. [4] is modified and shown in Fig. 2, an explanation for this change is given in Section-IV. A 5-nm thick control oxide (d_{fg}) is deposited over the nanocrystals to bury them. An n+ poly is used as the top contact and n+ guard ring is implanted around the MOS capacitor, which facilitates feeding the inversion layer with current at a fast rate. The area of the capacitor is 1mm^2. Detailed description of fabrication of these structures can be found in Ref. [6, 7].

III. MODEL

A detailed model to describe the I-t and I-V characteristics should include both energy level computation of the quantum

1-4244-0267-0/06/$25.00 ©2006 IEEE

dots, energy level balance across the structure for various gate voltages and the mechanisms of carrier transport through the tunnel oxide to and from to the quantum dots. The mechanism of transport is assumed tunneling. Details of the model are as follows.

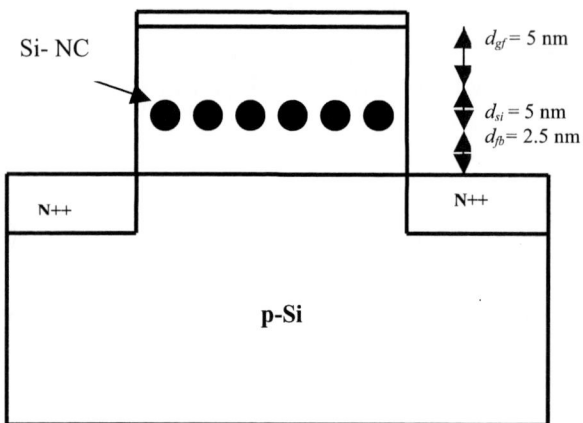

Fig.1: Schematic cross section of the silicon nanocrystal flash memory device

A. Quantum Energy levels calculation

Quantum confinement effects are very important for this device since the size of the nanocrystals is in the range of 0-12nm. The allowed energy levels in the conduction band of Si nanocrystals are not continuous, but discrete with the energy difference between subsequent levels varying with the size of the crystal. These energy levels are obtained by solving the Schrödinger's equation given by:

$$\frac{\hbar^2}{2m_{si}}\left(\frac{\partial^2}{\partial x^2}+\frac{\partial^2}{\partial y^2}+\frac{\partial^2}{\partial z^2}\right)\psi(r)+V(r)\psi(r)=E_r\psi(r) \qquad (1)$$

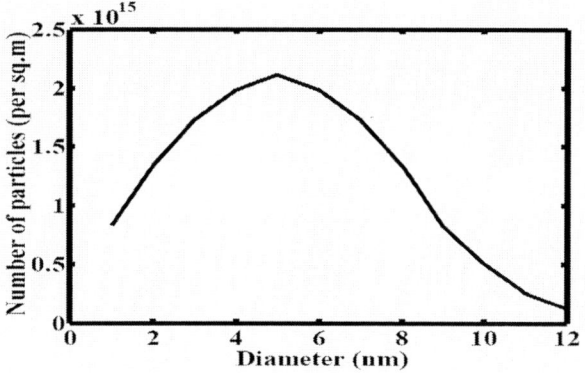

Fig. 2: Size distribution of silicon nanocrystals embedded in oxide.

for a spherical quantum dot surrounded by a finite barrier, where $\hbar=\dfrac{h}{2\pi}$ and h is Planck's constant, m_{si} is the effective mass of electron in silicon, $V(r)$ is electrostatic potential energy, $\psi(r)$ is the wave function and E_r is the Energy with

index r indicating carrier confinement along the radial direction. Due to the spherical symmetry of the nanocrystal, the potential and also the wave function are expected to be spherically symmetric. For the analysis of the energy levels of the quantum dot, a numerical program supplied in Ref. [8], is used. Once a nanocrystal is charged with an electron, the energy levels are shifted to higher levels because of the increase in charging energy determined by self capacitance. The shift in energy level is calculated to a first order approximation using electrostatics and is given by:

$$\Delta E = \frac{i^2 q^2}{4\pi\varepsilon_o\varepsilon_{ox}d_{si}} \qquad (2)$$

where i is the number of electrons in the nanocrystal, q is charge of an electron, ε_o is the permittivity of free space, ε_{ox} is the relative permittivity of oxide and d_{si} is the diameter of the nanocrystal. Energy values for the first, second and third levels as a function of diameter of the Si nanocrystals is shown in Fig. 3. **E2** and **E3** represent energy levels when a nanocrystal is filled with one and two electrons respectively. It is to be noted that they do not represent the higher order energy levels in the nanocrystals but will be referred later as second and third level subbands respectively.

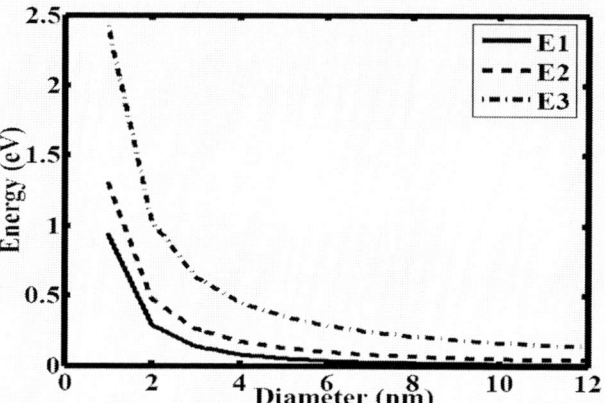

Fig 3: Energy values for the first, second and third level as a function of diameter of the Si nanocrystals obtained using Eq. 1.

B. Feeding current

The dominant charge transport mechanisms in the NC memory device are tunneling of electrons from (3-D) bulk Si into (0-D) quantum dots and Fowler-Nordheim tunneling of carriers from quantum dots to bulk Si. The forward tunneling current from the Si bulk to the $(i+1)^{th}$ energy level of the quantum dot, J_{f_i}, is given by :

$$J_{f_i} = \sum_{i=0}^{n}\frac{4\pi m_{si}qkT}{h^3}\int_{Ei\min}^{Ei\max}T(E_{i+1},F_i)\ln(1+e^{(E_f(F_i)-E)/kT})dE \qquad (3)$$

where q is the charge of an electron, k is the Boltzmann's constant, T is the temperature in °K, F_i is the electric field under the nanocrystal with i number of electrons in the

1-4244-0267-0/06/$25.00 ©2006 IEEE

nanocrystal and E_f is the position of the Fermi level in the quasi neutral bulk of Si substrate. The term $T(E_{i+1}, F_i)$ is the transparency factor and is given by [9]:

$$
\begin{cases}
T(E_{i+1}, F_i) = \exp\left[-b\dfrac{\phi_b^{3/2} - (\phi_b^{3/2} - F_i d_{fb})^{3/2}}{F_i}\right] \\[2mm]
b = \dfrac{8\pi\sqrt{2m_{ox}}}{3hq}
\end{cases}
\tag{4}
$$

where, m_{ox} is the effective mass of electron in oxide and $\phi_b = 2.85 eV$ is the barrier energy between Si and SiO$_2$ at the interface. The effects of image force lowering and quantum barrier lowering due to the quantized energy subbands in the surface potential well is taken into account while obtaining the barrier energy [10]. The reverse current of electrons tunneling back into the Si- bulk from the quantum dot energy levels, $J_{r_{i+1}}$, is given by the Fowler- Nordheim tunnel current as [11]:

$$
\begin{cases}
J_{r_{i+1}} = \displaystyle\sum_{i=0}^{n} A F_{i+1}^2 \exp\left[-b\dfrac{\phi_b^{3/2} - (\phi_b^{3/2} - F_{i+1} d_{fb})^{3/2}}{F_{i+1}}\right] \\[2mm]
A = \dfrac{q^3 m_{si}}{8\pi m_{ox}\phi_b}
\end{cases}
\tag{5}
$$

C. Energy band balance to obtain tunnel oxide electric fields

Calculation of electric fields under the nanocrystal is important as the fields vary based on the presence and absence of electrons in nanocrystals and their sizes. These fields are evaluated based on energy band balance as follows. The voltage drop on the oxide, V_{ox}, (Fig. 4.) is obtained as:

$$
V_g = V_{ox} + V_{gFB} + \psi_b(F_i)
\tag{6}
$$

where V_g is the applied gate bias voltage, V_{gFB} is the flat band voltage and $\psi_b(F_i)$ is the surface band bending. Since the surface band bending in the n+ poly silicon contact is negligibly small due to heavy doping, it is ignored in estimating the flat band voltage. The relation among V_{ox}, F_i and the local 2D charge density, Q_i, is given by:

$$
V_{ox} = F_i\left(d_{fb} + d_{fg} + \frac{\varepsilon_{SiO_2}}{\varepsilon_{Si}} d_{Si}\right) + Q_i\left(\frac{d_{Si}}{\varepsilon_{Si}}(1-a) + \frac{d_{fg}}{\varepsilon_{SiO_2}}\right)
\tag{7}
$$

where ε_{Si} and ε_{SiO2} are the dielectric constants for bulk silicon and silicon dioxide, respectively. The electronic charge in the nanocrystals is assumed to be located at distance $a.d_{Si}$ in a plane in the nanocrystal from the Si-tunnel oxide interface, with a satisfying $0 < a < 1$. The charge density, Q_i, is given by:

$$
Q_i = \frac{iq}{S_{d_{Si}}}\Big|\, i = 0..,n
\tag{8}
$$

where ($S_{d_{Si}} = \dfrac{\pi d_{si}^2}{4}$) is the cross sectional area of the spherical nanocrystal with diameter d_{Si}. It is noted that for the same number of charge, nanocrystals with different diameters will have different charge density given by Eq. 7. The electric field term, F_i, is evaluated by solving Eqs. (6-8).

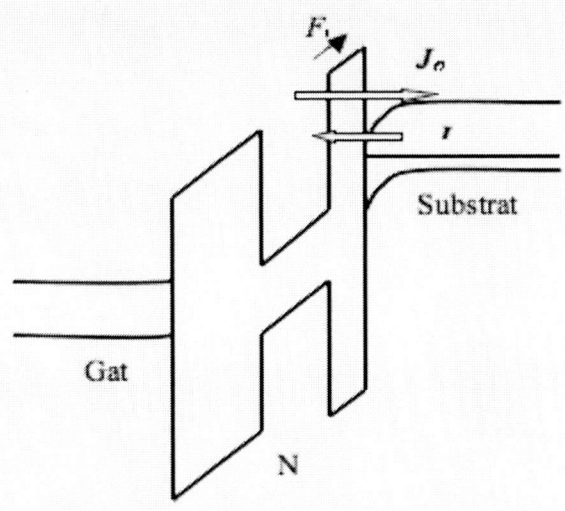

Fig. 4: Band scheme inside the nanocrystal memory device after the application of a bias voltage.

D. Carrier Dynamics

The model employed to obtain the charge dynamics in the nanocrystals due to tunneling processes is as follows. Let the initial total number of neutral (with state "i=0") nanocrystals be N. Let a gate voltage be applied at t = 0, At time t, let the nanocrystals charge be denoted as n_i (with state "i") and the neutral nanocrystals be denoted as n_o. The total number of nanocrystals will always satisfy the condition:

$$
N = \sum_{i=0}^{n} n_i
\tag{9}
$$

In the time interval between time t and $t + dt$, the number of crystal switching from state "i" to state "$i+1$" is given by:

$$
dn_{i \to i+1} = \sum_{i=0}^{n} \frac{1}{q} J_{f_i} S_{d_{Si}} n_i dt
\tag{10}
$$

Similarly, in the time interval dt, the number of crystals switching from state "$i+1$" to "i" is given by:

$$
dn_{i+1 \to i} = \sum_{i=0}^{n} \frac{1}{q} J_{r_{i+1}} S_{d_{Si}} n_i dt
\tag{11}
$$

The total number of nanocrystals switching states in time interval, dt, are given by:

$$\begin{cases} dn_0 = dn_{1\rightarrow 0} - dn_{0\rightarrow 1} \\ dn_i = \sum_{i=1}^{n} dn_{i-1\rightarrow i} - dn_{i\rightarrow i-1} \end{cases} \qquad (12)$$

The net charging current for subband energy i at time t, can be calculated as:

$$j_i = \sum_{i=1}^{n} dn_i \left(\frac{q}{dt} \right) \qquad (13)$$

E. Procedure for calculating current

The position of various subband energy levels from the bottom of the conduction band corresponding to a diameter of the nanocrystal is obtained from Fig. 2. Similarly for any diameter, the density of nanocrystals is obtained from Fig.3. Thus for a diameter of the nanocrystal, the position of subband energy levels and also its density are known. Data from Figs. 2 & 3 are used to generate the plot of the number of subband energy levels available for every energy level as shown in Fig. 5. The feeding currents are due to filling up of these energy states and they are computed as follows. Initially, all the nanocrystals are assumed to be neutral. For a gate voltage, Vg, the drop on the oxide is determined using Eq. 6. Electric field F_0, is obtained from Eq. 7, with $i=0$ in Eq. 8, because the nanocrystals are neutral. During the first time step, dt, the forward current, J_{f_0}, is the current due to electrons flowing into the neutral nanocrystals from the inversion layer and is obtained from Eq. 3. Thus the first energy level of some of the nanocrystals is filled with an electron. The number of nanocrystals that have an electron in the first energy level is obtained by multiplying the feeding current density with the number of energy levels available, cross section area of the nanocrystal and time per unit electron charge. The reverse current, $J_{r_{i+1}}$, due to electrons tunneling from the nanocrystals into the inversion layer is obtained from Eq.4. The number of nanocrystals that became neutral is computed using a similar procedure as mentioned above, except that it is multiplied by the number of occupied energy levels instead of the available energy levels. After this, the number of nanocrystals having one electron dn_1 is computed using Eq. 12. The number of neutral nanocrystals is updated by taking the difference between the initial number of nanocrystals and the number of nanocrystals having one electron. Electrons do not remain in the subband energy levels and drop down immediately into the available trap states located in the energy gap of the nanocrystal instantaneously [12]. As a result the same energy level is available to get filled with an electron. But the presence of an electron in the nanocrystal increases its charging energy. This results in shift of the energy levels to higher levels with respect to the bottom of the conduction band and is updated using Eq.2. During the next time interval, current is not only from the remaining neutral nanocrystals but also from the nanocrystals occupied with one electron. The corresponding electric field, F_1, for these nanocrystals is

computed using Eq. 7, with $i=1$ in Eq. 8. A procedure as mentioned above is used to compute the feeding currents and also the number of nanocrystals occupied with two electrons. After the second time step, nanocrystals that have two electrons can be filled and the current computation is carried out in a similar fashion. The total current at any time instance is the sum contribution from all the subband energy levels.

IV. RESULTS AND DISCUSSION

In Fig. 3, the subband energy levels (1st, 2nd and 3rd) obtained using Eq. 1 are shown as a function of diameter of the nanocrystals. The energy level values (with respect to the bottom of the conduction band) decrease with the diameter of the nanocrystal for all three energy subbands. Also, it is observed that the difference between the subband energies decrease with increasing diameter. Using the experimental data of nanocrystal size distribution and the data shown in Fig. 3, the number of subband energy levels available, are obtained as a function of energy and are shown in Fig. 5. For all three subbands, the distribution is same, except that the higher subband curves are shifted to higher energies. The minimum energy separation between subbands occurs at the lowest energies.

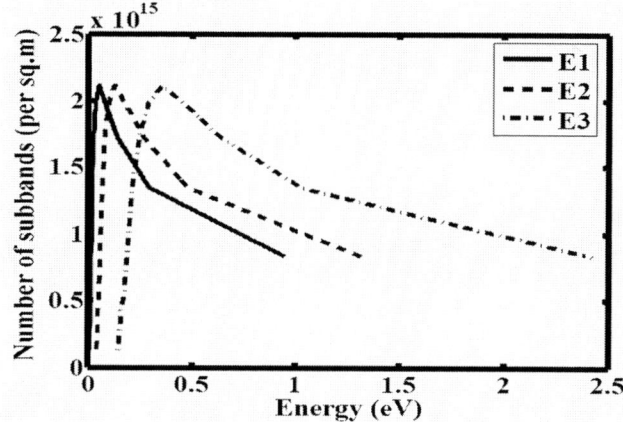

Fig. 5: Number of energy levels available for charge storage as a function of diameter of the silicon nanocrystal.

Current contribution as function of diameter of the nanocrystal is shown for the three subband energies at t = 0.2s is shown in Fig. 6. It is observed that the peak contribution of current results from different diameter particles for the three subbands. It is also observed that the peak contribution of the higher subbands is smaller due to higher energies. The total current contribution from all the three subbands as a function of diameter of the nanocrystal is shown for t = 0.2 sec and 1.5 sec respectively in Fig. 7. As expected the currents are smaller for longer times, but interestingly, the diameter corresponding to peak current contribution has decreased from

7.5 nm at 0.2 sec to 5 nm at 1.5 sec. This is due to the fact that as time progresses, the lower energy levels are getting occupied and only higher energy levels corresponding to lower diameter nanocrystals are available for occupation. The current magnitude for the higher energy levels is expected to be low as the number of carriers available for tunneling is low.

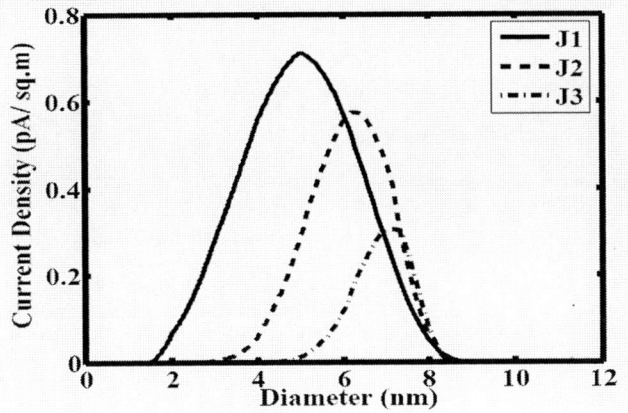

Fig. 6: Current contribution from the first three energy levels as a function of diameter after 0.2 s.

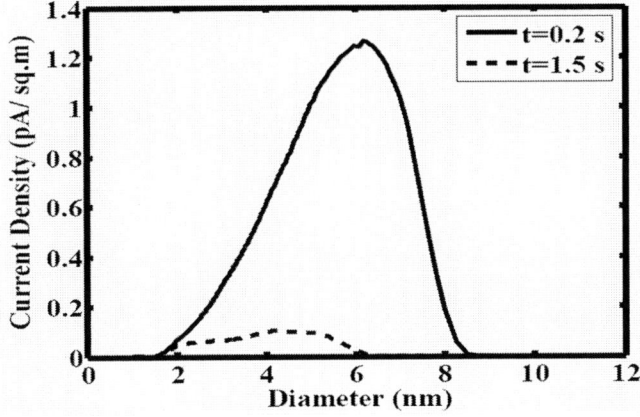

Fig. 7: Total current contribution from the first three energy levels as a function of diameter after 0.2 s and 1.5 s, respectively.

Current versus time plots for the subband energies are shown in Fig. 8. As expected, the lowest subband contributes to most of the current. As the subband number increases, their contribution decreases due to higher energy values. All the three plots show an exponential behavior at higher energy levels and subbands 2 and 3 shown an initial increase from zero at lower energy levels and then an exponential behavior. This behavior due to initial occupation of subband 1 and hence making states available for subband 2. Similarly, initial occupation of subband 2 makes states available for subband 3. Even though the first subband contributes most of the current, the second and third subband contribute significantly, approximately 15% and 6%, respectively.

Total current versus time plots for Vg= 0.7V and

Vg=1.5V, are shown in Fig. 9. The characteristics are exponential with time as expected and observed in experiments. The area under the curve in Fig.9, represents the total number of electrons present in the nanocrystals which is also related to the density of nanocrystals. The area obtained from experimental results from Ref. [4, 5] has more number of states than the reported value of density of nanocrystals. Hence we increased the reported density of crystals by a factor of 3.0. Our results agree with the experimental results reported in Ref. [4, 5].

This model can be employed to study the current versus voltage characteristics reported in Ref. [4]. Currently, this work is underway. It is speculated that the additional peaks observed in the current at various gate voltages are due to the charging of nanocrystals with more than one electron into quantum levels, shifted to higher energy levels due to the increase in charging energy, which was not allowed in the work of Ref. [4, 5].

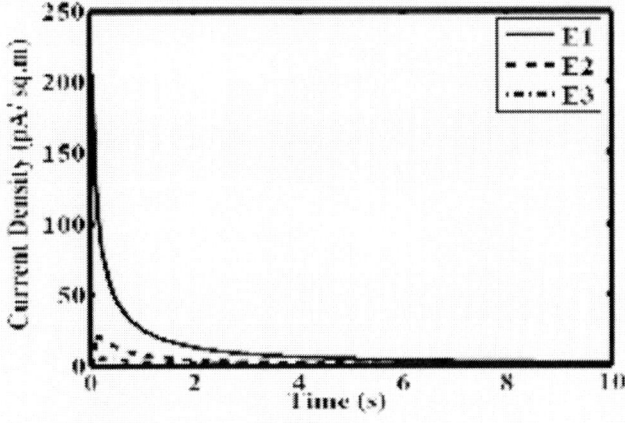

Fig. 8: Current density from the three subband energy levels as a function of time.

V. CONCLUSION

In this paper, we presented a model extending the work of Busseret et al [4] by including the various subband energies and their influence on the charging dynamics of nanocrystal flash memory devices. Even though, the first energy level contributes most of the current, the shifted energy levels contribute approximately 15% and 6 %, respectively. Current contributions from the energy levels and the total current exhibit an exponential behavior with time at longer times. The diameter of nanocrystals contributing maximum current is higher for higher subbands. This model can be employed to study the current versus voltage characteristics reported in Ref. [4]. Currently, this work is underway. It is speculated that the additional peaks observed in the current at various gate voltages are due to the charging of nanocrystals with more than one electron into quantum levels shifted to higher energy levels due to the increase in charging energy, which was not allowed in the work of Ref. [4, 5]

1-4244-0267-0/06/$25.00 ©2006 IEEE

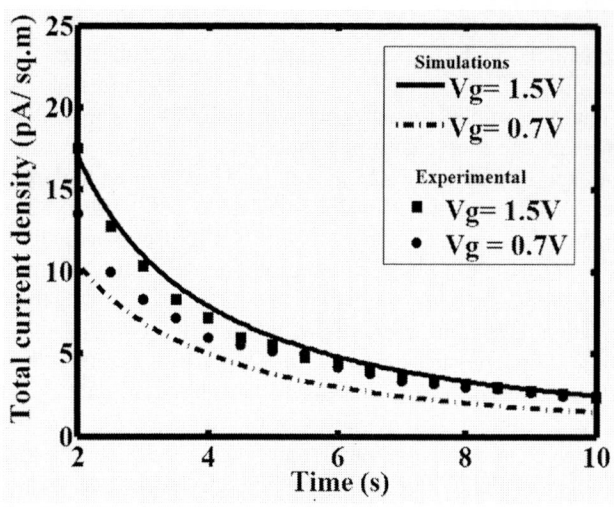

Fig. 9: Total current contribution from the first three energy levels as a function of time for Vg=0.5V and Vg=1.5V.

APPENDIX

Table I: List of parameters used in simulation

Physical Constant	Symbol	Value
Free electron mass	m_o	$9.1 \times 10^{-31}\,Kg$
Effective mass of electron in silicon	m_{si}	$1.06\,m_o$
Effective mass of electron in oxide	m_{ox}	$0.5 m_o$
Permittivity of free space	ε_o	$8.85 \times 10^{-12}\,F/m$
Permittivity of silicon	ε_{si}	$11.9\varepsilon_o$
Permittivity of oxide	ε_{ox}	$3.9\varepsilon_o$
Planck's constant	h	$6.625 \times 10^{-34}\,J\text{-}s$
Boltzmann's constant	k	$1.38 \times 10^{-23}\,eV/°K$
Charge of electron	q	$1.6 \times 10^{-19}\,Coul$
Barrier for electron	ϕ_b	$2.85\,eV$
Temperature	T	$300\,°K$

REFERENCES

[1] P. Cappelletti, C. Golla, P. Olivo, E. Zanoni, Flash Memories, Kluwer Academic Publishers, Dordrecht, MA, 1998.

[2]. S. Tiwari, F. Rana, H. Hanafi, A. Hartstein, E.F. Crabbe, K. Chan , "A silicon nanocrystals based memory", *Appl. Phys Lett.*, vol. 68, pp.1377-1379, 1996.

[3] B.V. Kashvan *et al.*, "MONOS memory element" in *IEDM Tech. Dig.*, Oct. 1968, p.140.

[4] C. Busseret, S. Ferraton, L. Montes and J. Zimmermann , "A three charge states Model for silicon nanocrystals nonvolatile memories", *IEEE Trans. Electron Devices*, vol.53, No.1, pp-14-22, Jan 2006.

[5] C. Busseret, S. Ferraton, L. Montes and J. Zimmermann, "Granular description of charging kinetics in silicon nanocrystal memories", *Solid State Electron*, vol.50, Issue 2, pp- 134-141, Feb 2006.

[6] T. Baron *et al.*, "Nucleation control of CVD growth silicon nanocrystals for quantum devices," *Microelectron. Eng.*,vol. 61–62, pp. 511–515, 2002.

[7] S. Ferraton, L. Montès, A. Souifi, and J. Zimmermann, *Microelectron.Eng.*, vol. 67–68, pp. 858–864, 2003.

[8] P. Harrison, "Quantum Wells, Wires and Dots - Theoretical and Computational Physics", John Wiley & Sons, Ltd., New York, 2000

[9] A. T. Fromhold Jr, *Quantum Mechanics for Applied physics and Engineering*. New York: Academic, 1981

[10] W. Quan, D.M. Kim, M.K. Cho, "Unified compact theory of tunneling gate current in metal-oxide-semiconductor structures: Quantum and image force barrier lowering" *J. Appl. Phys.*, vol 92, No.7 pp 3724-3729, 2002.

[11] M. Linzlinger and E. H. Snow, "Fowler-Nordheim tunneling into thermally grown SiO₂," *J. Appl. Phys.*, vol 40, pp 278-283, 1969.

[12] A. Camperra *et al*, "Modelling and Simulation of Charging and Discharging Processes in Nanocrystal Flash Memories during Program and Erase Operations" in *ICMTD-2005 – Special issue of Solid-State Electronics*

Signal Enhancement of Time-resolved Magneto-optic Measurements on Individual Nanomagnets

Suqin Wang, Naser Qureshi, Mark A. Lowther[†], Aaron R. Hawkins[†], Sunghoon Kwon[‡], Alexander Liddle[‡], Jeffrey Bokor[‡], and Holger Schmidt*

School of Engineering, University of California Santa Cruz, 1156 High Street, Santa Cruz, CA 95064

*Corresponding author. email: hschmidt@soe.ucsc.edu

[†]ECEn Department, Brigham Young University, 459 Clyde Building, Provo UT 84602

[‡]Molecular Foundry, Lawrence Berkeley National Laboratory, Berkeley, CA 94720

ABSTRACT

The sensitivity of magneto-optic Kerr measurements of nanomagnetic structures is significantly improved using cavity enhancement to increase the Kerr signal by depositing dielectric layers on the magnetic surface under study. We discuss different strategies to optimize the use of cavity enhancement for observation of nanomagnets. We show that maximization of the Kerr rotation from magnetic structures results in higher sensitivity for magnetic films, but does not yield the highest sensitivity for nanomagnets. Instead, the use of an anti-reflection coating on the substrate surrounding the magnet in conjunction with index optimization of the enhancement layer can increase the Kerr signal by more than two orders of magnitude in the deep nanometer range, and enables detection of individual nanomagnets using conventional far-field Kerr microscopy.

Nanomagnetic structures have received increasing attention over the past decade. This trend is driven partly by the rapid developments in the magnetic recording industry towards higher storage density, and partly by the emergence of sophisticated nanofabrication methods that allow for a large variety and degree of control of magnetic materials and their properties [1-4]. Nanoscale magnets have numerous applications ranging from the aforementioned patterned media for data storage [5] to magnetic memory (MRAM) and magnetic sensors [5] to biomedicine [6]. The magnetic properties of nanomagnet arrays can be determined with standard methods including magnetometry and magneto-optic Kerr measurements. However, to fully understand their behavior and properties, techniques with single magnet resolution are required in order to eliminate averaging and ensemble effects. High-resolution microscopy methods such as TEM and MFM have extensively been used to study magnetostatic properties on the nanoscale [7, 8]. Magnetization dynamics that can occur on timescales from pico- to nanoseconds, however, require a different approach. Optical methods based on the magneto-optic Kerr effect (MOKE) show the greatest promise for providing simultaneously high spatial and temporal resolution. Due to the small signal levels associated with time-resolved magnetization measurements on individual nanomagnets, it is necessary to optimize the experimental apparatus and the detected MOKE signal itself for high sensitivity.

The possibility of altering a MOKE signal with dielectric layers has been known [12,13]. It was recently demonstrated that cavity enhancement can be used to substantially increase magneto-optic signals from nanomagnetic structures [9, 10]. This technique was used to demonstrate magnetization switching in individual single-domain nanomagnets using both far-field and near-field Kerr microscopy [10]. The basic concept is shown in Fig. 1. The magnetic element under study is coated with one or more dielectric layers (EL) that lead to multiple reflections of an incident beam from the magnetic surface. The advantage of using dielectric coatings to form the cavity is that their deposition is compatible with the nanofabrication techniques used for making the magnetic structures. For a properly chosen thickness, the polarization rotations from each partial reflection add up in phase and the overall MOKE signal in the reflected beam can significantly exceed the signal from the uncoated magnet. The intensity profile of the incident beam is described by a Gaussian for far-field measurements using microscope objectives (see Fig. 1) or determined by the aperture of a near-field probe. Also shown in Fig. 1 are the reflection coefficients of the coated nanomagnet (R_{mag}) and the surrounding (possibly coated) substrate (R_{sub}). It was previously shown that the measured Kerr angle can be described very well with a geometric model that takes into account the relative amounts of light reflected from the nanomagnet and the substrate, respectively. For far-field microscopy in polar geometry, the total Kerr angle α_{tot} is given by [10]

$$\alpha_{tot} = \alpha_0 \frac{1}{1 + \dfrac{R_{sub}}{R_{mag}} \cdot \dfrac{1}{e^{\frac{D^2}{2w_0^2}} - 1}} \qquad (1)$$

,where α_0 is the enhanced Kerr angle from the magnet, D is the magnet diameter and w_0 is the width of the Gaussian beam.

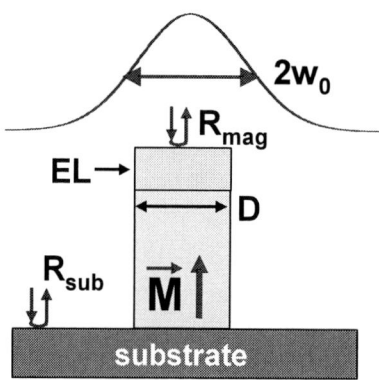

FIG 1: Nanomagnets coated with enhancement layer (EL) (reflectivity R_{mag}, diameter D) on substrate (reflectivity R_{sub}) are excited with a Gaussian beam in the far-field (spot size w_0).

Here, we discuss two approaches based on eqn. (1) to maximize α_{tot} for the observation of individual nanomagnets in the deep nanometer range (D<<100nm). We will show that the best choice of dielectric enhancement coating depends on the magnet diameter D and that careful optimization can lead to an increase in Kerr angle by more than two orders of magnitude. As a result, the sensitivity of Kerr measurements of individual nanomagnets can be pushed to diameters below 10nm, even using conventional far-field MOKE microscopy. All calculations and experiments are based on nickel as the magnetic material (films or cylinders of 150nm thickness) and an optical wavelength of 780nm. However, the principal conclusions are valid for other configurations as well.

When considering eqn. (1), the most obvious path to increasing α_{tot} appears to be an increase in α_0. Previous demonstrations used plasma deposition of single dielectric layers SiO_2 (enhancement factor

EF=2.3) or SiN (EF=5) which are lossless at 780nm. Larger factors can be observed by increasing the refractive index of the dielectric material, for example, amorphous silicon (EF=12). However, such materials tend to become lossy at optical wavelengths, which affects the detectable enhancement [9].

Maximizing α_{tot} by maximizing α_0 with high-index coating materials works well for magnetic films or structures with micron-sized dimensions. On the nanoscale, however, a different strategy must be pursued. The first important realization is that both α_0 and R_{mag} depend strongly on the index n of the enhancement layer, i.e. $\alpha_0=\alpha_0(n)$ and $R_{mag}=R_{mag}(n)$. The maximum Kerr enhancement is accompanied by a minimum in reflection intensity [9]. This is not of concern for magnetic films or micro-magnets where the excitation beam is entirely reflected by the magnetic material as long as the intensity exceeds the detector limit. On the nanoscale, however, a significant fraction of the detected light results from the illuminated substrate because the laser spot is larger than the nanomagnet. The relative contribution from the magnet under study is reduced, resulting in smaller α_{tot}. Fig. 2 shows the expected Kerr angle from nickel nanomagnets as a function of coating index for various magnet diameters D. It can be seen that an optimum index n_{opt} exists for each diameter. n_{opt} decreases with D which is also shown explicitly in the inset to Fig. 2. This means that the reduction in reflected intensity outweighs the increase in α_0 and some part of the enhancement has to be sacrificed for intensity. Fig. 3, finally, shows the expected effect on the measurable Kerr angle α_{tot} versus nanomagnet diameter D. The solid line represents the previously measured dependence for nickel cylinders coated with SiN [10]. The Kerr angle decreases for smaller magnets due to the decreasing contribution of the magnet reflection to the total signal (exponential term in eqn. (1)). The dashed line shows the result for identical conditions but choosing an optimized index based on the interplay of α_0 and R_{mag} for each D value. For microscopic dots, the influence of α_0 dominates and the signal can be increased by increasing the index. Once the diameter becomes smaller than the excitation spot size, an optimized index exists and improvements by a factor of two can be achieved.

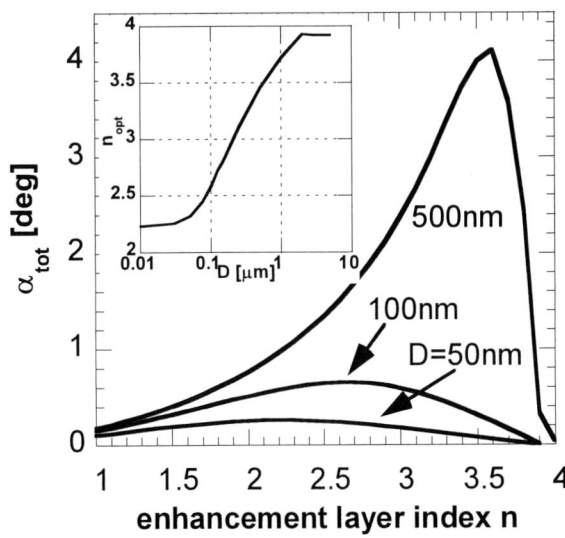

FIG.2: Kerr angle versus enhancement layer index for various magnet diameters; inset: optimum layer index versus magnet diameter.

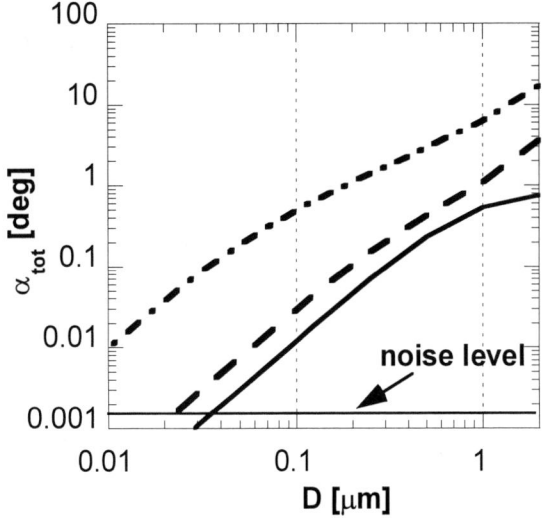

FIG.3: Kerr angle versus magnet diameter. Solid line: SiN-coated Ni magnets; dashed line: optimized coating index; dash-dotted line: index optimization and 0.1% AR coating on substrate

The final realization is that α_{tot} is maximized if the second term in the denominator of eqn. (1) tends towards zero. During sample design, this can be achieved by minimizing R_{sub} with an antireflection coating. This minimizes the unwanted contribution from the substrate to the total detected intensity. This effect is rather dramatic as can be seen by the dash-dotted line in Fig. 3 where an AR coating with 0.1% reflectivity was assumed instead of the measured 4.6% reflectivity for the solid line. It can be seen that the combined effect of index optimization and AR coating improves the signal by more than two orders of magnitude, especially for very small magnets below 100nm. In particular, the calculated signal for magnets of 10nm diameter exceeds the experimental noise level that was previously observed. This should allow for MOKE sensitivity to 10nm nanomagnets using conventional far-field Kerr microscopy without the need for an immersion lens objective.

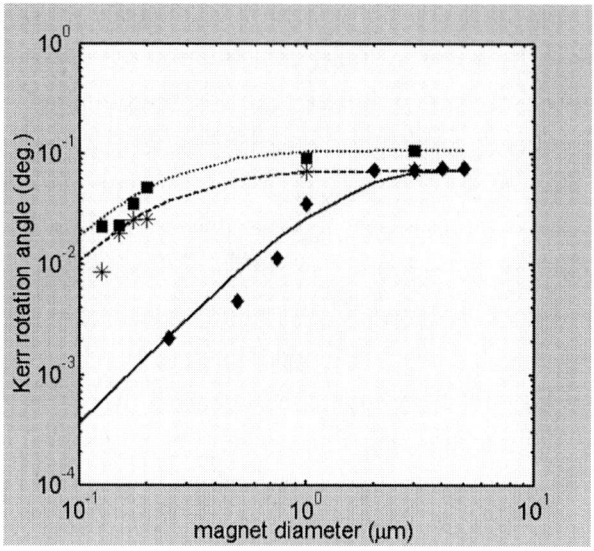

Fig. 4 Comparison between experiment data and theory fitting: 1% AR coating off the magnet, 58nm Ge layer on Ni magnet. symbols are experimental data, lines are theory fitting. diamond and solid line: Bare Ni everywhere; Asterisk and dashed line: Bare Ni magnet with AR coating on substrate; square and dotted line: AR coating on substrate, and 58nm Ge coating on Ni magnet

Fig. 4, shows our experimental results for the AR coating improvement. Comparing with bare Ni (diamond), there is a dramatic increase of the Kerr signal, especially for smaller magnets, due to the AR coating (asterisk). We then evaporated one layer of Ge (58nm) on the magnet to get further cavity enhancement. The reason we used Ge is that it can be deposited at low temperatures so that we can put it on nanomagnet without sacrificing the AR coating underneath. The square data show that the Ge coating does enhance the Kerr signal by a factor of 2~3. Resulting in the largest MOKE signals for D=125nm magnets to date. The Kerr angle can be further increased by optimizing the Ge thickness.

To increase data storage density and the speed of next generation magnetic recording devices, it is fundamentally important to study the ultrafast magnetization switching dynamics of individual nanomagnet. We have used a femtosecond laser and time-resolved magneto-optic Kerr effect (TR-MOKE) technique to study the magnetization dynamics in single nanomagnets down to 125nm [11]. In [11], we presented the results from Ni magnets with and without the enhancement layer coating and compared the dynamics. The experimental data shows that the cavity enhancement allows for improvement of the sensitivity from 500nm to 125nm without affecting the dynamics.

In summary, we have analyzed the optimization of magneto-optic Kerr enhancement for studies of nanomagnetic structures. Optimized dielectric layers increase MOKE signals by more than an order of magnitude, but are most effective on the microscale or for films. The most effective approach for obtaining large Kerr signals from nanoscale structures is the combination of an antireflection coating surrounding the magnets with a proper enhancement layer. For optimized parameters, large improvement in sensitivity is possible that should enable far-field MOKE of individual nanomagnets with dimensions as small as 10nm. This perspective is especially promising for time-resolved magnetization dynamics studies where small magnetization signals are prevalent. The experimental data proves the validity of this expectation.

The authors thank Anjan Barman for helpful discussions. This work was supported by the National Science foundation under grants ECS-0245425 and ECS-0216155 and the Molecular Foundry at Lawrence Berkeley National Laboratory.

REFERENCES

[1] S. Y. Chou, M. Wei, P. R. Krauss, and P. B. Fischer, *J. Vac. Sci. Technol. B* **1994**, 12, 3695.

[2] S. Sun, C. B. Murray, D. Weller, L. Folks, and A. Moser, *Science* **2000**, 287, 1989.

[3] C. A. Ross, M. Hwang, M. Shima, J. Y. Cheng, M. Farhoud, T. A. Savas, H. I. Smith, W. Schwarzacher, F. M Ross, M. Redjdal, F. B. Humphrey, *Phys. Rev. B* **2002**, 65, 144417.

[4] Nielsch K, Wehrspohn RB, Barthel J, Kirschner J, Gosele U, Fischer SF, Kronmuller H. *Applied Physics Letters*, vol.79, no.9, 27 Aug. 2001, pp.1360-2.

[5] Kirk KJ. *Contemporary Physics,* vol.41, no.2, March-April 2000, pp.61-78

[6] Koltsov D, Perry M, *Physics World*, vol.17, no.7, July 2004, pp.31-5.

[7] Dunin-Borkowski RE, Newcomb SB, McCartney MR, Ross CA, Farhoud M. Electron Microscopy and Analysis 2001. Proceedings. *IOP Publishing*. 2001, pp.485-8

[8] M. C. Abraham, H. Schmidt, T. A. Savas, H. I. Smith, C. A. Ross, R. J. Ram, *J. Appl. Phys.* **2001**, 89, 5667.

[9] N. Qureshi, H. Schmidt, A. R. Hawkins, *Appl. Phys. Lett.* 2004, 85, 431-433.

[10] N. Qureshi, S. Wang, M. Lowther, A.R. Hawkins, S. Kwon, B. Hartleneck, and H. Schmidt, *Nano Letters*, **5**, 1413, (2005).

[11] A. Barman, S. Wang, N. Qureshi, M. Lowther, A.R. Hawkins, S. Kwon, A. Liddle, J. Bokor, and H. Schmidt, *IEEE International Magnetics Conference*, San Diego, California, May8-12, 2006.

[12] A. V. Sokolov, *Optical Properties of Metals* (Blackie, London, 1967), p.311.

[13] K. Nakamura, T. Asaka, S. Asari, Y. Ota, and A. Itoh, *IEEE Trans. Magn.* **21**, 165 (1985).

1-4244-0267-0/06/$25.00 ©2006 IEEE

Integrated ARROW Waveguides for Molecule Specific Surface-enhanced Raman Sensing

Philip Measor[1], Leo Seballos[2], Dongliang Yin[1], John P. Barber[3], Aaron R. Hawkins[3],
Jin Zhang[2], and Holger Schmidt[1]

School of Engineering[1], Department of Chemistry[2], and ECEn Department[3]

University of CA Santa Cruz[1,2]

1156 High Street, Santa Cruz, CA 95064

Brigham Young University[3]

459 Clyde Building, Provo, UT *84602*

Abstract— **Planar integrated liquid-core ARROW waveguides for molecule specific surface-enhanced Raman (SERS) sensing are discussed. We demonstrate SERS detection from rhodamine 6G molecules bound to silver nanoparticles in picoliter volume ARROW waveguides and present waveguide designs for integrated Rayleigh-peak filtering.**

I. INTRODUCTION

Molecule sensing devices are of great interest to a wide array of disciplines including physics, biology, and biochemistry. Fully planar integrated liquid-core antiresonant reflecting optical waveguide (ARROW) devices have already established their potential as picoliter volume molecular fluorescence sensors [1], but fluorescence lacks the molecule specificity that Raman scattering spectroscopy has to offer [2]. Surface-enhanced Raman scattering (SERS) has been used to increase weak Raman cross sections and recently demonstrate single molecule detection [3,4]. Liquid-core ARROWs also have potential for Raman sensors with single-particle sensitivity. In this work, we demonstrate rhodamine 6G (R6G) SERS detection from a microfluidic ARROW waveguide using silver nanoparticles as SERS enhancers [5], and discuss ARROW design optimization for an important Raman signal wavelength range.

II. EXPERIMENT AND WAVEGUIDE DESIGN

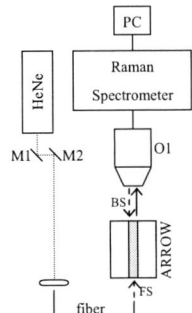

Fig. 1. SERS experimental setup with a red HeNe laser, Raman spectrometer, mirrors (M1, M2), objective lens (O1), and delineating forward scatter (FS) and backscatter (BS) geometries.

Liquid-core ARROW waveguides (core volume approx. 160 picoliters) were filled with a R6G concentration of 0.86 mM (about 10^{10} molecules), silver nanoparticles, and ethylene glycol dilute. The setup in Fig. 1 was used for SERS demonstration. In forward scatter (FS) geometry, a HeNe laser (632.8nm) was coupled into a single-mode fiber via two mirrors (M1, M2). The fiber was aligned to the ARROW core and the output was collected by a 100x/0.85NA objective lens (O1). In backscatter (BS) geometry the HeNe laser was coupled through O1 onto the sample and collected by O1. The spectra were analyzed by a Jobin Yvon Horiba HR800 Raman spectrometer.

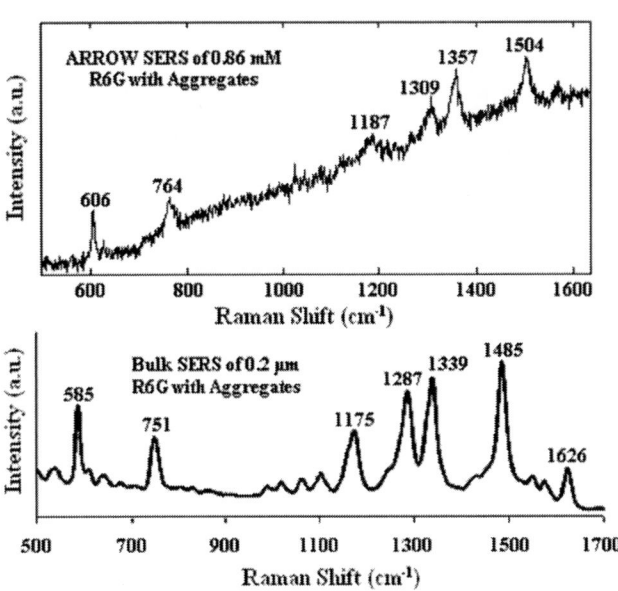

Fig. 2. (top) Forward scattered SERS spectra of R6G bound to silver nanoparticles inside ARROW waveguide; (bottom) SERS spectra of bulk R6G bound to gold nanoparticles from [8].

FS SERS spectra of R6G in an ARROW are compared to the SERS spectrum of R6G using Au nanoparticles from [8] in Fig. 2. The major Raman peaks of R6G are clearly identified, demonstrating SERS detection in integrated liquid-core waveguides for the first time.

1-4244-0267-0/06/$25.00 ©2006 IEEE

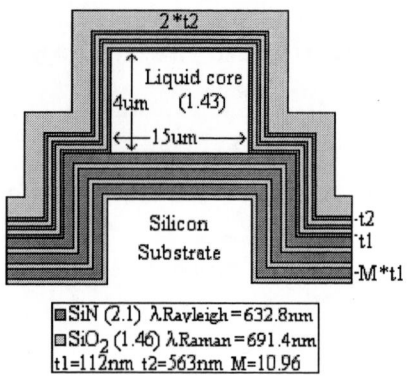

Fig. 3. Waveguide cross section showing optimization parameter M and other design parameters.

ARROWs use specifically chosen thicknesses of alternating dielectric layers to confine light in a liquid core [6], thus can be tailored for wavelength-dependent properties [7]. This can be exploited in SERS detection by designing the waveguide for high loss at the pump wavelength and low loss in the range of the Raman peaks. We achieve this by scaling the thickness of the first cladding layer, t_1, by $M = Q \lambda_1 / (\lambda_1 - \lambda_2)$ [7] with a scaling factor $Q = 0.929$ (λ_1: pump, λ_2: SERS signal; see Fig. 2). Altering only the bottom layers, as shown in Fig. 3, is sufficient to achieve the desired wavelength characteristics and allows for designing the layers above the hollow core independently. A region of interest for R6G Raman spectra is 700-1900 cm^{-1} (corresponding to 660-720nm with 632.8nm excitation).

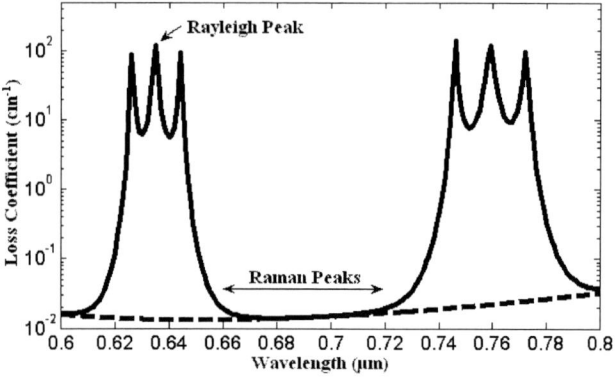

Fig. 4. Calculated loss for a liquid ethylene glycol ARROW core designed for high loss at excitation (632.8nm) and low loss at Raman peaks (660 to 720 nm).

The solid line in Fig. 3 shows wavelength dependent loss optimized for SERS detection of R6G with HeNe laser excitation (cladding materials SiN and SiO$_2$). Compared to a more weakly dispersive unoptimized structure (dashed line) the design shows the desired high loss at the Rayleigh wavelength (632.8nm) and low loss in the Raman peak range (660-720nm). This design will be particularly useful for a beam geometry with perpendicularly intersecting waveguides [7] and can be used to eliminate the need for external filters.

III. CONCLUSION

We have shown SERS detection in integrated liquid-core ARROW waveguides in planar beam geometry, and presented a waveguide design for integrated suppression of the Rayleigh peak. With further improvements such as filtering of individual Raman peaks and use of an intersection geometry [7], this approach has the potential for integrated Raman sensors with single-molecule sensitivity.

ACKNOWLEDGMENT

We acknowledge A. Shakouri, X. Wang, and A. Wolcott for help with the Raman spectrometer and chemical supplies, respectively. This work was supported by the W.M. Keck Foundation.

REFERENCES

[1] D. Yin et al., "Highly efficient fluorescence detection in picoliter volume liquid-core waveguides," Appl. Phys. Lett., **87**, 211111 (2005).
[2] A. Campion and P. Kambhampati, "Surface-Enhanced Raman Scattering," Chemical Society Reviews, **27**, 241 (1998).
[3] H. Nie et al., "Probing Single Molecules and Single Nanoparticles by Surface-Enhanced Raman Scattering," Science, **275** , 1102 (1997).
[4] K. Kneipp et al., "Single Molecule Detection Using Surface-Enhanced Raman Scattering (SERS)," Phys. Rev. Lett., **78**, 1667 (1997).
[5] C. Lee, D. Meisel, "Adsorption and Surface-Enhanced Raman of Dyes on Silver and Gold Sols," J. Phys. Chem., (1982) **86**, 3391.
[6] D. Yin et al., "Integrated optical waveguides with liquid cores," Appl. Phys. Lett., **85**, 3477 (2004).
[7] H. Schmidt et al., "Hollow-Core Waveguides and 2-D Waveguide Arrays for Integrated Optics of Gases and Liquids," IEEE J. Sel. Topics Quantum Electron., **11**, 519 (2005).
[8] A.M. Schwartzberg et al., "Unique Gold Nanoparticle Aggregates as a Highly Active Surface-Enhanced Raman Scattering Substrate," J. Phys. Chem. B, **108**, 19191 (2004).

Phonon Confinement in Germanium Nanowires

Xi Wang, Ali Shakouri,

Baskin School of Engineering, University of California Santa Cruz, CA 95064, USA

Bin Yu, Xuhui Sun, Meyya Meyyappan

Center for Nanotechnology, NASA Ames Research Center Moffett Field, CA 94035

ali@soe.ucsc.edu

Abstract: Raman spectra for different size Ge nanowires were measured with different excitation laser powers and wavelengths. By eliminating the heating of the sample under illumination, the phonon confinement effect for small size nanowires was clearly identified.

One-dimensional crystalline structures such as nanowires and nanotubes have been studied extensively during the past several years. Semi-conducting nanowires promise applications in future generation electronic and optoelectronic devices. Raman microscopy is a useful tool in the study of quasi-1D materials; it provides information about the surface and volume phonon modes and lattice vibrations, including how those vibrations are affected by extreme small dimensions. Recently, several papers [1, 9, 10] have analyzed the Raman peak shifts and the shape of the Raman spectrum for Si nanowires. However, the reported shifts and asymmetric broadenings vary depending on the experimental conditions. Studies show that the optical phonon peaks of Silicon nanowires are dependant on the excitation laser power and independent of wavelength. Thus low laser power is essential in order to examine the phonon spectrum of different size nanowires [1].

Self-assembled single crystalline Germanium nanowires can allow researchers to observe relatively strong one-dimensional confinement effects for both carriers and phonons. Compared to Si, Ge has smaller electron and hole effective masses and a lower dielectric constant; therefore, nanowires made of Ge should have stronger confinement characteristics than Si nanowires with the same diameters.

The samples considered in this paper were synthesized on lithographically patterned Au catalyst arrays, with sizes ranging from 5nm to 20 nm, by the Vapor-Liquid-Solid (VLS) method. The details of the process can be found in Ref [2]. As a reference, a piece of bulk Germanium wafer was also examined under same conditions. The ambient temperature was kept at typical room temperature, 22°C. Raman spectra for different size Ge nanowires were measured with different excitation laser powers and wavelengths. By eliminating the heating of the sample under illumination, the phonon confinement effect for small size nanowires can clearly be identified. Fig. 1 shows the scanning electron microscopy (SEM) images of the as-synthesized germanium nanowire.

Fig.1 A SEM image of as-synthesized GeNW sample.

1-4244-0267-0/06/$25.00 ©2006 IEEE

Fig. 2 shows the evolution of Raman spectrum as a function of wire diameter. All the four samples were excited by 514.532nm, $500\,\mu W$ Kr$^+$ laser line and examined the scattering light. We are mainly interested in the Strokes peaks in the Ge range (~300 cm^{-1}). Comparing these to the Strokes peak of bulk Ge, we observed obvious position shift-downs, broadenings and increases of asymmetry from GeNWs. As the wire size decreases, these features become more significant. Earlier paper [8] reported these phenomena as a entire contribution of scaling- induced phonon confinement effect. However, the pure confinement effect should be looked at only after carefully calibrating and removing the environmental contributions. In this experiment, we kept the ambient temperature stable and the excitation times for each measurement equal. We also used a clean room for taking measurements in order to reduce contaminations. But localized heating at the excite point can not be totally eliminated, since this is inevitable when focusing extreme small targets with lasers.

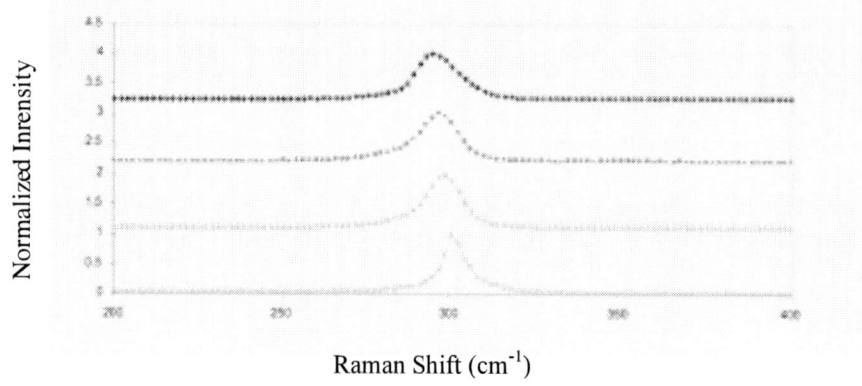

Fig.2 Typical evolution of Raman spectrum as a function of wire diameter.

In order to separate the influence of local heating from that of the pure phonon confinement, we took another set of Raman spectrums on these four samples. All experimental conditions and parameters were kept the same, except the excitation laser power was reduced to a very low level, $50\,\mu W$. Results are shown in Fig. 3(A). The spectrums obtained with the $500\,\mu W$ excitation were superimposed and shown in Fig. 3(B). Comparing Fig. 3 (A) with (B), we clearly obtained much smaller $\left|\Delta shift/\Delta size\right|$, and $\left|\Delta broadening/\Delta size\right|$ ratios.

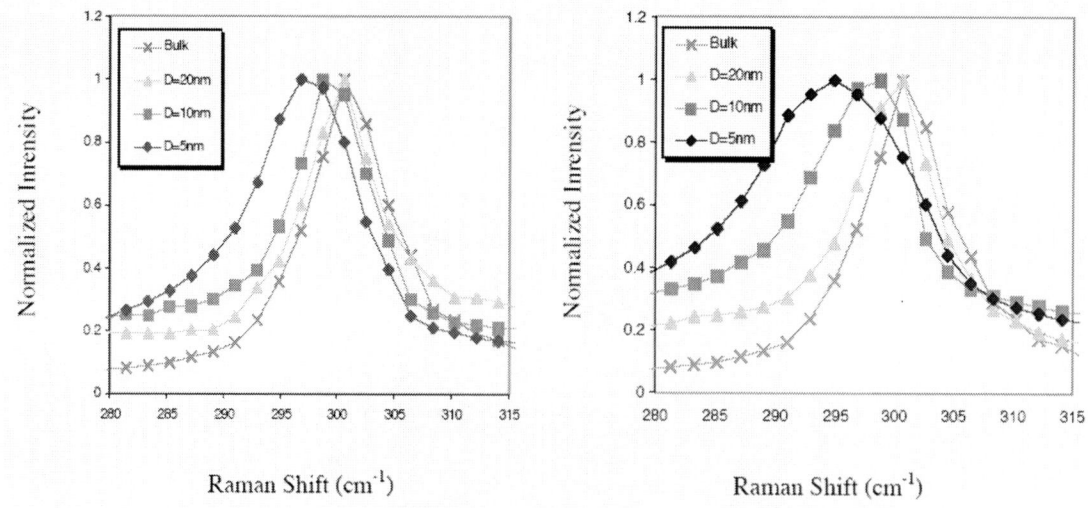

Fig.3 (A) Different size GeNWs' Raman Strokes peak at low excitation power;
(B) Different size GeNWs' Raman Strokes peak at high excitation power;

To further understand how the local heating affects Raman spectra, each of the differently sized Ge nanowire samples was excited by three different laser powers. See Fig. 4 (A-C). Neutral Density Filters (NSFs) were added in-between the sample and laser source to indicate the different excitation powers. From a $500\,\mu W$ source, the D0, D0.3 and D0.6 filters, allow, respectively, $500\,\mu W$, $250\,\mu W$ and $125\,\mu W$ laser powers to pass through. As the power increases, the Raman Strokes peaks of GeNWs generally move to lower frequencies, broaden and become less symmetric, while no significant change can be observed from those of the bulk Ge. This is consistent with what has been found for Si and Si nanowires [1]. Another interesting find is the difference between changing speeds of differently sized wires. Thinner wires respond to excitation power change much more obviously than thicker ones. This trend is valid for all different size nanowire samples, until there is no local heating effect, as approximately what happens with the bulk material.

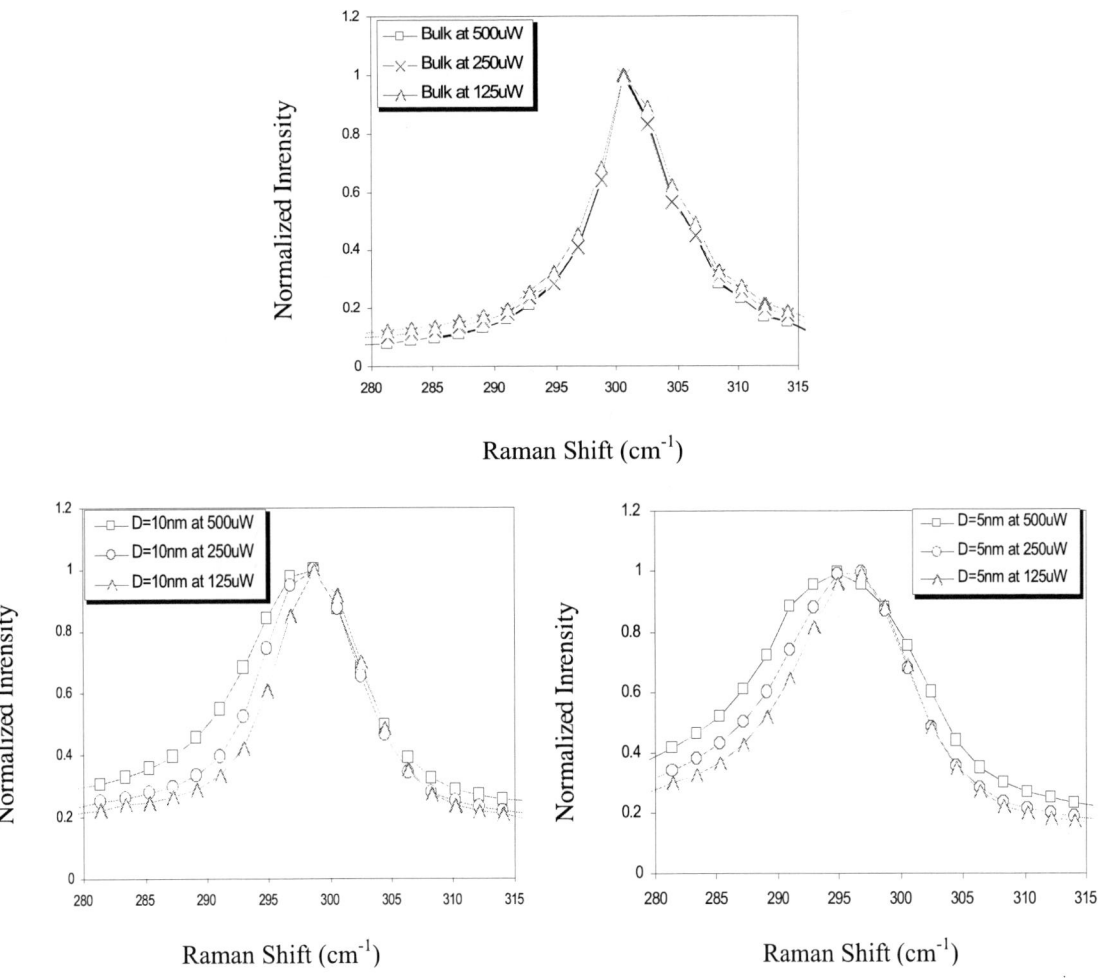

Fig. 4 (A) Raman Strokes peak of bulk Ge with different excitation powers;
(B) Raman Strokes   peak of 10nm GeNWs with different excitation powers;
(C) Raman Strokes peak of 5nm GeNWs with different excitation powers;

A quantitative analysis can help us distinguish the scaling-induced phonon confinement effect from the gross. Take the spectra of 5nm size GeNW as an example: if the downshift at low power excitation, $\sim 3.1\,cm^{-1}$, is considered as a pure local heating contribution, the downshift accompanying the higher power excitations can be estimated by doing a first order approximation and should be very close to the measurement value. To estimate the Strokes peak downshift with high power excitation, with calculated the ratios between magnitudes of Stokes and Anti-Stokes peaks. $I_{AS}/I_S = \gamma \exp(k_B T/h\nu)$, where T is the local temperature, ν is phonon energy and γ is a coefficient relates to peak position and FWHM. [11, 12, 13]. We found out that the local temperature range is 305-380K, as the power varied from 50uW~250uW. Then, as a rule of thumb, for the 5.6cm^{-1} downshift excitated by $500\,\mu W$ power, a local heating of ~440K should be found in the case of excluding all other contributions rather than local heating. S. Piscanec *et al.* have used the similar method in their study of Si nanowires [1]. However, only 405K was found based on the peak features. It is much smaller. The only issue that may cause inaccuracy in this approximation is that the shape of the peaks could not be perfectly fitted by ideal Lorentzian pulses. After investigation, the Least-Mean-Square fitting error was found only gives less than 5% of the intensity ratio difference and less than 0.1cm^{-1} of the position variance, which respond to downshift uncertainties of ~0.35% and ~0.5%, respectively. Therefore, we are confident of the participation of optical phonon confinement.

The small physical dimension of the scattering crystalline nanowire leads to downshifts and broadening of the first –order Raman line, known as the scaling-induced phonon confinement effect. More specifically, in one-dimensional structures, such as nanowires, a zero momentum exists along the axis directions, and phonons with momentums close to the Brillouin zone center allow the Raman scattering to occur at 300.6 cm^{-1}. However, in directions perpendicular to the axis of the wire, momentum level is discrete and spaced by $2\pi/d$, where d is the physical dimension and indicates the size effect. When phonon modes are confined, the degeneration of longitudinal optical mode (LO) and transversal optical mode (TO) disappears, while the momentum k=0 in infinite structures allows the degeneration to exist. This causes the general broadening. In this study, different speeds of shifting and broadening as a function of excitation power for different size wires are considered the results of the size distribution of wires on each sample. Although the size of catalysts can control the size of synthesized nanowires, a slight variance of the wire size is inevitable due to the nature of self-assembling growth. For instance, if the sizes of wires range from 4nm to 6nm, averaged and centered at a catalyst size, 5nm; while another set of wires has a size distribution from 9nm to 11nm, centered at 10nm, then both of the samples have an internal variation $\Delta d = \pm 1nm$. However, the same size variations that exist in different samples do not exhibit the same changes. More specifically, it broadens the shape and shifts the central position of the Raman peaks more with thinner wires, while less with thicker wires. The local heating enlarges this phenomenon, since the thermal conductivity of semiconductor nanowires is more than approximately two orders of magnitude lower than the bulk value and it decreases as the size scales down [3]. The laser power more easily accumulates in thin nanowires.

As a supplemental study, two different excitation laser lines: a 514.523 nm Ar$^+$ laser line and a 633.817 nm Kr$^+$ laser line, were used on all four samples to examine the wavelength independency. Fig. 5 (A-B) shows these results. Power used for both of the laser lines were 500uW. The absorption coefficients of Germanium are 600 cm^{-1} for Ar$^+$ laser and 150cm^{-1} for Kr$^+$ laser. Therefore, with the different absorption coefficients the excitation powers match what have been used above: $125\,\mu W$ (Kr$^+$) and 500uW (Ar$^+$). Comparing these spectra to Fig. 4 (B, C), features of spectrums—position, FWHM, and asymmetry level— almost stay the same. In other words, the spectrum changes with the different wavelengths used are not due to resonant Raman selection of different size wires, but the results from the absorbed power difference.

In conclusion, we showed that the Raman spectra for different size Ge nanowires were measured with different excitation laser powers and wavelengths. Study shows the excitation power dependency and wavelength independency of Raman spectrum evolution. By eliminating the heating of the sample under illumination, we can clearly identify the phonon confinement effect for small size nanowires.

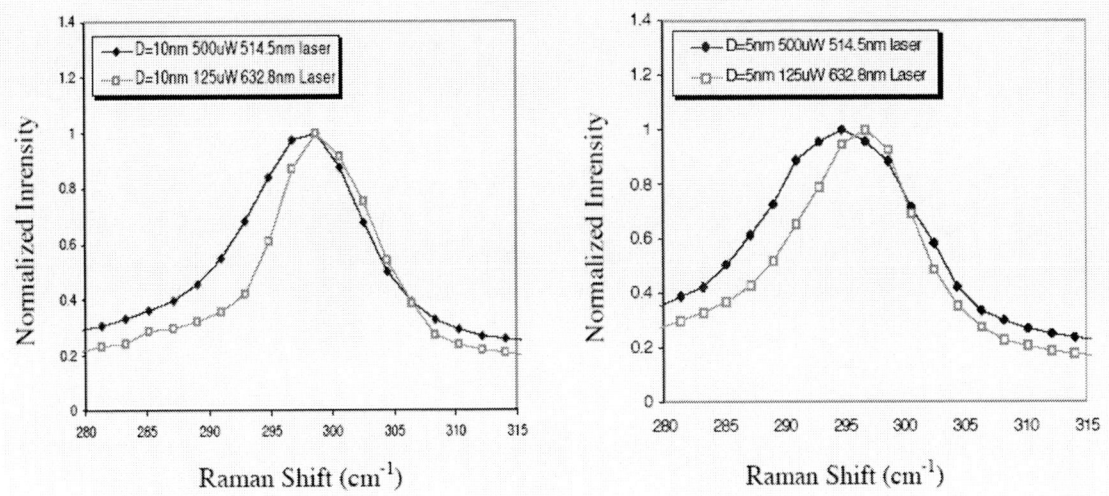

Fig. 5(A) Raman Strokes peak of 10nm GeNWs with different excitation wavelength;

(B) Raman Strokes peak of 5nm GeNWs with different excitation wavelength;

References

[1] S. Piscanec, A. C. Ferrari, M. Cantoro, S. Hofmann, J. A. Zapien, Y. Lifshitz, and S. T.Lee, J. Robertson, Phys. Rev. B 68,241312(R) (2003)

[2] B. Yu, G. Calebotta, K. Yuan, and M. Meyyappan, NTSI (2005)

[3] D. Li, Y. Wu, P. Kim, L. Shi, P. Yang, and A. Majumdar, "Thermal Conductivity of Individual Silicon Nanowires," Appl. Phys. Lett., Vol. 83, p. 2934, 2003

[4] A. C. Ferrari, S. Piscanec, S. Hofmann, M. Cantoro, and C. Ducati, J. Robertson, Proc. of IWEPNM, AIP, Melville, NY, 2003.

[5] R. P. Wang, G.W. Zhou, Y. Liu, S. Pan, H. Zhang, D. Yu, and Z. Zhang, Phys. Rev., B 61 (2000) 16827.

[6] H. Richter, Z. P. Wang, and L. Ley, Solid State Commun. 39 (1981) 625.

[7] I. H. Campbell and P. M. Fauchet, Solid State Commun. 58 (1986) 739.

[8] Y. F. Zhang, Y. H. Tang, N. Wang, C. S. Lee, I. Bello, and S. T. Lee, Phys Rev, B61 7 (2000)

[9] N. Fukata, T. Oshima, K. Murakami, T. Kizuka, T. Tsurui, S. Ito, Appl. Phys. Lett., Vol. 86, 213112, 2005

[10] J. Qi, J. M. White, A. M. Belcher, Y. Masumoto, Chem. Phys. Lett., 372 763-766, 2003

[11] M. Malyj and J.E. Griffiths, Appl. Spectrosc. **37**, 315 (1983)

[12] F. LaPlant, G. Laurence and D. Ben-Amotz **50**, number 8, (1996)

[13] B.J. Kip and R.J. Meier, Appl. Spectrosc. **44**, 707 (1990)

1-4244-0267-0/06/$25.00 ©2006 IEEE

The Design of MOS-BJT-NDR-Based Cellular Neural Network

Dong-Shong Liang ,Yaw-Hwang Chen, Chun-Min Wen, Chun-Da Tu, Kwang-Jow Gan, And
Cher-Shiung Tsai

Department of Electronic Engineering, Kun Shan University, Tainan, Taiwan 710, Republic of China
Lecturer, suln@ mail.ksu.edu.tw; assistant professor, yhchen@mail.ksu.edu.tw; graduate student;
takoglans@yahoo.com.tw; graduate student, kukoc0306@yahoo.com.tw; Professor, gankj@ms52.hinet.net; and
assistant professor, e5040@mail.ksu.edu.tw

Abstract—The cell of cellular neural network (CNN) studied in this work is realized by negative differential resistance (NDR) devices. The NDR device is composed of metal-oxide semiconductor field-effect transistor (MOS) devices and bipolar transistor (BJT). Therefore, we can fabricate the cellular neural network by standard CMOS or BiCMOS process.

Key words : Cellular neural network(CNN), Negative differential resistance(NDR).

I.   I.INTRODUCTION

THE cellular neural network was invented by Chua and Yang in 1988[1,2]. The definition of cellular neural network is n-dimensional array of mainly identical dynamical systems, called cells, which satisfies two properties: most interactions are local with a finite radius r, and all state variables are continuous valued signals. CNN is a high nonlinear neural network. It contains many interesting phenomena and applications. The stable solutions of CNN are determined by some conditions [2]. It is easy to find the stable solutions.

In order to realize the CNN by hardware, it must use the electronic devices which have the nonlinear characteristics. Recently, the resonant tunneling diode (RTD)-based CNN cell was proposed [3]. In this paper, metal-oxide semiconductor field-effect transistor (MOS) negative differential resistance (NDR)-based and MOS-BJT-NDR-based CNN cells are proposed [4,5]. They are monostable-bistable transition logic elements (MOBILE) [6]. A MOBILE is a first order CNN cell, which acts as a flip-flop with only two possible states

II. MOS-NDR AND MOS-BJT-NDR DEVICES

The MOS-NDR and MOS-BJT-NDR devices consist of the metal-oxide-semiconductor field- effect- transistors (MOS) and bipolar-junction transistor. A MOS-NDR device is composed of three n-channel MOS, and one p-channel MOS. Transistor mn1 act as a load resistor with the gate shorted to the drain, which is used to modulate the input gate voltage of transistor mn3, and transistor mn2 behave as an active switch with the gate electrode connected to the gate of driver transistor mn3 and the drain electrode of driver

transistor mp4. The substrate of load transistor mn1, the substrate and source of transistor mn2, the source of transistor mn3 which are connected together with the gate and source of transistor mp4. The circuit representation of the MOS-NDR device is shown in figure 1. The output characteristics of the MOS-NDR device simulated by HSPICE program are shown in figure 2. The structure of MOS-BJT-NDR device is similar to MOS-NDR device, the only difference is mn3 is replaced by a BJT. Its voltage of I_p(peak current) is lower than V_p of MOS-NDR.

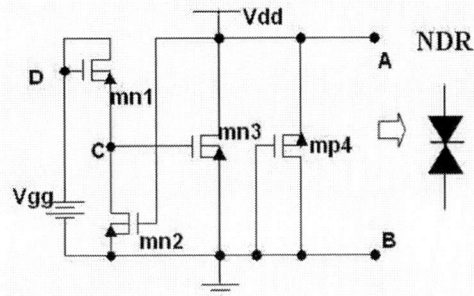

Fig.1. A N-type NDR device circuit is composed of three NMOS and one PMOS.

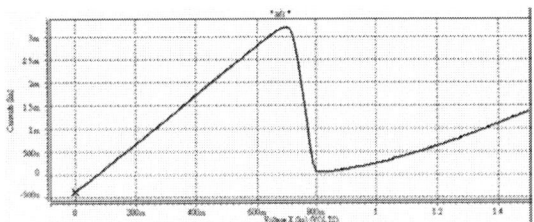

Fig. 2. The I-V characteristic of MOS-NDR device.

III. THE NDR-BASED CNN'S CELLS

The use of MOS-NDR devices in applications to CNN's cell was proposed. The schematic circuit is shown in figure 3. The output of cell is determined by V_{in} and V_s according to the MOBILE (monostable-bistable transition logic elements) theory. We use this MOS-BJT-NDR-based CNN's cell to compose of a 5x5 CNN, it is shown in figure 4. The negative differential resistance devices are suitable for the performance of characteristics of cellular neural network. The resonant tunneling diode is the famous NDR device. But it is less flexible than the MOS-NDR device. The MOS-NDR device can easily be modulated by width and gate voltage of MOS transistor. We have proposed a cell of cellular neural

1-4244-0267-0/06/$25.00 ©2006 IEEE   189

network which is composed of MOS and BJT transistors and a 5×5 network has been constructed. It can be modulated by widths of MOS transistors and I-V characteristics of MOS-BJT-NDR devices.

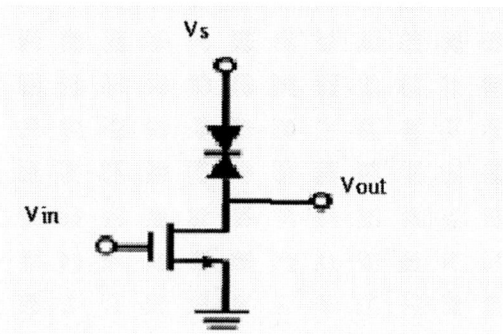

Fig.3. It is the schematic circuit of MOS-NDR based CNN's cell.

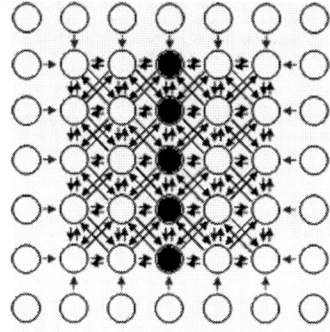

Fig.4. It is the pattern "1".

IV. CONCLUSIONS

The cellular neural network has many applications. It has been realized by RTD-based cells. The negative differential resistance devices are suitable for the performance of characteristics of cellular neural network. We have proposed MOS-NDR based and MOS-BJT-NDR based CNN's cells. The MOS-NDR and MOS-BJT-NDR devices can easily be modulated by width and gate voltage of MOS transistor. We will continuously study MOS-NDR-based N×N cellular neural network and realize the integral circuit of MOS-NDR and MOS-BJT-NDR based CNN.

Acknowledgements

This work was supported in part by the National Science Council of Republic of China under Grants Nos. NSC94-2215-E-168-001 and NSC94-2112-M-006-015.

References

[1] L.O. Chua and L. Yang, "Cellular Neural Networks : Theory," *IEEE Transactions on circuits and systems*, vol.35, no.10, pp.1257 -1272, 1988.

[2] R. Dogaru and L.O. Chua, "Universal CNN Cells," *International Journal of Bifurcation and Chaos*, vol. 9, no. 1, pp1-48, 1999.

[3] M. Itoh, P. Julian, and L.O. Chua, "RTD-based Cellular Neural Networks with Multiple Steady States," *International Journal of Bifurcation and Chaos*, vol. 11, no. 12, pp2913-2959, 2001.

[4] Dong-Shong Liang, et. al., "Novel Voltage-Controlled Oscillator Design by MOS-NDR Devices and Circuits," *The 5th IEEE International Workshop on System-On-Chip for Real-Time Applications*, Banff, Alberta, Canada, 20-24 July 2005, pp78-81.

[5] K.J. Gan, "Hysteresis phenomena for the series circuit of two identical negative differential resistance devices," *Japanese Journal of Applied Physics*, vol. 40, no. 4A, pp2159-2164, 2001.

[6] T. Akeyoshi, K. Maezawa, and T. Mizutani, "Weighted sum threshold logic operation of MOBILE (monostable-bistable transition logic elements) using resonant-tunneling transistors," *IEEE Electron Device Lett.*, vol. 14, pp.475-477, 1993.

Leakage Current in DRAM Memory Cell

Jonathan Yu and Koorosh Aflatooni

Department of Electrical Engineering
San Jose State University
One Washington Square - San José, California USA, 95192

Abstract - Retention time is a critical characteristic in dynamic random access memory (DRAM) design. In order to improve DRAM retention time characteristics, leakage current must be reduced and various solutions are proposed. The major leakage paths in a DRAM cell stem from reverse junction leakage from the storage node, and gate induced drain leakage (GIDL) current. Empirically it is known that the junction leakage is affected by the lateral electric field near the storage node, which is enhanced by an increase in substrate doping due to threshold adjustment. The DRAM cell becomes more susceptible to GIDL when the storage node stores a "1" and a negative bias is applied to the gate. The voltage drop across the gate oxide creates a vertical electric field that leads to a higher leakage current. In this paper, these leakage paths are investigated with device simulation. Tradeoffs between substrate doping, gate thickness and leakage are explored using Silvaco Athena/Atlas. The DRAM cell was modeled using a 0.24μm NMOS transistor. Substrate doping was varied to analyze its effect on depletion region, lateral electric field, and reverse current. Device simulation has shown that a lower substrate doping yields better results on lateral field and reverse current. As the substrate doping was reduced from $5.8 \times 10^{17} \text{cm}^{-3}$ to $9 \times 10^{16} \text{cm}^{-3}$, the lateral field decreased from $7.25 \times 10^5 \text{V/cm}$ to $5.6 \times 10^5 \text{V/cm}$. Subsequently, the reverse current decreased from 3.1pA to 0.5pA. The GIDL was explored by varying LDD implantation energy and gate oxide thickness. As oxide thickness increased, the vertical field decreased because the voltage drop occurred over a larger distance. The vertical field was reduced from $1.25 \times 10^6 \text{V/cm}$ to $6.9 \times 10^5 \text{V/cm}$ when the oxide thickness was increased from 11.7 to 21.6nm.

I. INTRODUCTION

Since the invention of the one-transistor/one-capacitor cell in the 1960's, DRAM technology has developed rapidly, with chip density quadrupling every three years [1]. Although DRAM is among the smaller memory cells, it will lose its stored information when the power is turned off (volatile memory), and must be constantly refreshed or else the stored charge will leak away [2]. A DRAM cell is composed of a MOSFET connected in series with a storage capacitor. The wordline is connected to the gate of the transistor, while the bitline is connected to the n-type diffusion opposite of the storage capacitor node as shown in Figure 1. Data is stored by applying a low voltage to the wordline and trapping the charge on the storage capacitor. Subsequently, the data is accessed by applying a high voltage to the wordline and sensing the voltage difference on the bitline.

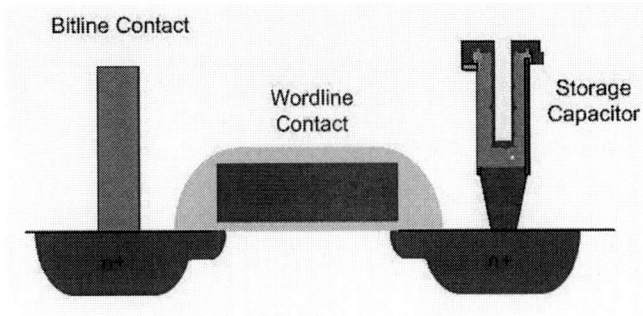

Fig. 1 DRAM memory cell.

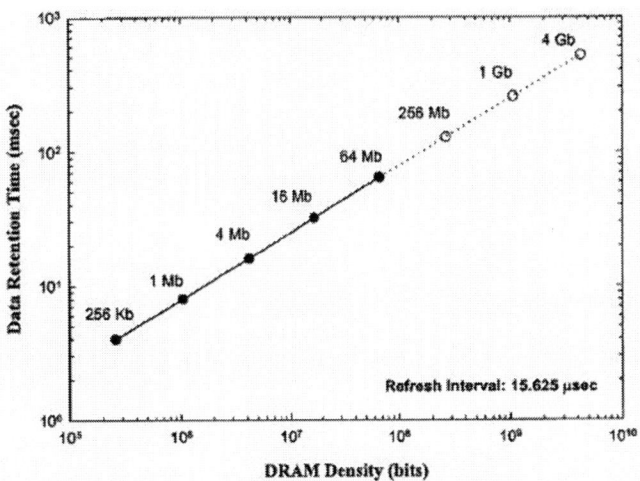

Fig. 2 Trend of retention time as a function of DRAM density.

The time from when the capacitor stores the data until it is read out is known as retention time. As shown in Figure 2, the retention time specifications have become more demanding for each successive DRAM generation [3]. One reason is due to battery-operated equipment requiring longer retention times to reduce power consumption from refreshing the data [4]. Retention time is dependent on the weakest cell, generally in the tail distribution, which loses the stored data due to leakage currents before it can be refreshed. The electric field in memory cells is becoming stronger, increasing leakage current, and causing poor retention characteristics. The major leakage paths in a DRAM cell are reverse junction leakage from the storage node and gate induced drain leakage (GIDL) current. In this paper, these leakage paths are investigated with device simulation.

1-4244-0267-0/06/$25.00 ©2006 IEEE

TABLE I
PROCESS AND BIAS CONDITIONS

Channel length	0.24μm
Gate oxide thickness	10nm
Boron doping concentration	$4.1 \times 10^{17} cm^{-3}$
Junction depth of n-layer	93nm
Concentration of n-layer	$3.91 \times 10^{19} cm^{-3}$
Substrate voltage	0V
Wordline voltage	-1V
Storage node voltage	4V

II. EXPERIMENT

Leakage current was explored using Silvaco Athena/Atlas. The DRAM cell was modeled in Athena using a 0.24μm NMOS transistor. Table I summarizes the structure investigated in this paper. The electrical device simulations in Atlas focused on junction leakage and GIDL. Leakage is examined in the worst case when the DRAM cell is storing a "1." Data is stored by applying -1V to the wordline and trapping the charge on the storage capacitor. The applied voltage on the storage node is 4V for signal "1." Since this paper studied leakage due to electric fields, accurate electric field and tunneling models are needed. Band-to-band tunneling and trap-assisted tunneling models were added to the default Atlas MOS models. In the band-to-band tunneling model, for high electric fields, local band bending would allow electrons to tunnel, by internal field emission, from the valence band into the conduction band. Additional electron hole pairs would be formed. In the trap-assisted tunneling model, high electric fields would cause the tunneling of electrons from the valence band to the conduction band through trap or defect states; therefore, there would be a great effect on current.

III. JUNCTION LEAKAGE

Junction leakage occurs when the device is off and current is not expected to flow. As the reverse voltage increases, leakage current increases exponentially. This characteristic is explained by thermal field emission (TFE) current [5]. Hamamoto et al. reported that while diffusion and generation-recombination (G-R) current is inversely proportional to substrate doping, TFE current increases as substrate doping increases. An enhanced thermal emission rate is due to a strong electric field in the depletion region. The electric field in the depletion layer increases when the substrate doping is increased. It results in the enhancement of the tunnelling probability of electrons and the increase in TFE. Therefore, one method to reduce junction leakage is to decrease the electric field in the depletion layer by lowering substrate doping.

The substrate doping was decreased incrementally from $5.8 \times 10^{17} cm^{-3}$ to $9 \times 10^{16} cm^{-3}$ to study its effects on:
1) Depletion region
2) Lateral electric field in the depletion region
3) Storage node to substrate junction leakage

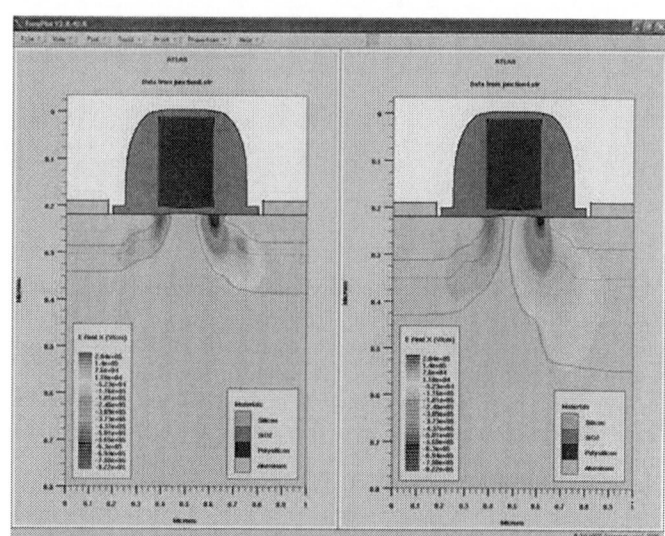

Fig. 3 Effects on depletion region and lateral electric field as substrate doping is decreased from $5.8 \times 10^{17} cm^{-3}$ (left) to $9 \times 10^{16} cm^{-3}$ (right).

After multiple simulations, it was evident that substrate doping had an effect on depletion region and electric field, as shown in Figure 3. As the substrate doping concentration decreased, the depletion region increased and the junctions became deeper. These factors contributed to a decrease in the lateral electric fields. The purpose of the LDD is to reduce the electric field near the channel region. Since the LDD is becoming larger, the field is becoming weaker. A deeper junction also reduces the field because the junction curvature is becoming more rounded. Simulations show that for a substrate doping of $5.8 \times 10^{17} cm^{-3}$, the maximum lateral field in the storage node is $7.25 \times 10^5 V/cm$ and the junction leakage is 3.1pA. As the substrate doping decreases, the lateral fields and leakage current also decrease as shown in Figure 4. When the substrate doping decreases to $9 \times 10^{16} cm^{-3}$, the electric field reduces to $5.6 \times 10^5 V/cm$ and the junction leakage is 0.5pA.

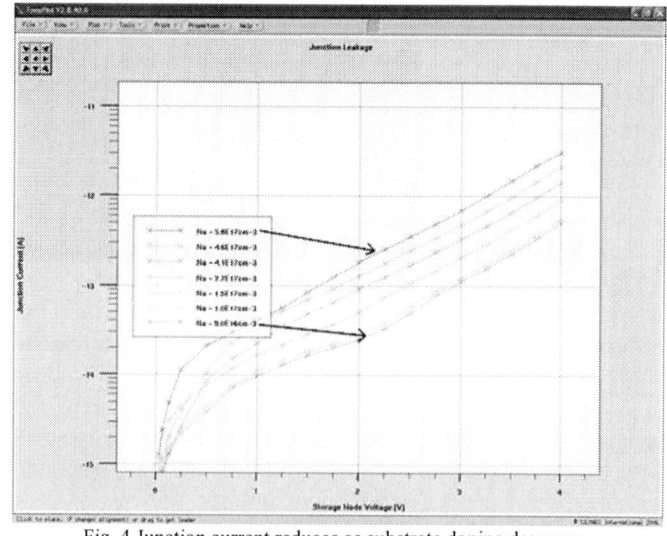

Fig. 4 Junction current reduces as substrate doping decreases.

1-4244-0267-0/06/$25.00 ©2006 IEEE

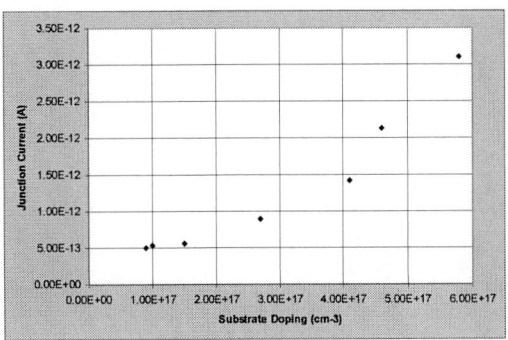

Fig. 5 Junction current (at storage node voltage = 4V) reduces as substrate doping decreases.

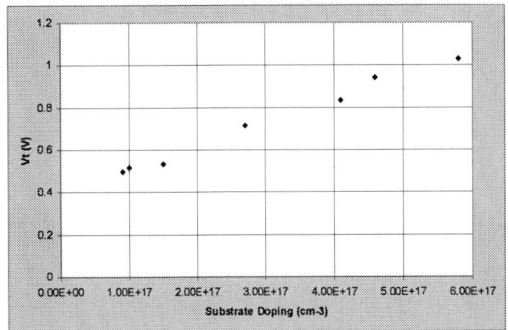

Fig. 6 Threshold voltage reduces as substrate doping decreases.

Figures 5 and 6 display the tradeoff considerations that must be taken into account when substrate doping is adjusted. A low substrate doping is desirable for low junction current, but it has a negative effect on threshold voltage. It is difficult to reduce substrate doping without degrading the subthreshold characteristics of the transistor. DRAM cells usually require a high threshold voltage (~0.8V) to reduce short channel effects and prevent subthreshold conduction of charge from the storage node to the bitline when the wordline is turned off [1].

IV. GIDL LEAKAGE

To remedy the subthreshold conduction due to low substrate doping, a negative voltage is commonly applied to the wordline. A wordline voltage ranging from -0.5V to -1V is applied to assist in further turning off the transistor [1]. Although the subthreshold characteristic improves, the GIDL problem becomes serious. When the capacitor stores a value "1," the n-type region around the capacitor is also at a high level, which causes a voltage drop across the gate oxide in the overlap region as shown in Figure 7. This creates a vertical electric field which could induce leakage current by causing band-to-band tunnelling in regions where the bandgap voltage is dropped across a short distance. Methods to reduce GIDL include increasing the LDD implantation energy and increasing the oxide thickness [6].

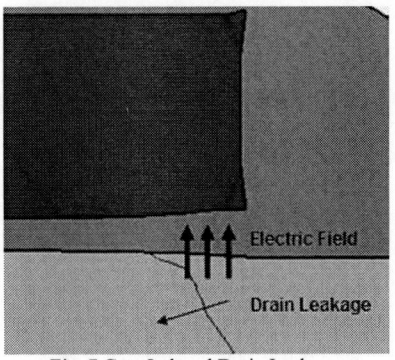

Fig. 7 Gate Induced Drain Leakage.

TABLE II
LDD IMPLANTATION RESULTS

LDD Implantation Dose/Energy	Max Vertical Electric Field (V/cm)	Junction Current (A)
$2 \times 10^{13} cm^{-2}/25 KeV$	1.25×10^{6}	1.07×10^{-12}
$2 \times 10^{13} cm^{-2}/30 KeV$	1.10×10^{6}	6.03×10^{-13}
$2 \times 10^{13} cm^{-2}/35 KeV$	1.00×10^{6}	4.65×10^{-13}
$2 \times 10^{13} cm^{-2}/40 KeV$	9.60×10^{5}	4.13×10^{-13}

Simulations show that as LDD implantation energy increases, vertical electric field and junction leakage decrease (Table 2). For implantation energy of 25KeV, the junction leakage is 1.07pA. The implantation energy was increased to 40KeV, and the junction leakage was reduced to 0.413pA. As the LDD implantation energy increases, the n-type area under the gate increases. Since the LDD region is moving further away from the gate, the vertical fields have less influence on that region.

Gate oxide thickness was increased to examine the effects on vertical field and junction leakage. As the oxide thickness increases from 11.7nm to 21.6nm, the maximum field decreases from 1.25×10^{6}V/cm to 6.9×10^{5}V/cm respectively (Figure 8).

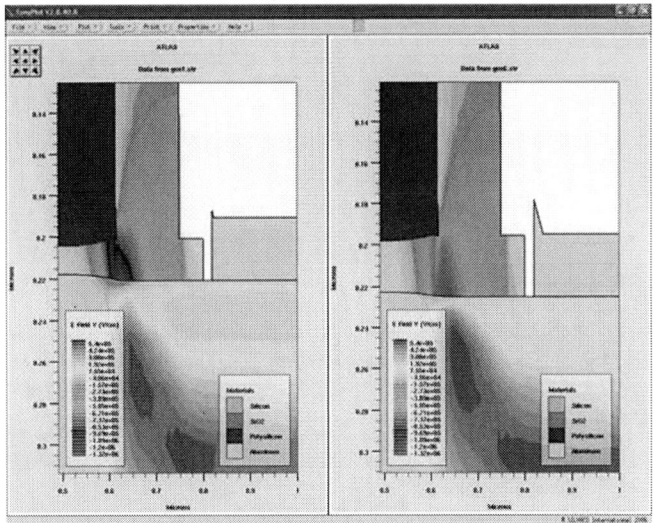

Fig. 8 Vertical electric field decreases as gate oxide increases from 11.7nm (left) to 21.6nm (right).

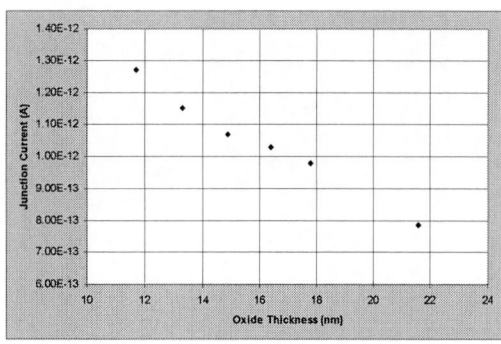

Fig. 9 Junction current (at storage node voltage = 4V) reduces as oxide thickness increases.

GIDL is reduced by increasing the gate oxide thickness as shown in Figure 9. Since E = V/d, an increase in oxide thickness will relax the vertical fields.

V. CONCLUSION

The purpose of this paper was to explore leakage current in DRAM cells: reverse junction leakage from the storage node and gate induced drain leakage. The proposed solutions were examined and simulated using Silvaco Athena/Atlas. It was determined that substrate doping had a great effect on leakage current. This was observed by studying the effects on depletion region, junction depth, and lateral electric fields. Careful consideration must be taken into account because a reduction in substrate doping can severely degrade the subthreshold performance. It was also determined that LDD implantation energy and gate oxide thickness had a great effect on GIDL. As newer DRAM generations emerge, innovations in process design will be needed to reduce leakage current.

REFERENCES

[1] J. A. Mandelman, et al, "Challenges and future directions for the scaling of dynamic random-access memory (DRAM)," *IBM J. RES. & DEV.*, vol. 46, no. 2/3, pp. 187-212, March/May 2002.

[2] M. T. Bohr, "Nanotechnology Goals and Challenges for Electronic Applications," *IEEE Transactions on Nanotechnology*, vol. 1, no. 1, pp. 56-62, March 2002.

[3] J. Lee, D. Ha, and K. Kim, "Novel Cell Transistor Using Retracted Si_3N_4-Liner for the Improvement of Data Retention Time in Gigabit Density DRAM and Beyond," *IEEE Transactions on Electron Devices*, vol. 48, no. 6, pp. 1152-1158, June 2001.

[4] E. Adler, et al, "The evolution of IBM CMOS DRAM technology," *IBM J. RES. & DEV.*, vol. 39, no. 1/2, pp. 167-188, January/March 1995.

[5] T. Hamamoto, S. Sugiura, and S. Sawada, "On the Retention Time Distribution of Dynamic Random Access Memory (DRAM)," *IEEE Transactions on Electron Devices*, vol. 45, no. 6, pp. 1300-1309, June 1998.

[6] K. Saino, et al, "Impact of Gate-Induced Drain Leakage Current on the Tail Distribution of DRAM Retention Time," *IEDM Tech. Dtg*, pp. 837-840, 2000.

1-4244-0267-0/06/$25.00 ©2006 IEEE

A Gain Control Low Power CMOS Power Amplifier for Ultra Wideband Applications

Jack C. Reed, Houshang Amir Aghahassan, Jane Chi, and Albert Yen

UMC, Sunnyvale, CA, USA

Abstract – This paper presents a gain control, low power, wideband power amplifier operating in the lower band frequency of 3.1 GHz to 4.8 GHz for UWB applications. This design was implemented in UMC 0.13 μm MMRF CMOS technology with an on-chip transformer and biasing circuitry. This PA works in the class A region and has high linearity with power consumption as low as 32 mW from a 1.2 V supply voltage. The differential common-source with single ended output topology provides a good isolation and good matching to 50 Ω at the output port. This PA delivers a total gain of 16 dB with a 2 dB step level for on-chip gain control circuitry.

Index Terms: CMOS, power amplifier (PA), UWB, low power, Radio Frequency (RF), RF transceivers.

I. INTRODUCTION

Since the FCC allocated large contiguous blocks of spectrums for unlicensed applications, ultra wideband (UWB) technology has drawn increased industrial attention because of its short-range high data rate, spectrum reuse and low power consumption. In addition, the low transmission power avoids interference with narrowband RF traffic and enables new functions like Wireless USB. Although technologies like SiGe or GaAs has been used for transceiver realization, the CMOS technology is used here to achieve a low cost SoC solution [1]. However, the features of low power and wide bandwidth also complicate the implementation of key RF components, such as the low noise amplifier (LNA), power amplifier (PA) and mixer.

Three design approaches were commonly used for the broadband circuit [2]. Distributed amplifiers achieve a wideband of frequency through high linearity and good matching [3]. However, the trade off with this approach is increased power consumption and area. Another approach is to use a shunt-feedback network with a resistor, but this is limited by the headroom of voltage supply [2]. Finally, the filter theory, which was used in many narrowband applications, can be extended for the wide band situation [4]. The down side is that the LC stages occupy significant area.

This paper presents a CMOS power amplifier that overcomes the area issue in filter theory technique by using a transformer for matching the circuit and output load. This design uses UMC 0.13 μm CMOS technology and focuses

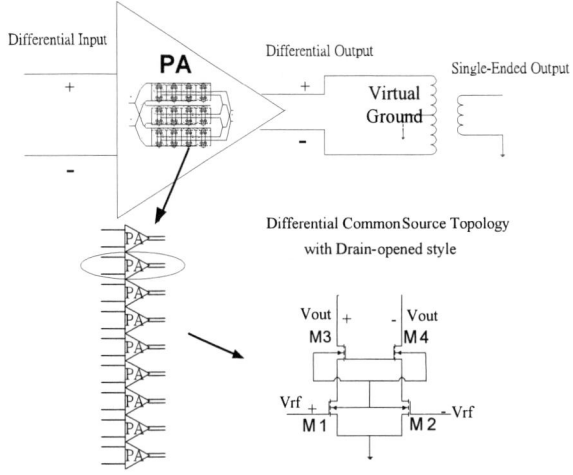

Figure 1. Gain Control Power Amplifier with Tranformer

on band group number one of the IEEE standard (from 3.1 GHz to 4.8 GHz). The higher frequency design can wait until CMOS technologies mature for frequencies above this band group [5]. This paper also includes the control voltage block providing 0 -16 dB attenuation with a 2dB step. The layout and simulation results of the proposed PA are shown at the end of this paper.

II. CIRCUIT DESIGN

Class A amplifiers operate linearly across the full input and output range. Although only a theoretical maximum of 50% efficiency is obtainable, this waste of power is extremely small for small signals, and can be easily tolerated [6]. For an application like UWB system, the EIRP is constrained to -41.3 dBm/MHz by FCC Part 15. The advantage of linearity and wider operating range in class A outweighs the increased power consumption. Therefore, the proposed PA uses a differential cascode class-A model (Fig. 1) to achieve high linearity, high speed and low power.

The common source topology was used in our design methodology, since it provides the highest efficiency among

1-4244-0267-0/06/$25.00 ©2006 IEEE

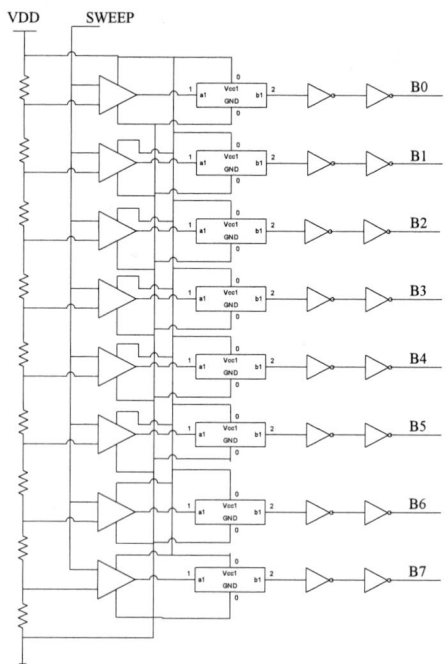

Figure 2. The Control Circuitry of PA

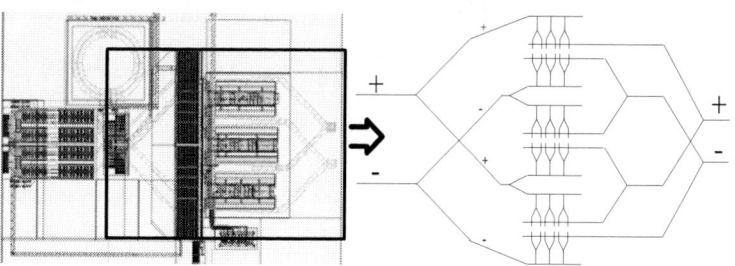

Figure 3. Layout of the proposed UWB PA with transformer

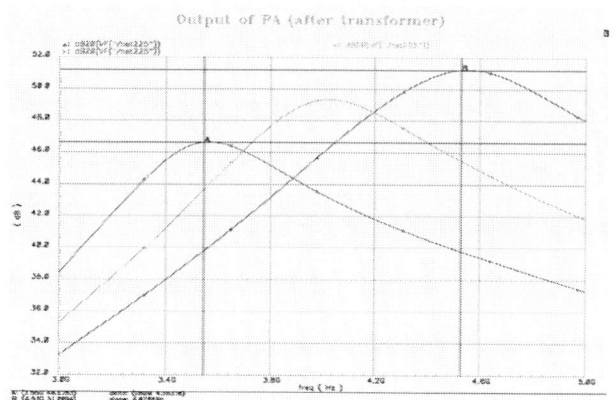

Figure 4. The AC plot of PA in three different center frequency, covering the range of interest.

the three one-transistor amplifier configurations [6]. The transistors M1 and M2 perform the input gain amplification, while M3 and M4, the cascode stage, are used to improve the signal isolation between input and output. The drain-opened style design was intended to leave the output port open for transformer loading. Figure 1 depicts a diagram of the transformer model showing the virtual ground on the left side when the circuit is tuned to perfectly match with the output impedance. The inductors on the side of the virtual ground can then be seen as the load of PA. Moreover, the transformer transforms the differential signal from PA to a single output signal for the antenna stage.

There are a total of 9 such models assembled in the presented PA (Fig. 1), and each model were designed in sized to provide 2dB power gain and controlled by a control circuitry. Fig. 2 reveals the control circuit diagram to turn on and off when the input signal sweeps from low to high. The fundamental idea of the control circuitry is to use the op-amp to compare the SWEEP signal with different voltages and to deliver a high/low voltage signal to turn on and off the corresponding PA.

III. IMPLEMENTATOIN AND LAYOUT

A gain control, low power CMOS PA operating frequency range from 3.1 GHz to 4.8 GHz was designed based on the described model above. The active transistors M1, M2 and cascode transistors M3, M4 are sized identical with gate finger widths of 4 μm, gate lengths of 120nm, and a finger number of 4. The transformer was tuned to a 2-to1

ratio to match the differential impedance. The model was connected to the UWB mixer and transformer differentially for design validation purpose. Because the insignificant imaginary component of impedance, there is no matching circuit used in the design. Fig. 3 shows the complete circuit layout view.

There are numerous transistors are used to equally distribute the total amplifier current in the layout, so we can avoid heat and other problems. IBM provided a wiring distribution layout style to improve the power distribution among a plurality of parallel heterojunction bipolar transistors. The basic structure of this embodiment is using several symmetrical H branches path. The input signal line first distributes into four sub-branches forming H style layout, while each sub-branch connects to another four sub-branches [7]. Although this structure can deliver well-distributed signal to each transistor, the output signal is hard to collect. This is because the transistors spread out by the H style layout, and the output lines need to be emerged by relatively long path. This situation is especially hard to be feasible in differential circuit. In the layout shown in Fig. 3, we proposed a Y type layout style to reach similar results in H style. However, the advantage of using Y style is that the concept can be applied to both differential and single ended.

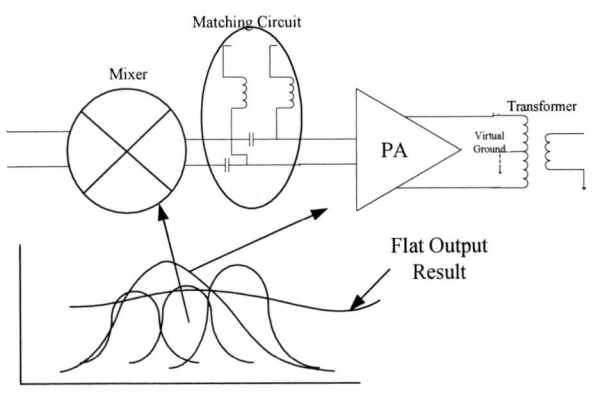

Figure 5. Tuning at low operating frequency compensate out the gain variation from two design block resulting in steady output power.

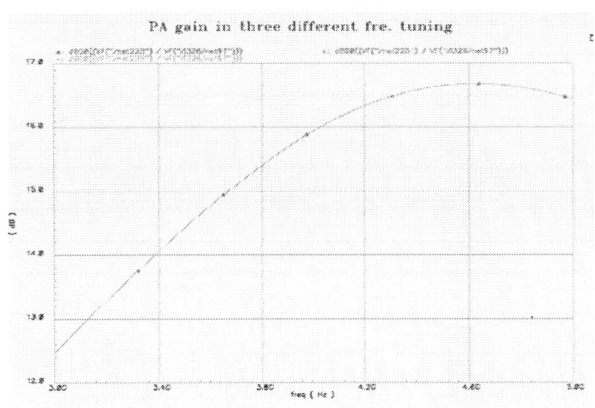

Figure 6. PA gain plot in three center frequencies.

In our design, there are totally of six arrays of active transistors to drive the amplifier current, and every two arrays form a differential pair device. To drive the three differential pairs in Fig. 3 equally, signal first travels through two input lines, which are divided into two branches and cross-connected to a pair of matching circuits. Then the signal propagates through each matching circuit where are divided into three branches in the end of the path line and connects to one set of transistors in one array. In each array the paths are divided into two until the path connects to the active transistors. This methodology can equally distributed input signal to each active transistor, and can be extended to the output path connection. The simulation results proof that this layout style simplifies the layout work and helps the post layout simulation to meet the design specifications.

IV. SIMULATION RESULTS

The post-layout was simulated using Cadence SpectreRF with United Microelectronics Corporation (UMC) 0.13 μm MMRF CMOS models and Process Design Kit (PDK). Fig. 4 shows AC characteristics of PA tuned in three different center frequencies 3.43 GHz, 3.96 GHz, and 4.49 GHz covering the frequency range of interest.

The output signal shows frequency dependency over the band because the center frequency was tuned by an external LC circuit providing additional load to the previous stage. (Fig. 5) This load incorporates a Q value from inductor and affects the previous stage's gain to delivery the frequency dependent output signals. The gains of PA block in three different frequency bands shown in Fig. 6 reveals the consistent gain performance over the desired frequency range.

The UMC process ensures that the gain variation was managed within 1dB in the corner cases. Due to many manufacture process issues, the transconductance factor (gm) of each transistor may vary about 60% making the output

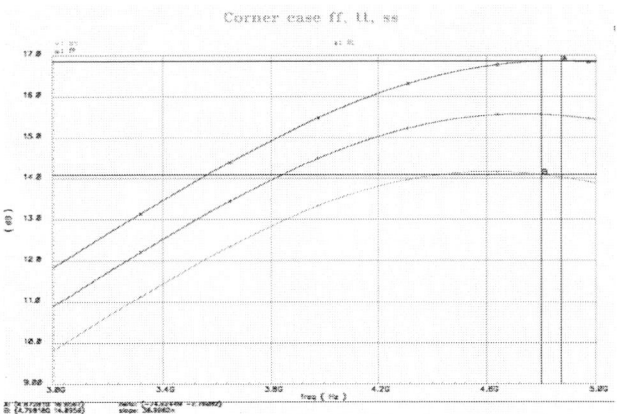

Figure 7. The worst and best scenario plot compare to the typical case.

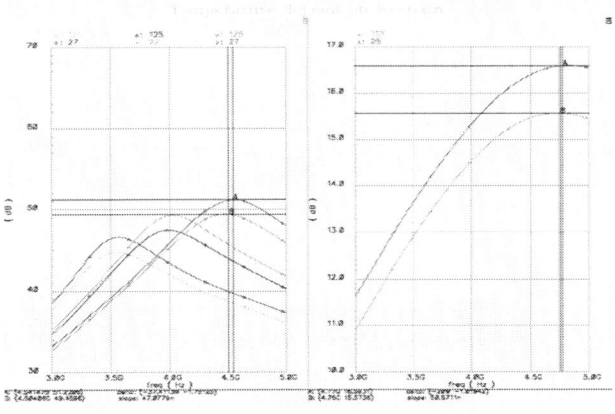

Figure 8. PA characteristics in temperature variation

1-4244-0267-0/06/$25.00 ©2006 IEEE

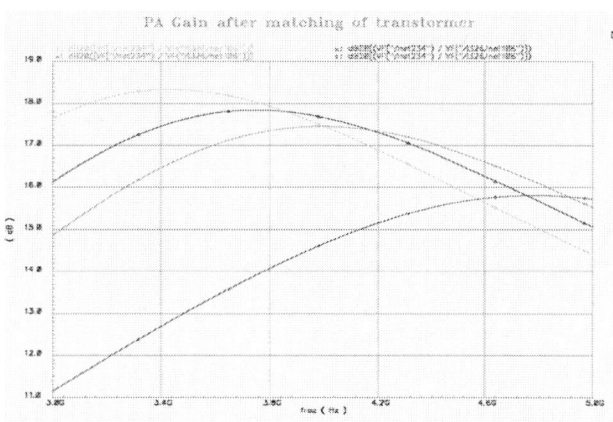

Figure 9. PA gains in different matching of transformer. (transformer resonances from 5G Hz to 3G Hz)

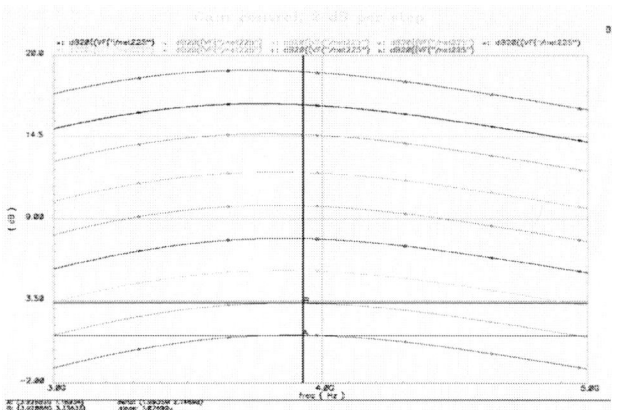

Figure 10. Gain control plot form.

results of the manufactured device hard to manage. The simulation using UMC PDK shows that, in Fig. 7, gains from both extreme cases (fast-fast, slow-slow) shift 1 dB away from that of the typical case.

Temperature is another key factor in output results measurement and analysis. Fundamental circuit elements are highly dependent on the temperature. For example, the mobility of transistors changes for different temperatures. The deviation is 1dB in from -40 degree Celsius to 125 degree Celsius (Fig. 8) the PA block. The information on temperature coefficients was found in UMC Topological Layout Rule (TLR) and Electrical Design Rule (EDR) documents.

Due to the unmatched transformer, the gain difference over the operating range is about ±2 dB. If a properly designed transformer was provided, the circuit permits the delivery of a flatter gain performance. Fig 9 shows the PA block delivering 16.5 ±1.25dB gains over the frequency range. Moreover, if the mixer block is taken into consideration, center frequency tuned at lower frequency

Parameter	Distributed Amp [3]	Shunt Feedback [2]	Filter Theory	Units
Supply Voltage	2	1.8	1.2	V
Frequency Range	2 G ~ 11 G	3.1 G ~ 4.8 G	3G ~ 5G	Hz
Power Consumption	2 V * 50 mA = 100	25	1.2 V * 27m A = 32	mW
Gain	17	19 +/- 1	16	dB
Output P1 dB	3.5	-4.2	6	dBm
S22	<-5	<-8	< -10	dB
Technology	0.13 μm	0.18 μm	0.13 μm	CMOS

TABLE I.   COMPARISON ON THREE TYPES OF PAS

becomes preferable. Because the mixer raises the gain as frequency increases, having a higher gain at lower frequency in PA can compensate for the mixer resulting in a flat power output from PA (Fig. 5). Finally, the gain control 2dB per step plot is shown in Fig. 10. Table 1 lists performance summary and comparisons of this work to other two types recent research reports. The power consumption of this PA is only one-third of that in distrusted amplifier structure, but it is able to achieve the same or better design parameters. On the other hand, this PA uses a little more power than shunt feedback structure, but the output P1 dB performance is much better while other design parameters are similar. Therefore, combining all the design issues together, the filter theory PA model proved by this paper is one of the ideal candidates for UWB application.

V. CONCLUSION

This concludes the presentation of a gain control low power CMOS PA designed in UMC 0.13μm technology for UWB application. The frequency range of interest is 3.1 GHz to 4.8 GHz. In this frequency range, this PA consumes only 32 mW of power from a 1.2 V supply providing 16dB gain in 2dB step. An on-chip transformer was implemented with the design block for matching and frequency tuning purpose. If the system adjusts operating frequency to 3.4 GHz with proper modification on transformer, better performances in gain and matching can be expected. The UMC 0.13μm MMRF CMOS technology assures the gain performance floating within 1dB over temperature variation and worst-case scenario. Since this design consumes very low power, it is an ideal candidate for UWB application and the design methodology can be easily extended to higher frequency applications in the future.

REFERENCES

[1] W. Lawrence, "2004 Ansoft Worldwide Technical Workshops" Ansoft Corporation

[2] S. Jose, Lee Hyung-Jin, Ha Dong, S.S. Choi, "A Low-power CMOS Power Amplifier for Ultra wideband (UWB) Applications", 2005.

ISCAS 2005. IEEE International Symposium on Circuits and Systems 23-26 Page(s):5111 – 5114, May 2005

[3] C. Grewing, K. Winterberg, S. van Waasen, M. Friedrich, G.L. Puma, A. Wiesbauer, C. Sandner, "Fully integrated distributed power amplifier in CMOS technology, optimized for UWB transmitters" Radio Frequency Integrated Circuits (RFIC) Symposium, 2004. Digest of Papers. 2004 IEEE Page(s):87 – 90, June 2004

[4] Bo Shi and Michael Yan Wah Chia, "Design of a SiGe low-noise amplifier for 3.1 – 10.6 GHz ultra wide band radio", ISCAS May 2004.

[5] TI Physical Layer Proposal for IEEE 802.15 Task Group 3a, May 2003.

[6] B. Razavi, "RF Microelectronics" Upper Saddle River, NJ: Prentice Hall PTR, 1998, pp. 298 -302

[7] David R. Helms, and Phillip Antognetti, "Mask Layout for Sidefed RF Power Anplifier", United States Patent, Patent No. US 6,448,858 B1, Sep, 2002.

1-4244-0267-0/06/$25.00 ©2006 IEEE

A 2.4-GHz, Wide Tuning Range GmC VCO Using a Novel Load Biasing Technique

Vivek Verma, *Member, IEEE*, Tamara Papalias, *Member, IEEE*

Abstract— **This paper presents a 2.4-GHz monolithic GmC fully differential voltage controlled oscillator (VCO) fabricated in 0.24-μm CMOS technology with a wide tuning range between 1.77GHz and 2.73GHz. This VCO can be used for data clock recovery applications in optical receivers for such standards as SONET OC-48 or European SDH STM-16. Two published techniques for quadrature generation are discussed for use in data clock recovery. The circuit implementation is purely transistor based OTA stages with the entire VCO covering 60μmx90μm. A novel technique for active load biasing is introduced which provides marginal temperature insensitivity and eliminates a need for external bias source. The power dissipation is less than 50mW at 2.0V power supply and the measured phase-noise is -93dBc/Hz at 2MHz offset.**

Index Terms—**CMOSFET oscillator, SONET OC-48, wide tuning range, GmC oscillator, ring oscillator**

I. INTRODUCTION

OPTICAL transmission standards such as synchronous optical networks (SONET) and synchronous digital hierarchy (SDH) require the transmitters to embed clock information with the data to reduce jitter susceptibility. To receive such a signal, optical receivers need a data and clock recovery (DCR) unit[1] to extract the clock edge placement information from the incoming signal, and use it for data recovery. The SONET OC-48 standard, and the European equivalent SDH STM-16, requires clock extraction at a bit-rate of 2.5 Gb/s[2]. Most all integrated DCR circuits are Phase Locked Loop (PLL) based[1], and consequently require an integrated oscillator to provide the spectrally pure high frequency signal for maintaining frequency lock with the incoming signal. Conventional wisdom is that a high degree of phase noise performance cannot be achieved with a ring oscillator design in a CMOS process, and that an inductor-capacitor (LC) based oscillator is required because of high

Manuscript received May 23, 2006. This work was supported in part by Cadence Design Systems with grants for the use of Spectre simulator at San Jose State University, by MOSIS with grants for fabrication services and Agilent/Verigy USA for supporting the test and characterization of these oscillators.

V. Verma was with San Jose State University, San Jose, CA 95192 USA. He is now with Agilent/Verigy USA, Santa Clara, CA 95054 USA (e-mail: vivek.verma@verigy.com).

T. A. Papalias was with Stanford University, Stanford, CA 94309 USA. She is now with the Department of Electrical Engineering, San Jose State University, San Jose, CA 95192 USA (e-mail: drpapalias@yahoo.com).

quality resonance and power concentration in a narrow band[3,4]. Other high performance, high-frequency operational transconductor amplifier (GmC OTA) based oscillators have been reported but use costly technologies that may not be as widely available[5,6]. We present a 2.4GHz, area efficient, wide tuning range GmC voltage controlled oscillator (VCO) with a high carrier to noise ratio, while dissipating less than 50mW of power using 0.24μm CMOS without using any inductors and without any extraordinary process requirements. The following sections attempt to explain the innovation, complexities, and the trade-offs necessary.

The OTA stage evolution is presented in Section II. A novel active load biasing scheme using the DC component of the oscillator feedback loop to create a DC bias and eliminate voltage reference generators is also explained in Section II. Possible circuit implementations for in-phase (I) and quadrature (Q) signal generation necessary for DCR are discussed in Section III. Experimental results for this VCO implementation are discussed in Section IV and a qualitative comparison with other reported similar oscillators is described in Section V. The layout considerations and physical effects encountered during the development cycle are discussed in Section VI and a brief conclusion is presented in Section VII.

II. CIRCUIT DESIGN

To achieve high frequency of oscillation (f_c) and noise performance, a high quality transconductor is required. To achieve the best possible results, every non-essential transistor or passive component or interconnect was eliminated to reduce losses caused by parasitic. This thought process kept the OTA stage complexity to a minimum[7]. Previously published reports and analyses of differential ring oscillators using a linear time variant phase noise model have shown that minimizing the number of stages can result in phase noise performance improvement[8]. Without reproducing the math here, the addition of every stage increases the phase noise by a multiple of two[8]. So, the key to achieving a high performance GmC ring-oscillator using CMOS is to keep the number of stages to a minimum. By reducing the number of stages, power can be efficiently focused to achieve a higher quality factor. GmC ring-oscillators have been reported ranging from 2-stages to the more commonly used 4-stage and 5-stage configurations[5,6,9]. A block diagram of the GmC ring-oscillator implementation being presented here is shown

1-4244-0267-0/06/$25.00 ©2006 IEEE

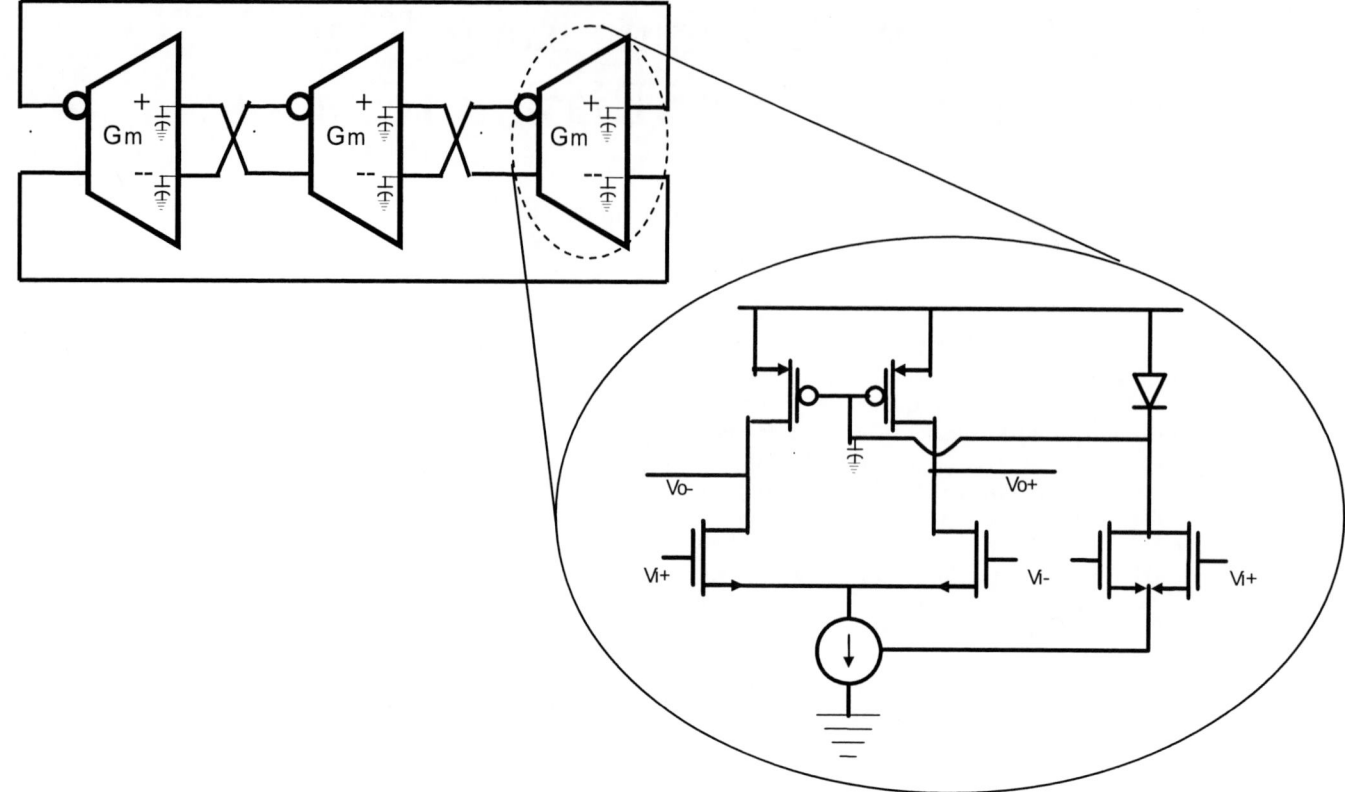

Figure 1 - 3-stage VCO implementation and the GmC OTA stage

in Figure 1. A 3-stage configuration was chosen for higher frequency of oscillation, area efficiency and lower power consumption.

To meet additional area efficiency and monolithic goals, a novel technique has been developed for biasing the active loads by using the complementary nature of the differential signals to create a dc voltage using an ac cancellation circuit. As illustrated in Figure 1, the source and drains of two weak transistors are tied together such that the input differential signals cancel out each other, leaving a dc residue. The voltage drop across a diode placed between the power supply and the drains of the two transistors is used as the dc bias and can be tuned for accuracy by varying the size of diode during design. As the dc output levels of the OTA stages drift with temperature and power supply noise, this novel technique allows the active load bias to relatively scale with the rest of the OTA. In addition to providing temperature insensitivity, it also helps phase noise performance as the OTA drive strength stays the same in every stage. Comparatively, an external stable reference would not be able to provide this noise rejection in the VCO. Additionally, the power consumed in generating this bias is relatively small as the dc current used to keep the diode in the forward biased region is very small compared to the main differential pair since weak transistors were used.

This biasing scheme however increases the input capacitance in the OTA stages, affecting the overall f_c and

introduces noise on the bias signal i.e. an ac signal generated at twice the oscillation frequency due to the switching action of the weak transistors, similar to the noise on the common source node of a differential pair. Albeit, this biasing noise is a small signal that can be attenuated by certain design choices. A common practice of a grounded MOS capacitor at the biasing node was used in this implementation to reduce the affect of the ac noise. Similar biasing noise attenuation can also be achieved by designing the active load transistors to be very large relative to the weak bias generating transistors.

III. QUADRATURE GENERATION

DCR relies on the availability of clean amplitude and phase balanced I and Q signals for a high image rejection ratio and low bit error rate data reconstruction. Most published techniques for I and Q signal creation focus on using RC-CR networks, or by extension poly-phase filters (PPF), in conjunction with limiters (a.k.a. Haven's technique)[10,11,12]. A second novel method of quadrature generation recently published uses bipolar devices as active current sourcing components in a two step process[13]. Here we discuss both techniques with a brief analysis of the theory and an overview of the implementation.

An RC-CR network, a splitter, comprises of two signal paths, one through a low-pass filter and the other through a high-pass filter, producing a $\pi/2$ phase difference between the forked signals. However, the two signals are only amplitude balanced in a narrow band at the corner frequency of the two filters. Besides this, a principle drawback of an RC-CR network when used by itself is susceptibility to temperature and process variations[13,14]. In an integrated implementation, layout optimizations can be made to limit the affect of such variations, yet the practical limitations of RC-CR network make it difficult to achieve gain and phase balance. Methodologies such as Haven's techniques have been proposed and widely used to make quadrature generation insensitive to the amplitude variations of the phase splitter to some extent by using amplitude limiters and variable capacitors[10,13,14].

Another technique recently reported for quadrature generation uses bipolar devices to implement all pass transconductance and transimpedance circuits. It eliminates the need for limiters, as required by other methodologies, and produces amplitude and phase matched I and Q signals by design. Effectively, this technique generates the I and Q signals in a two step process, first converting the LO voltage signal into current and then converting it back into voltage, generating two signals $\pi/2$ apart in phase, by adding and subtracting the currents generated in the first step with trans-impedance amplifiers. The limiters are not necessary here because the last stage trans-impedance amplifiers also act as buffers[13], and can be scaled to generate very high amplitude matching between the I and Q signals with optimized layout. Even though this implementation is more efficient and accurate in terms of phase splitting and matching, relative to the RC-CR networks, it does increase power consumption, and adds shot noise and flicker noise components to the signal because of its current mode operation.

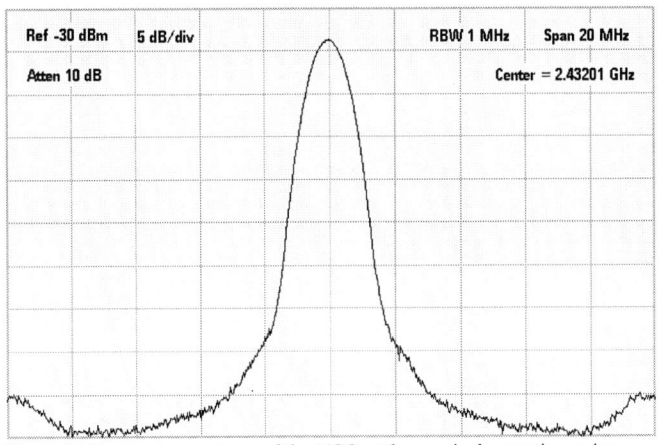

Figure 2 - Frequency spectrum of the VCO at the nominal operating point.

IV. IMPLEMENTATION AND EXPERIMENTAL RESULTS

An operational GmC voltage controlled oscillator has been realized in TSMC 0.24μm CMOS. A micrograph of the realized silicon is shown in Figure 7. The highlight shows the effective area covered by the VCO, approximately 60μm², and exists completely independent of any biasing circuits. All measurements reported here were made using an FR4/SMA

test rig with QFN type surface mounted packaged samples, and an Agilent E4440A spectrum analyzer. The f_c is approx 2.4GHz at the power supply voltage (V_{supply}) of 2.0V, and tuning voltage (V_{tune}) of 765mV. This is referred to as the nominal operating point for the rest of the report. A power spectrum plot of the freerunning VCO at the nominal operating point is shown in Figure 2. The power concentration at the fringes of the spectrum plot is the expected substrate and supply noise component beyond 10MHz[7,15]. The VCO can be tuned from 1.77GHz to 2.73GHz over a 300mV change in V_{tune}. Figure 3 shows a plot of the f_c vs. V_{tune} at different V_{supply} levels. The total tuning range is more than 40%, and is linear above V_{tune} = 725mV. During lab testing, the VCO demonstrated an even wider tuning range than reported, but performance degrades quite rapidly for V_{tune} below 664mV.

This wide tuning range, however, results in higher power dissipation due to a large standing dc current necessary to

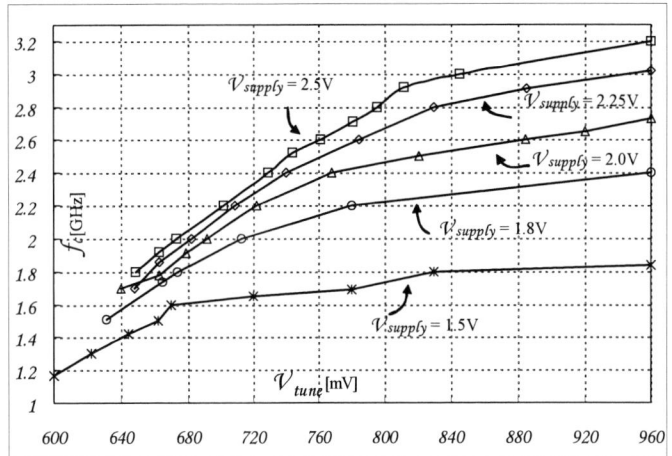

Figure 3 – A plot of the VCO tuning behavior at different power supply levels

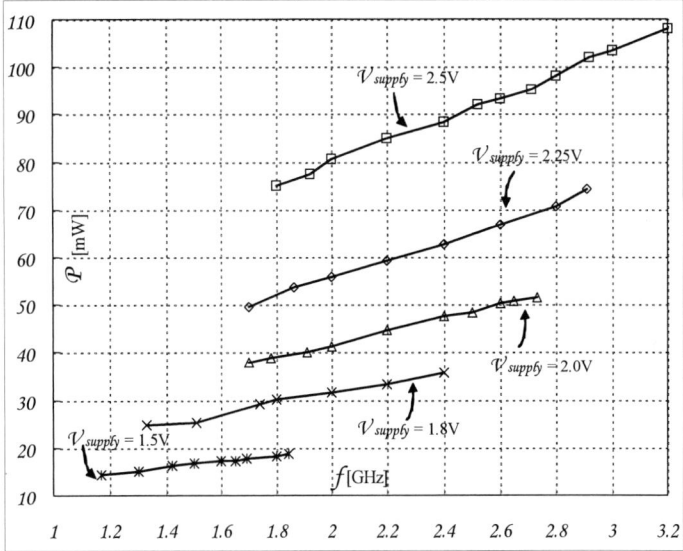

Figure 4 - A plot of the power dissipation over the tuning range at different power supply levels

support the wide band operation. At nominal operating point, the VCO dissipates about 47.5mW. A plot of the power dissipation over the tunable frequency range at different $\mathcal{V}_{supply}$ levels is shown in Figure 4. The power consumption behavior changes quite consistently over the tuning range at about 2.5mW for every 100MHz. The output power measurements were pessimistic due to apparatus limitations and high losses in the measurement setup. The output power measured at the nominal operating point is -33dBm, and changes 10dB over the tuning range. The noise floor in the tuning range is consistent at -90dBm.

A logarithmic plot of the phase noise performance of the freerunning VCO at three different frequencies is shown in Figure 6. The measurements were taken with the VCO selectively tuned to upper, mid, and lower band frequencies in the tuning range to show the consistency in performance. At nominal operating point, the resulting phase noise measured at 2MHz offset is -93dBc/Hz, and it improves at 10dB/dec between 1MHz - 10MHz. The three overlaid phase noise plots practically overlap each other in the decade of interest

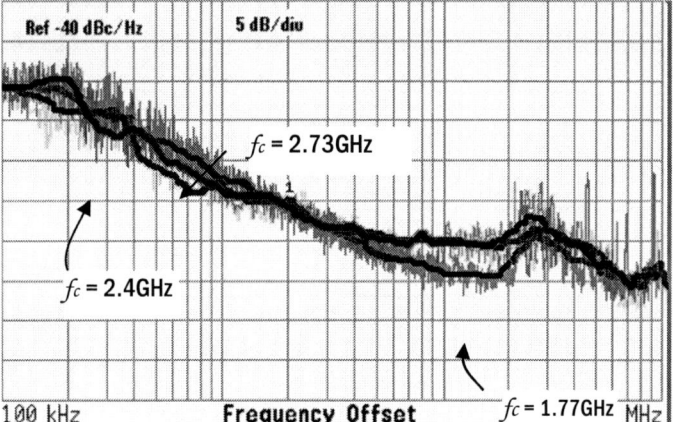

Figure 6 - A logarithmic plot of the phase noise measured at the three labeled frequencies with power supply at 2.0V. The consistency of the phase noise performance across the tuning range is evident.

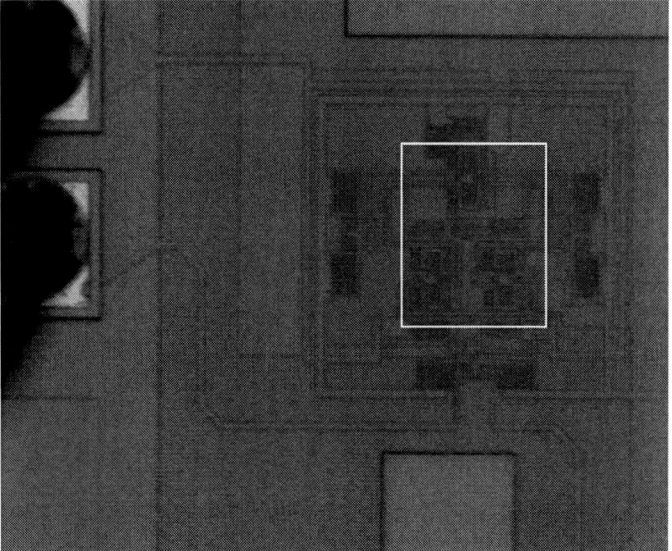

Figure 7 - A micrograph of the 3-stage VCO implementation. The active area highlighted is 60μm x 90μm

dominated by the 1/f noise component. There is 5dB phase noise degradation beyond 4MHz when tuning from low to mid band frequencies as the quality factor decreases[7]. The supply noise component noticed earlier in the power spectrum plot can also be seen here as the power roll-off diminishes beyond 10MHz.

Additional characterization data of note is the VCO robustness across $\mathcal{V}_{supply} \pm 500\text{mV}$ as seen in Figure 3 and 4. Changing $\mathcal{V}_{supply}$ from 1.5V to 2.5V, the phase noise performance stays consistent, but there is a pronounced trade-off between tuning range and power dissipation. A higher $\mathcal{V}_{supply}$ results in a wider tuning range and a higher power dissipation and vice versa.

V. QUALITATIVE COMPARISON

A wide range of VCO using similar topologies, and similar performance ranges have been reported. Even though they use a variety of different technologies such as CMOS, BiCMOS, double poly CMOS, un-silicided poly layers, etc, some technique must be developed to quantize the overall performance for comparison with this implementation. Previous publications have defined benchmarks such as figure of merit (FOM)[16] for doing similar comparisons. In addition to the parameters used in the past, other distinguishing attributes must be considered here. A large tuning range, as discussed earlier, is necessary for operation over process and temperature variations. Additionally, area required on silicon can be a practical limitation when considering cost and integration in an SOC. Temperature is ignored here since all oscillators are assumed to have been tested at room temperature. So, FOM for this comparison is defined as:

$$\text{FOM} = \frac{1}{P} + \frac{ft}{fc} - \frac{\zeta[fo]}{fo} - \log 10(\mathcal{A}_{eff}) \qquad (1)$$

where P is the power dissipated, ft is the total tuning range, fc is the frequency of oscillation, fo is the offset from the carrier where phase noise is measured, $\zeta[fo]$ is the phase noise measured at that offset, and $\mathcal{A}_{eff}$ is the effective area used by the VCO. FOM is a unit less quantity where a higher value indicates better performance.

All implementations chosen for comparison here are of inductorless GmC VCO topology and have been realized in silicon. Figure 8 shows a plot comparing [5], [7], [8], [15], [17], [18] and this implementation, based on (1). The only VCO that is superior to this implementation is [7]. A parameter that is not considered in this comparison is the ratio of fc/fT, where fT is process transition frequency. This is important when evaluating any VCO because it demonstrates its ability to exploit the process to its full extent. It was not included in the FOM because of lack of verifiable data.

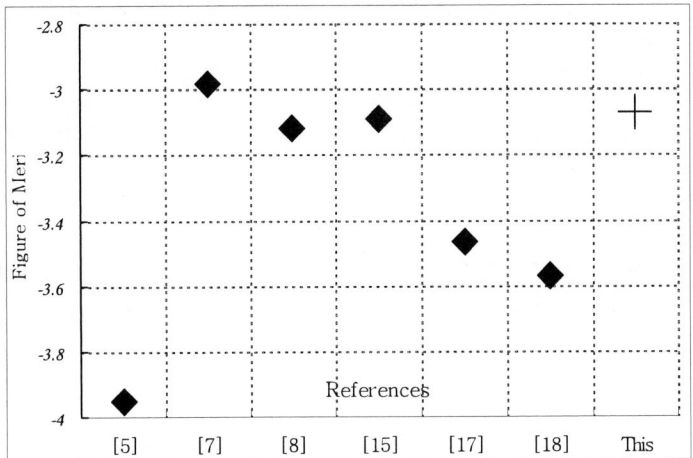

Fig. 8 - A qualitative comparison of six other VCO with this implementation
Note: [5] used a BiCMOS process
Note: [8] used un-silicided polysilicon for passive loads to reduce 1/f noise
Note: [18] used double poly layer CMOS process

VI. LAYOUT CONSIDERATIONS

For high performance oscillator design in CMOS, parasitics are the biggest impediment[14]. Even though design techniques can mitigate some of these affects; physical layout determines the final effectiveness of these techniques. Basically, layout becomes the most critical part of the design phase when trying to extract every last bit of performance. One of the biggest concerns in this implementation was matching differential signals, and matching different stages with minimal overhead. Any unwanted imbalance in the symmetry of a differential circuit, or a stage, can offset differential crossover levels, degrading phase noise performance[15], and spreading power away from f_c. Popular approaches such as common-centroid were prototyped, but its efficacy in implementation was less than impressive due to its inherent nature of splitting every component to achieve symmetry. The parasitic overhead caused by using common-centroid degraded f_c by almost 50%, also adversely affecting phase noise performance and power efficiency. Instead, more practical rules such as current direction matching, no ac signal cross over to increase signal isolation, and close physical proximity were followed in an attempt to match circuits[19,20]. Interconnects between stages can be an artificial power transfer bottleneck in multi-stage circuits due to high frequency phenomenon such as skin effect, and can also restrict the maximum power that can be driven off chip. To minimize these physical limitations, thicker metals higher up in the stack were used where ever possible with wide interconnects, and maximum number of vias within reason, on every layer change, to minimize the resistive and capacitive overhead.

VII. CONCLUSION

A low phase noise GmC VCO suitable for DCR circuits in optical receivers has been presented. The oscillator can be tuned between 1.77GHz – 2.73GHz. It does not require any passive components minimizing the area required, or any special process requirements, to achieve this performance, effectively a lower cost solution relative to other reported implementations. A novel active load biasing technique that utilizes the dc component of the oscillating signal, further reducing the area required, has been introduced.

At $\mathcal{V}_{supply}$ of 2.0V, the power dissipation is 47.5mW in the VCO core at f_c of 2.4GHz where the measured phase noise is -93dBc/Hz at 2 MHz offset. The entire oscillator covers an area less than 60μmx90μm. Its robustness is evident with a consistent phase noise performance over 40% of the tuning range.

ACKNOWLEDGMENT

The authors would like to thank Dr. David Parent of San Jose State University for his work in support of this project.

REFERENCES

[1] A. Buchwald and K. Martin, Integrated Fiber-Optic Receivers. Norwell, MA: Kluwer, 1995.
[2] Momtaz, A et al, "Fully-integrated SONET OC48 transceiver in standard CMOS," in *Solid-State Circuits Conference, 2001. Digest of Technical Papers. ISSCC. 2001 IEEE International*, no.pp.76-77, 433, 2001.
[3] T. I. Ahrens, A. Hajimiri, and T. H. Lee, "A 1.6 GHz, 0.5 mW CMOS LC low phase noise VCO using bond wire inductance," in *Proc. 1st Int. Workshop Design of the Mixed-Mode Integrated Circuits and Applications*, Cancun, Mexico, July 1997, pp. 69–71.
[4] A. Hajimiri and T. H. Lee, "Phase noise in CMOS differential LC oscillators," in *Proc. VLSI Circuits*, June 1998, pp. 48–51.
[5] Johan D. van der Tang, Dieter Kasperkovitz, and Arthur van Roermund, "A 9.8–11.5-GHz Quadrature Ring Oscillator for Optical Receivers," *IEEE Journal of Solid-State Circuits*, vol. 37, No 3, March 2002.
[6] Pottbaker, A.; Langmann, U., "An 8 GHz silicon bipolar clock-recovery and data-regenerator IC," *IEEE Journal of Solid-State Circuits*, vol. 29, No 12, Dec 1994.
[7] Razavi,B "A Study of Phase Noise in CMOS Oscillators," *IEEE Journal of Solid-State Circuits*, Volume 31, Issue 3, March 1996.
[8] A. Hajimiri et al., "Jitter and phase noise in ring oscillators," *IEEE Journal of Solid-State Circuits*, vol. 34, June 1999.
[9] Seog-Jun Lee, Beomsup Kim, and Kwyro Lee, "A Fully Integrated Low Noise 1-GHz Frequency Synthesizer for Mobile Communication Application," *IEEE Journal of Solid-State Circuits*, vol. 32, No 5, May 1997.
[10] Tsuneo Tsukahara, Masayuki Ishikawa, and Masahiro Muraguchi, "A 2-V 2-GHz Si-Bipolar Direct-Conversion Quadrature Modulator," *IEEE Journal of Solid-State Circuits*, vol. 31, No 2, February 1996.
[11] Rofougaran, A. et al., "A single-chip 900-MHz spread-spectrum wireless transceiver in 1-μm CMOS. I. Architecture and transmitter design," *IEEE Journal of Solid-State Circuits*, vol.33, no.4pp.515-534, Apr 1998.
[12] Ahola, R. et al., "A single chip CMOS transceiver for 802.11 a/b/g WLANs," in *Solid-State Circuits Conference, 2004. Digest of Technical Papers. ISSCC. 2004 IEEE International*, pp. 92 - 515 Vol 1, 15-19 Feb. 2004.
[13] MAT Sanduleanu, JP Frambach, "1GHz Tuning Range, Low Phase Noise, LC Oscillator with Replica Biasing Common-Mode Control and Quadrature Outputs," in *Solid-State Circuits Conference, 2001. ESSCIRC 2001*.
[14] Behzad Razavi, "Architectures and Circuits for RF CMOS Receivers," in *Custom Integrated Circuits Conference, 1998. IEEE 1998*.
[15] Manop Thamsirianunt and Tadeusz A. Kwasniewski, "CMOS VCO's for PLL Frequency Synthesis in GHz Digital Mobile Radio Communications," *IEEE Journal of Solid-State Circuits*, vol. 32, No 10, October 1997.
[16] D. Ham and A. Hajimiri, "Concepts and Methods in Optimization of Integrated LC VCOs," *IEEE Journal of Solid-State Circuits*, Volume 36, pp 896-909, June 2001.

1-4244-0267-0/06/$25.00 ©2006 IEEE

[17] Chan-Hong Park, Beomsup Kim, "A Low-Noise 900MHz VCO in 0.6pm CMOS," in *Symposium on VLSl Circuits Digest of Technical Papers.1998*

[18] Dong-Youl Jeong et al., "CMOS current-controlled oscillators using multiple-feedback-loop ring architectures," *E Solid-State Circuits Conference, 1997. Digest of Technical Papers. 44th ISSCC., 1997 IEEE International , vol., no.pp.386-387, 491, 6-8 Feb 1997.*

[19] A. A. Abidi, "High-frequency noise measurements of FETs with small dimensions," *IEEE Trans. Electron Devices*, vol. ED-33, pp. 1801–1805, Nov. 1986.

[20] Marcel J.M. Pelgrom, Aad C.J. Duinmaijer, Anton P.G. Welbers, "Matching Properties of MOS Transistors" *IEEE Journal of Solid-State Circuits*, vol. 24, No 5, May 1989.

Lithography Solutions for a 0.35µm 25V PDMOS Technology

Brett Williams [1], Mike Thomason [1], Chuck Belisle [2], Bruce Greenwood [1]
[1] Technology Research and Development, [2] Fab10 Process Engineering
AMI Semiconductor, Inc.
2300 Buckskin Rd., Pocatello, ID 83201

Abstract— Lateral extended-drain MOS transistors (DMOS) are very sensitive to the well and Reduced Surface Field (RESURF) implant critical dimensions (CDs) as well as the layer-to-layer alignment (overlay). The photoresist that is used for the well and RESURF implants of the DMOS was originally highly dependant on reticle transmission (RT). This caused significant variability in the P-channel DMOS (PDMOS) performance, while the N-channel DMOS (NDMOS) performance remained stable. The lithography solutions to improve the robustness of the DMOS structures are discussed in this paper.

I. INTRODUCTION

THE I3T25 technology was developed at AMI Semiconductor to address the market for automotive, industrial, and medical applications (e.g., pacemakers, neurostimulators, etc.), requiring breakdown voltages > 25V. This technology belongs to a series of intelligent interface technologies developed within the company over the past years [1]-[2]. The technology is built on a 0.35µm CMOS base and incorporates specific isolation rules to allow for high voltages. The spacing of key dimensions such as implanted wells and fields are critical in producing a reliable, manufacturable device.

The I3T25 technology is built on n-type epitaxial wafers. The RESURF [3]-[4] implant for the PDMOS transistor is referred to as a P-channel field (PFLD) implant. The PFLD dopant is Boron. In the development of this technology, a new testchip (19427) was created using a reticle for the PFLD implant that only had very small openings for the PDMOS only. Prior development work in I3T25 was done using a 0.35µm CMOS technology testchip (19234), which had a reticle for this implant that had significantly less chrome.

When the first wafers of 19427 came out of the fab, it was soon discovered that the PDMOS structure did not function the same as it did on the 19234 testchip. The breakdown voltages (BVDSS) were 7-8 volts lower in magnitude, the on-resistance (RON) was 50 ohms lower, and the threshold voltage (VT_GM) was 50mV higher in magnitude on the same device. Weeks of investigation into the cause of this shift produced several theories. However, the one theory that endured all scrutiny and was ultimately verified as the root cause was the sensitivity of the PFLD mask step to reticle transmission (RT). Due to a large difference in RT between the two testchips at the PFLD mask step and also due to the low resolution resist that is used, a significant difference in PFLD CD was found to be the culprit in the shift of PDMOS

II. BACKGROUND

Initially the Boron doped p-type well (PWELL) implant (used for the NMOS transistors) was used in the PDMOS structure as the RESURF implant in the I3T25 technology. However, it was found that a unique implant dose was needed in the PDMOS structure to meet the target electrical characteristics. A PFLD implant was introduced to avoid changing the n-channel CMOS (NMOS) structures. The 19234 testchip was designed without a separate PFLD mask, so in order to do initial work on the PDMOS structures, a process flow change was needed. Since this flow does not contain a PFLD mask or implant step, the PFLD reticle is used in place of the PWELL reticle with the standard PFLD dose in place of the PWELL and NMOS threshold adjust (VTN) implants. This allows the gathering of electrical data from the PDMOS structures only. The standard NMOS structures and the NDMOS device will not have standard functionality.

Fig. 1 shows a cross-section of the PDMOS device. The PFLD area is below the green field oxide (FOX) to the right. In Fig. 2 the complimentary NDMOS device cross-section is shown. Notice that the PFLD implant is not used in the NDMOS structure.

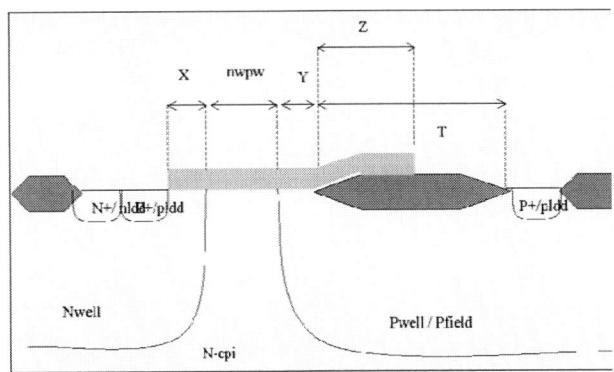

Fig. 1. I3T25 PDMOS Structure Cross-section. The RESURF implant is shown as Pwell/Pfield. The 19234 used the Pwell implant here while the 19427 used the Pfield implant. Note: the field oxide is green and the poly is gray.

1-4244-0267-0/06/$25.00 ©2006 IEEE

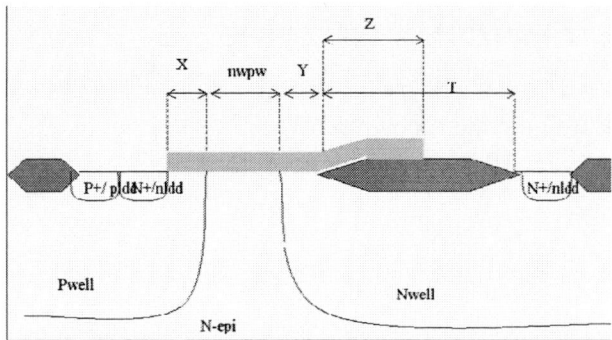

Fig. 2. I3T25 NDMOS Structure Cross-section. The RESURF implant is shown as Nwell.

Once the 19427 testchip was available, a new flow was developed, which has the PFLD implant module implemented so that both the NDMOS and PDMOS structures, as well as the CMOS, are functional on the same wafers. This certainly simplified the analysis process.

The PDMOS functionality difference was soon discovered on the initial wafer lots of 19427 that came out of fab. The data comparison with the 19234 testchip is shown in Table I. The 19427 BVDSS is < 25V in magnitude. For sufficient margin in this technology, it is desired to have the BVDSS > 30V in magnitude as in the 19234. As you can see, the RON and VT_GM values have also shifted. These results were thought to be consistent with a difference in dopant level under the field oxide (FOX) of the PDMOS device.

III. INVESTIGATION AND THEORIES

The known differences between the processing of the two testchips were the first to be investigated. Since each testchip needed a different process flow, these flows were carefully reviewed. All was in order.

The implant doses were checked to make sure that the new 19427 lots didn't receive a different amount of dopant at PFLD. The photoresist thickness of each lot was confirmed to verify that extra dopant was not bleeding through a thinner photoresist at PWELL implant. Secondary Ion Mass Spectrometry (SIMS) was done in an area of PFLD under FOX comparing the 19427 with the 19234 testchips to determine if there was a possible N-type epitaxial layer (N-EPI) or PFLD doping difference. The results of this SIMS can be found in Fig. 3. There is a slightly different dose level at the very surface that has not been fully explained.

The Etest data was also verified to make sure that this difference was not purely a testing issue. Additional structures were also reviewed. It was found that on a 19427 PDMOS with the n-type well (NWELL) intentionally pulled

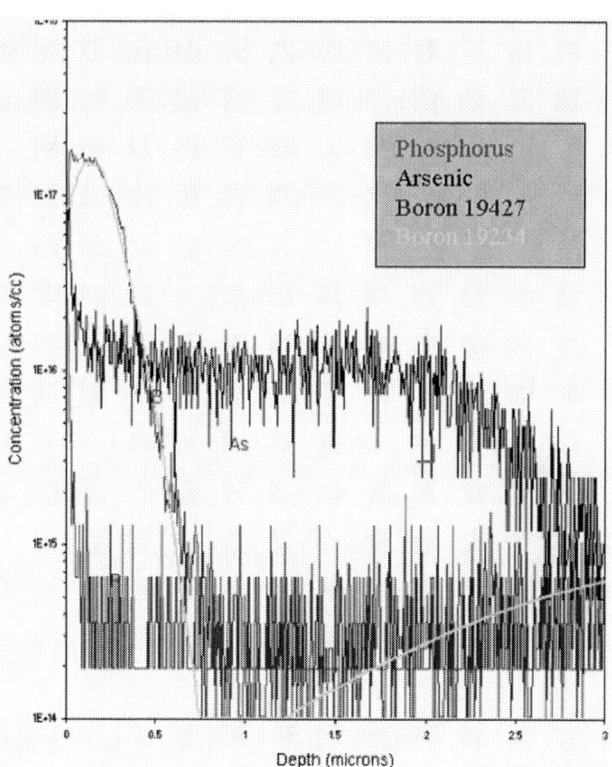

Fig. 3. SIMS comparison of Boron in the PFLD area under the PDMOS of the 19427 vs. 19234.

back to the left by ~ 0.2µm (see Fig. 1) the BVDSS and RON values seemed to match the 19234 standard device much better. This was an important clue to finding the root cause of this difference.

The reticle data (MEBES) from the two testchips was also reviewed and compared to see if there was a difference in the fracture that may have caused such a shift. Although the structures were not identical, since they were created at different times by hand, the critical measurements were as expected.

The PFLD opening was measured on MEBES for both testchips on the standard devices. Then shortloop wafers were processed in the fab for each of the masking steps in question (PFLD, NWELL, PWELL) for each testchip. These wafers were only masked, leaving the patterned photoresist. The wafers were then submitted for SEM cross-section. Once the data was reviewed, it was easy to see that there was a significant difference in the PFLD MEBES to photoresist SEM offset between the two testchips as seen in Table II.

The 0.4µm difference between the SEM and the MEBES for the 19427 PFLD openings means that the PFLD edge is actually 0.2µm further to the left (see Fig. 1) than the 19234. This would have a similar affect on the BVDSS as moving the

TABLE I
19234 AND 19427 TESTCHIP PARAMETRIC COMPARISON

Product	Structure	BVDSS	RON	VT_GM
19427	PDMOS18	-24.388V	481.211Ω	-0.674V
19234	PDMOS18	-31.759V	527.067Ω	-0.625V

TABLE II
COMPARISON OF PFLD OPENINGS IN THE TWO TESTCHIPS FROM SEM VS. DRAWN MEBES DATA.

Testchip	SEM1	SEM2	MEBES	Difference
19427	3.06um	3.13um	2.7um	~0.4um
19234	5.61um	5.73um	5.7um	~0.0um

NWELL edge to the left 0.2µm. It was stated earlier that a structure with the NWELL shifted had breakdowns that matched the 19234. It was apparent that the photo process at PFLD mask needed further investigation.

IV. PHOTO PROCESS REVIEW

The first thing that needs to be explained is that the PFLD reticle on the 19427 has a reticle transmission (RT) value of only 0.65%. This means that only 0.65% of the reticle is open and that 99.35% of the reticle is chrome. The PDMOS structure is the only place where this implant is needed. On the other hand the RT of the PWELL reticle that was used with the 19234 testchip is 74.52%. This means that there is a significantly higher amount of photoresist that needs to be exposed on the 19234. The exposed photoresist at the 19427 PFLD mask is very small and therefore it gets over-developed in comparison. This causes the PFLD opening to be much larger than expected as seen in Table II.

The photoresist being used for the PFLD and Ntub masking step is referred to as Prg-6. This is a very inexpensive resist with very little photoactive compound (PAC), resulting in a low-resolution resist that does not require much energy to expose. However, without much PAC the resist resin will actually develop slowly even if it is not exposed. Minimum resolution of this program is ~1.5um, with AMIS using it for only pad and large CD implant levels. The clearing energy (E0) of this program is ~110mJ. Resist thickness loss due to develop is ~700Å. This photoresist has been found to have a high variability in thickness uniformity and RT dependence.

The photoresist that has been proposed to improve this issue is Prg-3. This is a mid-grade resist that offers low exposure energy with decent resolution. This is our highest volume resist in the fab and is used for a majority of the transistor implant levels. Minimum resolution of this program is ~0.5µm, with AMIS using it for several metal levels at 0.6µm. The clearing energy (E0) of this program is ~90mJ. Resist thickness loss due to develop is ~450Å. The thickness used is only 1.6µm and has a much wider process window.

The Prg-3 photoresist is also much less susceptible to RT differences as seen in Fig. 4. This data matches well with

Table II in that at low RTs the difference in CD will be ~0.4µm between Prg-3 and Prg-6.

V. EXPERIMENTS

There have been four experiments, which have been used to verify both the short-term fix and the long-term solution to the problem addressed in this paper. The CDs referred to in this section are photoresist bars in the scribe line. Therefore, if the CD is larger, the photoresist opening for the implant is smaller. For example, since the PFLD photoresist opening on the 19427 testchip is too large (according to Table II) it will be necessary to increase the PFLD CD measurement.

Table III shows the design of the first two experiments. Both the 19234 and 19427 testchips were used. The TUBMSK process step is for the NWELL implant. The PTUBMSK is for PWELL implant. A photolithography focus exposure matrix (FEM) was included on these wafers in order to obtain various CDs. The details of the FEM is located in the comment section of Table III.

The most significant results of the experiment are that the PFLD CD does not as strongly affect the BVDSS on the 19234 as on the 19427 testchip (see Fig. 5). However, if the 19427 PFLD reticle were biased up by 0.4µm, the plots of the 19234 and 19427 BVDSS results in Fig. 5 would likely line up. Also, the 19234 data shows that if the PFLD CD is enlarged too much there will eventually be a reduction in BVDSS. Be aware that to obtain the highest photoresist CDs, the exposure energy is significantly reduced from 230mJ to just 90mJ. This causes scumming of the photoresist in the openings. The scumming may be the source of this decrease in BVDSS.

TABLE III
DESCRIPTION OF THE FIRST TWO EXPERIMENTS.

| | | | | | WAFER NUMBER | | | |
OP Name	Photoresist	Reticle	FEM	CD Target	1	2	3	Comment
TUBMSK	PRG 6	Rev A	Yes	2.05um	X			230mJ w/ 35mJ steps
	PRG 6	Rev A	No	2.05um		X	X	
PTUBMSK	PRG 6	Rev A	Yes	2.06um		X		230mJ w/ 35mJ steps
	PRG 6	Rev A	No	2.06um	X		X	
OP Name	Photoresist	Reticle	FEM	CD Target	1	2	3	Comment

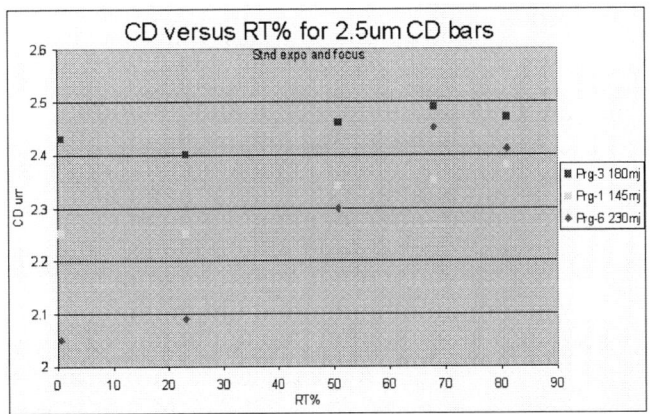

Fig. 4. Printed CD uniformity vs. reticle transmission % for 3 different photoresists. Prg-3 and Prg-6 are discussed in this paper.

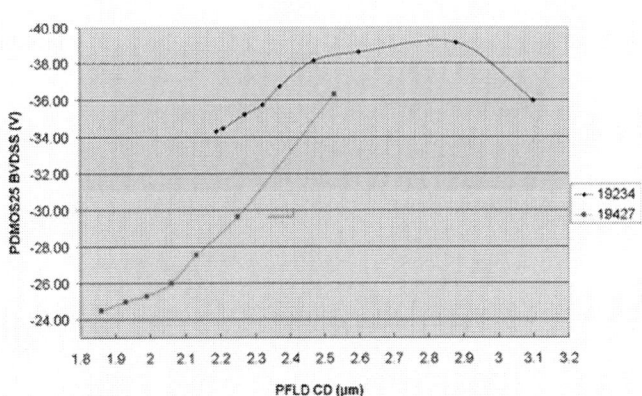

Fig. 5. PDMOS25 BVDSS results from 1st 2 experiments

Although there are > 30V BVDSS results from the experiment on the 19427 testchip, there is not enough resolution in the results. This is due to the fact that it is difficult to control the CD at such a low exposure energy (90mJ).

The third experiment used only the 19427 testchip. Table IV explains what was done. This time the new photoresist (PRG-3) was used as well as the current photoresist recipe (PRG-6). The PWELL mask was included in this experiment, which would affect the NDMOS structure (see Fig. 2). A FEM was repeated as described in Table IV.

In Fig. 6, the electrical results of this experiment indicate that using the PRG-3 photoresist at TUBMSK (NWELL mask) has a significant impact on PDMOS Vt and that at a

TABLE IV
DESCRIPTION OF THE THIRD EXPERIMENT.

U3V	19427-001			Purpose: Characterize new PR program for PFLD/I							
					WAFER NUMBER						
OP Name	Photoresist	Reticle	FEM	CD Target	1	2	3	4	5	6	Comment
TUBMSK	PRG 3	Rev A	Yes	2.35um			x	x		x	180mJ w/ 20mJ steps
	PRG 6	Rev A	Yes	2.06um						x	230mJ w/ 20mJ steps
	PRG 6	Rev A	No	2.06um	x	x					
PFLDMSK	PRG 3	Rev A	Yes	2.4um			x	x		x	180mJ w/ 20mJ steps
	PRG 6	Rev A	Yes	2.02um						x	230mJ w/ 20mJ steps
	PRG 6	Rev A	No	2.02um	x						
PTUBMSK	PRG 3	Rev A	Yes	2.45um	x		x				180mJ w/ 20mJ steps
	PRG 3	Rev A	No	2.45um						x	
	PRG 6	Rev A	No	2.35um			x		x	x	
OP Name	Photoresist	Reticle	FEM	CD Target	1	2	3	4	5	6	Comment

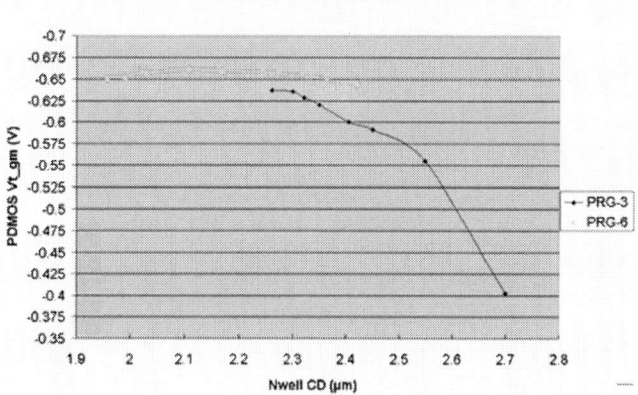

Fig. 6. 19427 PDMOS Vt_gm vs. Nwell Mask photoresist and CD

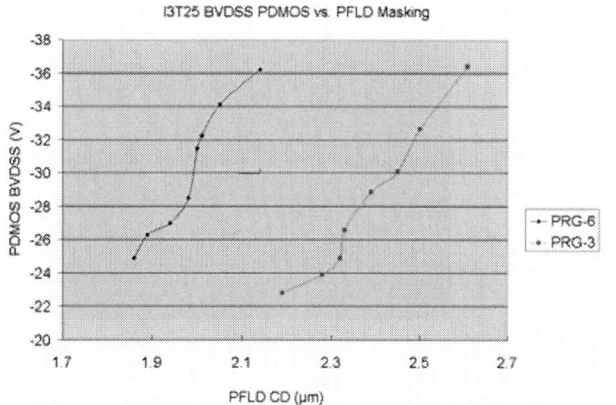

Fig. 7. 19427 PDMOS BVDSS vs. PFLD Mask photoresist and CD

TABLE V
DESCRIPTION OF THE FOURTH EXPERIMENT.

U3V	19427-001			Purpose: Characterize PFLD mask with ne							
					WAFER NUMBER						
OP Name	Photoresist	Reticle		FEM	CD Target	1	2	3	4	5	6	Comment
TUBMSK	PRG 3	Rev A		Yes	2 35um	x			x			
	PRG 3	Rev A		No	2 35um						x	
	PRG 6	Rev A		Yes	2 06um		x	x				
	PRG 6	Rev A		No	2 06um					x		
PFLDMSK	PRG 3	Rev A top/ Rev B bottom of wafer	No	2 6um			x			x	
	PRG 3	Rev B		Yes	2 6um				x			
	PRG 6	Rev A top/ Rev B bottom of wafer	No	2 4um	x			x			
	PRG 6	Rev B		Yes	2 4um		x					
OP Name	Photoresist	Reticle		FEM	CD Target	1	2	3	4	5	6	Comment

2.4μm CD, the Vt_gm matches the 19234 results in Table I. This is very promising. The results also show that with an increase of PFLD CD, the > 30V |BVDSS| results can be obtained on either photoresist program (see Fig. 7). The data also shows that there is an offset between the CD values of the PRG-6 photoresist and the PRG-3.

The BVDSS results from 2 wafers in the experiment were significantly lower than the rest. This can partially be explained by the BVDSS result from two PDMOS structures that have a 180° difference in orientation. The results are found in Fig. 8. The rotation increased the BVDSS by 4V on the average. Since the BVDSS is orientation dependent, there must be an overlay issue involved as well. The most likely overlay problem is between NWELL and PFLD implant for the PDMOS structure. This may have also caused the offset seen in the data from Fig. 7 as well.

The fourth experiment also used the 19427 testchip. Table V explains what was done. Both the new photoresist (PRG-3) as well as the current photoresist recipe (PRG-6) were used. The PFLD reticle for the 19427 testchip has been revised to bias down the openings by 0.2μm per side. This is referred to as 'Rev B' in Table V. By using the Rev B pfield reticle we were able to see the result of smaller PFLD openings without scumming.

Using the result of this experiment, we were able to model the behavior of the PDMOS BVDSS. From Fig. 1, the length of free N-EPI (no NWELL, no PFLD) is key in determining the drain diode breakdown. This in turn is determined by the relative NWELL and PFLD edge positions. The BVDSS is well modeled by the relation in (1).

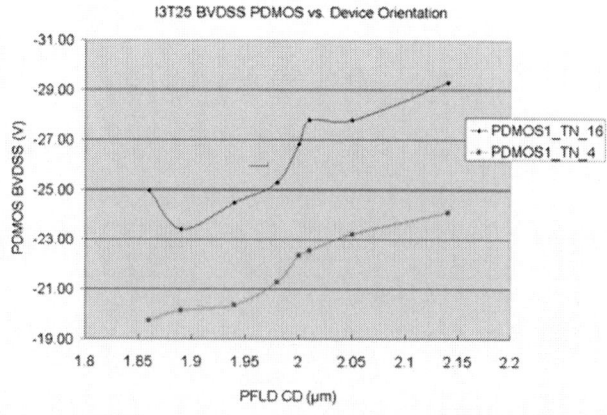

Fig. 8. 4V Difference in BVDSS depending on device orientation

$$BVDSS = 6.963*nwellcd-16.5*pfldcd-2.58 \quad (1)$$

Actual BVDSS data is plotted against the model prediction in Fig. 9. Note that the model saturates at –30V BVDSS, however, for the data in this regime, the NWELL resist is starting to scum. This results in lower n-tub doping and even larger bvdss, but randomly due to the variable nature of the resist scumming.

Also note that as the NWELL edge moves toward the source edge, the VT_GM of the device becomes lower in magnitude (see Fig. 10). This is due to lower doping at the NWELL edge. At the onset of scumming, the VT_GM drops markedly due to reduced dopant because of resist scum blocking the implant. (compare to Fig. 6).

NDMOS behavior was strongly influenced by NWELL CD (see Fig. 11). At the onset of scumming, NWELL CD no longer predicted BVDSS for the NDMOS. The new resist system made obtaining CDs in the 2.4µm range manufacturable, thus assuring good BVDSS control for the NDMOS.

VI. CONCLUSIONS

Selection of proper resist systems is critical in obtaining good process control of DMOS devices implemented in existing CMOS processes. Resist systems, which are adequate for CMOS processing, are not adequate when DMOS devices are added. Both NDMOS and PDMOS devices are shown to have high sensitivity to the NWELL and PFLD CD control. Implementation of PRG-3 resist system was done for PFLD, NWELL, and PWELL layers. PFLD reticles were adjusted to a higher CD to ensure process margin for the PDMOS. CD and overlay measurements at these layers were implemented, along with statistical process controls (SPC). The result of this was a manufacturable DMOS/CMOS process.

REFERENCES

[1] P. Moens, D. Bolognesi, L. Delobel, D. Villanueva, H. Hakim, S. Trinh, K. Reynders, F. De Pestel, A. Lowe, E. De Backer, G. Van Herzeele, M. Tack, "I3T80: a 0.35um Based System-on-Chip Technology for 42 Volt Battery Automotive Applications", *Proceedings of the ISPSD 2002 Symposium*, pp. 225-228.

[2] F. De Pestel, P. Moens, H. Hakim, H. De Vleeschouwer, K. Reynders, T. Colpaert, P. Colson, P. Coppens, S. Boonen, D. Bolognesi, M. Tack, "Development of a Robust 50V 0.35um Based Smart Power Technology Using Trench Isolation", *Proceedings of the ISPSD 2003 Symposium*, pp. 182-185.

[3] A. W. Ludikhuize, "Review of RESURF Technology", *Proceedings of the ISPSD 2000 Symposium*, p. 11.

[4] J. A. Appels, H. M. J. Vaes, "High Voltage Thin Layer Devices (RESURF DEVICES)", *IEDM*, p. 238-241, 1979.

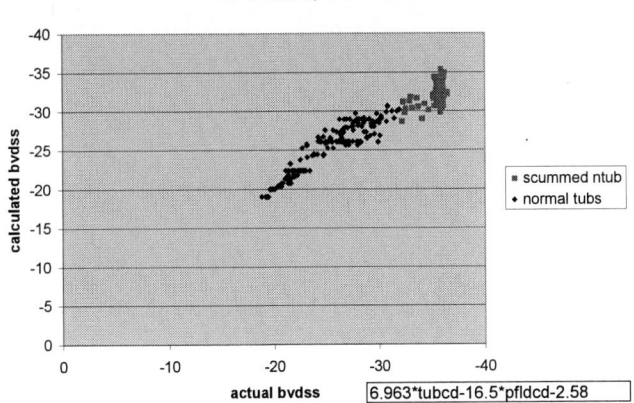

Fig. 9. Calculated vs. Actual BVDSS using Equation 1.

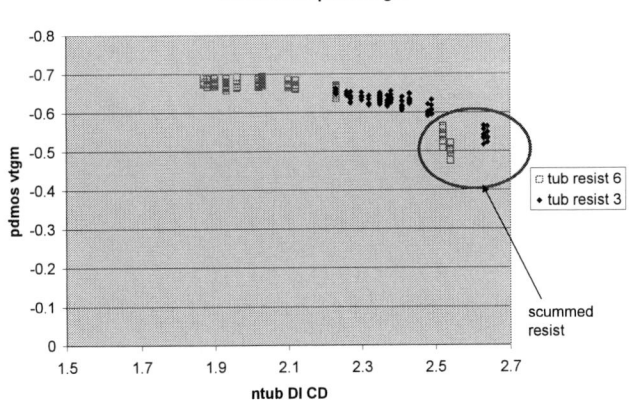

Fig. 10. PDMOS VT_GM vs. NWELL CD

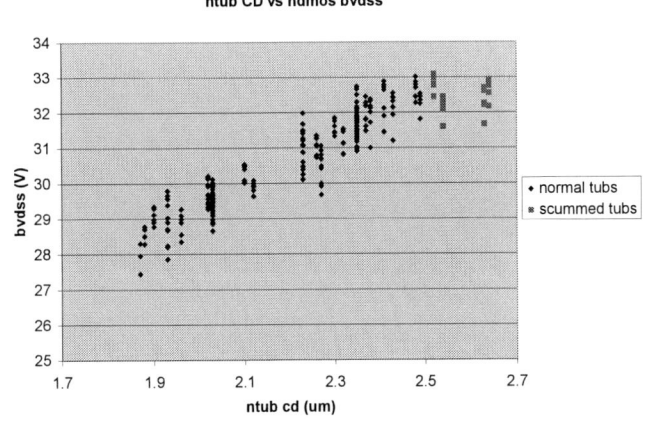

Fig. 11. NDMOS Bvdss vs. NWELL CD

1-4244-0267-0/06/$25.00 ©2006 IEEE

Advanced Process Simulation - Laser and Flash Annealing

Mark E. Law

Department of Electrical and Computer Engineering

University of Florida, Gainesville, FL 32611

Continuing a trend that began with RTA, anneals continue to shrink in time and increase in temperature. The latest developments are with either flash lamp anneals in the millesecond regime or laser anneals which can be as short as nanoseconds. Both of these anneal technologies use the wafer as a heat sink for the energy pulse provided to achieve rapid cooling. There can be no serious modeling of laser annealing until more complete details are known about the time-temperature profile during the laser anneal. Without this, any process simulation of defects, dopants, or strain is impossible. It is a critical first step for any further investigations.

From a modeling perspective, there are three key challenges in accomplishing of accurately modeling these anneal cycles. First, whenever the substrate is used as a heat sink, the spatial variation of temperature is important. Most process simulators do not allow for this. We have addressed this and it has been fixed in FLOOPS over the last two years. I will discuss techniques employed to minimize computational overhead in doing this. Second, the simulator must solve the heat flow equation to help generate the spatial and temporal variation of temperature. FLOOPS' scripting language is well suited to this task, and we have built preliminary models that help predict the onset of melting, for example. Finally, these models need to be calibrated. Can equilibrium models of heat transport be used at this time and length scale? How critical is carrier transport to the heat removal?

After the simulation capability is constructed, we then need to explore how these anneals interact with dopants and defects. Is the temperature variation critical? Do models breakdown at short times? Many of the existing models were developed based on longer time furnace anneals. For example, we have tuned boron cluster models to fit primarily the cluster dissolution kinetics. These fits and work tell us little in detail about how the clusters form, which is more likely to be important at short times. This also extends to extended defects. These extended defects are interstitial agglomerations that can control TED and activation. Preliminary experimental and simulation results will be presented from this regime that indicate where the pressing problems exist.

1-4244-0267-0/06/$25.00 ©2006 IEEE

6 Bit Decimation Filter in Sub-threshold Region

Ritu Jain, Pratibha Guttal, and D. W. Parent

Department of Electrical Engineering, San Jose State University, California

Abstract—We show the design of a 6-bit decimation filter with a decimation factor 4 operating in sub threshold region. We find that the optimum Wp/Wn ratio for an inverter operating in the sub-threshold was 24 as measure by equalizing NMOS and PMOS transistors drives. At the maximum achievable operating frequency of 50 KHz the circuit uses 2.5nW of power and occupies an area of 47μm X 100μm. TSMC 0.18um technology, with a supply voltage of 300mV was used.

Index Terms—CMOS, Subthreshold, Low Power

I. INTRODUCTION

In battery operated medical and wireless sensors applications, reduced power consumption is more important than the speed of the circuit. Sub-threshold operation is emerging as a promising technology for ultra low power applications [1].

Decimation filters are placed after a sigma-delta modulator to select a desired channel in the presence of channel interference and quantization noise from the digitization process [2]. The decimation filter is also used to reduce sampling rate. The decimation filter samples the incoming data until the sum of k inputs has been accumulated. Then, the sum is dumped into an output register. Through this process, the filter reduces (decimates) the output frequency by K times. Even though the sampling rate of a sub-threshold circuit is quite low, some of the metrics that remote sensor networks measure are even slower.

In this paper, a decimation filter under sub-threshold region is designed, which is substantially different from strong inversion operation of digital circuits. In sub-threshold speed requirements of the circuit are relaxed and the supply voltage is scaled down to well below the threshold voltage to minimize switching energy of the circuit. However, at low supply voltages and low clock frequencies, leakage energy dissipation can exceed active energy of the circuit, which leads to an optimal solution for the operating frequency and the supply voltage that minimizes energy consumption in the circuit.

This paper discusses an approach for sizing the circuit and the effect of reducing the supply voltage in sub threshold region. A comparison study is done for inverter, flip-flops and full adder (which forms the basic blocks) in static and dynamic logic. Individual blocks are choosen based upon the performance and the ease of design. In this paper spice simulation and layout is also shown.

II. ANALYSIS

In the sub-threshold region, choosing the transistor widths is substantially different from strong inversion operation. In the sub-threshold region, V_{DD} is less than the V_T of the transistor, gate leakage current is insignificant in comparison to sub threshold current and total current can justifiably be equal to sub threshold current. The basic equation (1) used to model sub-threshold current and total current is [3]

$$I_{SUB} = I_0 \exp((Vgs-V_T)/nV_{th}) \qquad (1)$$

I_0 is the drain current when $V_{GS}=V_T$, n is the sub threshold slope factor; V_TH is the thermal voltage, kT/q.

A. Modeling of Wp and Wn for inverter

For minimum energy operation, Wn is taken as 270n, minimum width possible in TSMC 0.18u technology. In sub threshold static inverter, due to low supply voltage V_{DD}, at Vin=0 current in PMOS is comparable to the leakage current in NMOS and inverter works more like a ratio circuit. By using a wider PMOS, we allow the PMOS current to overpower the NMOS for lower supply voltages.

In (Fast NMOS, Slow PMOS) FS process corner, PMOS is weakest and there for WP needs to be oversized to maintain proper operation [4]. To achieve the optimum width of Wp for fastest operation with respect to supply voltage, switching threshold is provided as $V_{DD}/2$. For Vin= Vout= $V_{DD}/2$, for a Wp, when Ip=In (for the circuit shown in Fig. 1), gives optimum Wp for which circuit operates at highest speed [1]. From our results (Fig 2), Wp/Wn ratio is coming as 24 for highest speed, while the results from the paper by Benton H. Calhoun et. al. show the ratio as 12 [1].

1-4244-0267-0/06/$25.00 ©2006 IEEE

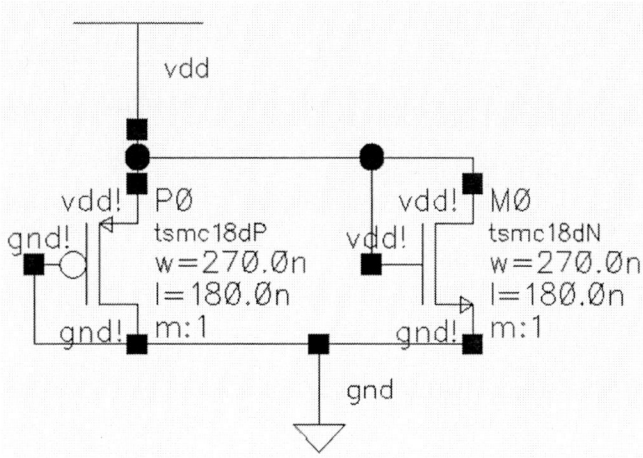

Fig. 1 Equivalent circuit of an inverter, when switching threshold is Vdd/2

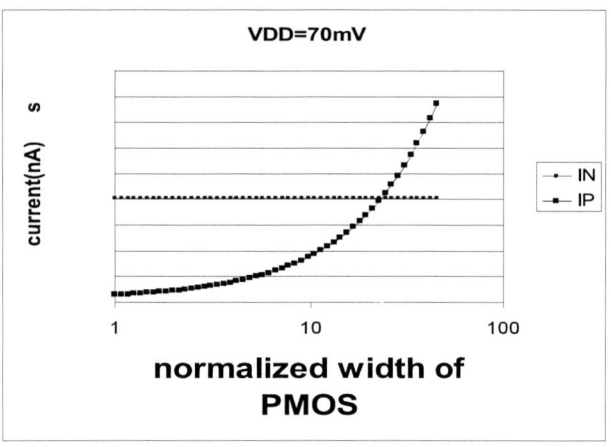

Fig. 2 Sizing of inverter for highest speed operation, when, switching threshold Vdd/2 and PMOS and NMOS are balanced (Ip=In). Vdd = 70mV.

In Figure 3, WP was swept from 270nm to 12um for steps in $V_{DD}/2$ from 70mV to .9V. The normalized WP value where $I_N=I_P$ was found taking the minimum value of the absolute value of the difference of the two currents. The W_P values were then normalized to 270nm. The observed steps are probably due to the BSIM model boundaries. V_{TN} was .34 Volts and V_{TP} was -.4 Volts. We can see that the W_P value that should produce an inverter switching threshold voltage of $V_{DD}/2$ becomes relatively constant as $V_{DD}/2$ approaches the threshold voltage of the devices. Using the results in Figure 3 to size W_P to have a switching threshold voltage of $V_{DD}/2$ results less than 6% error when extracting switching threshold voltage from a DC V_{IN} vs. V_{OUT} simulation with VDD varied from .14mV to 1.8V.

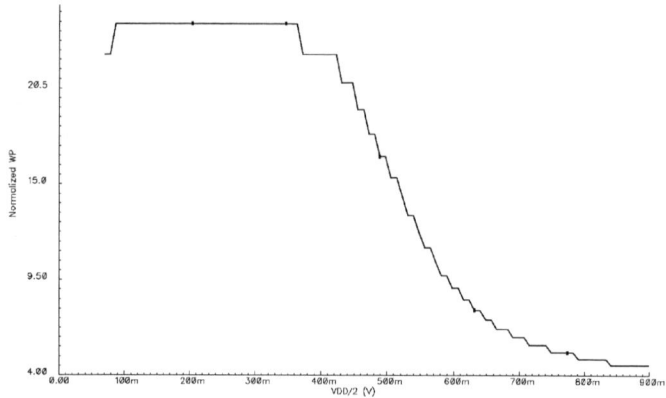

Fig. 3 Analysis of optimum Wp with respect to supply voltage for an inverter

B .Analysis of Wp/Wn Ratio vs. Vdd

If the supply voltage V_{DD} is scaled to the point where V_OH is only 90%of V_{DD} then we consider the circuit to be non-functional for that Wp/Wn ratio V_{DD} pair. Results are shown for Wp/Wn=4, where minimum possible V_{DD} is 120mV (Fig. 4). For safer operation we are using V_{DD} as 300mV.

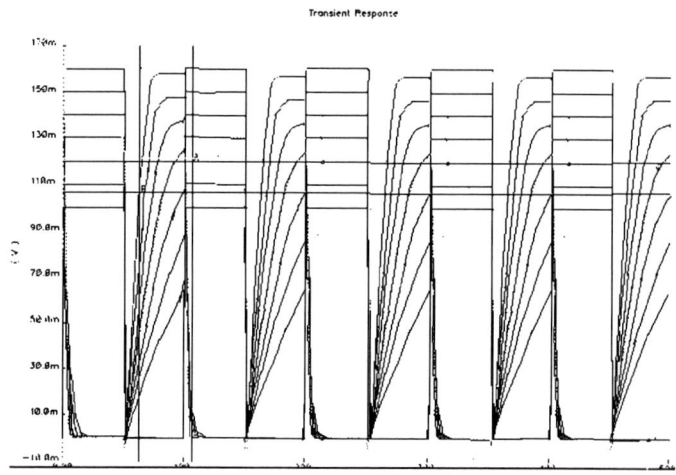

Fig. 4 Analysis of Vout vs. Vin for Wp/ Wn =4

III. COMPARISON OF STATIC VS DYNAMIC SUBBLOCKS

A. Comparison of Static Vs Dynamic Inverter in Sub-threshold

By comparing the average power dissipated in static vs. dynamic inverter for various ratios of Wp/Wn (Fig. 5), static inverter is better choice for minimum energy operation. If we compare, the propagation delay for static vs. dynamic inverter, τ_{phl} (propagation delay for high to low) for static inverter is much lower than dynamic inverter for various Wp/Wn values. τ_{plh} (propagation delay for low to high) for dynamic inverter is lower than static inverter. Since our aim is to reduce the average power consumed, static inverter makes a better choice.

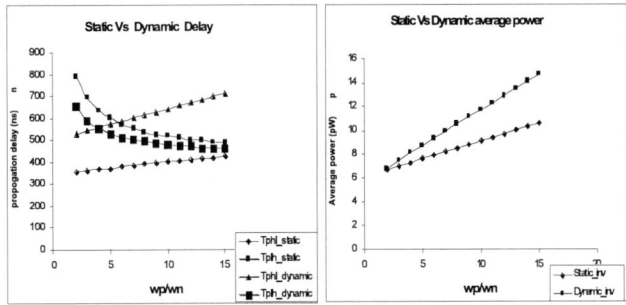

Fig. 5 Comparison of static & dynamic inverter.

B. Comparison of Static Vs Dynamic Full Adder

Two full adder blocks are compared, Static full adder using XOR and pass transistor gate [5] vs. Dynamic Full Adder using Pipelined NORA CMOS logic in Table 1[6]. Based on better performance and less area consumed, Static Full Adder with XOR and Pass Logic (figure 6) is used in the circuit [5]. This 18-transistor full adder alleviates the problems of threshold voltage loss and a non-zero standby power consumption. This circuit provides a full voltage swing (0 to V_{DD}) at all nodes and the signal level at the outputs are seen to be correct in all cases at a low supply voltage [5].

	Static	Dynamic
No of transistors	18	24
τ_{phl}/τ_{plh}(ns))	1030/322	1487,/8295
Avg Power (pw)	269	1692
Block/Logic	XOR, pass transistor	Pipelined /NORA CMOS

Table 1.Comparison of static vs. dynamic Full Adder.

C. Comparison of Flip Flops

A Static D Flip Flop with two transparent latches and a feedback transistor is compared with a Dynamic TSPC (True Single Phase clock) D Flip flop [6] in Table 2. Since the TSPC flip flop has better performance and uses less area, we used a TSPC flip flop in our design. Average power consumed for both types of flip-flops is hard to measure for only one gate.

	Static	**Dynamic**
No of transistors	20	11
τ_{phl}/τ_{plh}(ns)	1130/241	210/260
Block/Logic	Two transparent latches, weak feedback transistor	True single phase TSPC

Table 2 Comparison of Static vs. Dynamic D Flip-Flop

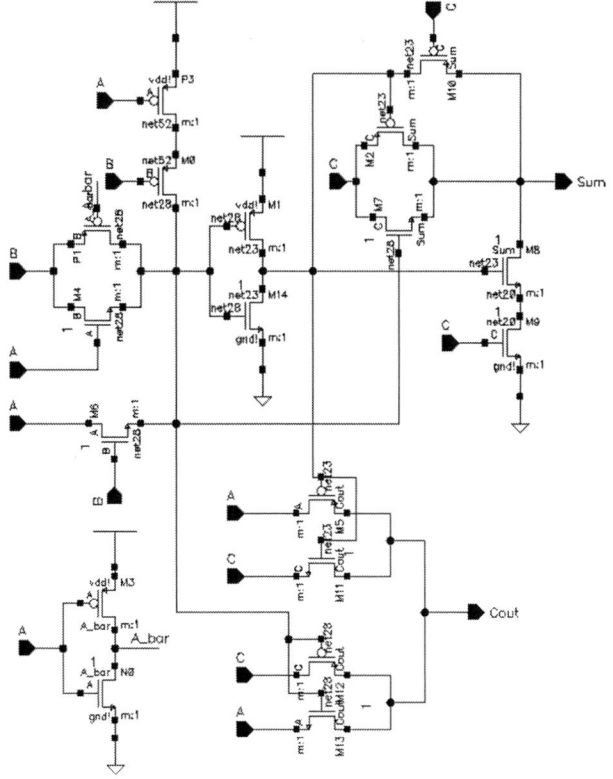

Fig. 6 Full Adder circuit with XOR and Pass gate logic.

IV. DECIMATION FILTER IMPLEMENTATION

The block diagram of 6-bit decimation filter for decimation factor of 4 is shown in figure 7. The main sub blocks used in the decimation filter are a full adder, TSPC D flip flop and TSPC D flip flop with reset. This decimation filter uses ripple carry adder to add 6-bit input data for 4 clocks and then sends the data to flip flops connected to output. the clock is routed to the output flip flops and has period 4 times than the period of clock at which input data is coming in. Once the output data is sent, reset pulse resets the flip-flops at the accumulator output. Hence it accumulates input data for 4 clocks and sends it to the output with a divided by 4 input frequency.

The reset pulse is delayed by using buffers from the F/4 clock going to the output Flip-Flops to prevent hold time violations.

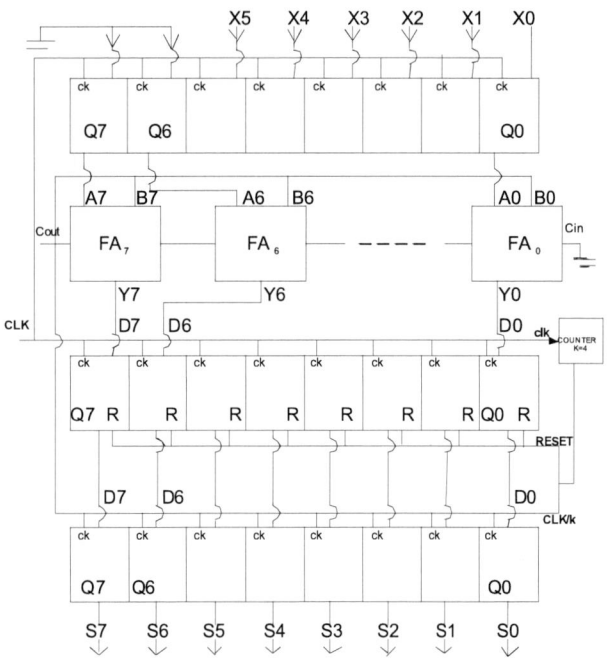

Fig. 7 Schematic of 6 bit Decimation Filter with decimation factor 4.

V. SIMULATION RESULTS

Simulation result is shown for the 6-bit decimation filter in Figure 8 for set up conditions as Wp/Wn ratio as 4 and power supply as 300mV. Worst case for the addition is considered when all the inputs are tied high and Cin is tied low with reset held high for the first clock cycle. Result for the longest path S7 is shown in the figure 8. Total average current drawn by the filter is 20.69 nA. The average power for the first 10 clock cycles is 2.5nW. (The power is measured using the technique described by Kang and Leblebici [8].)

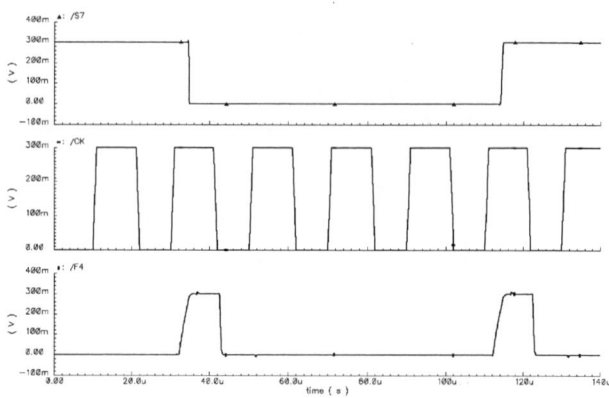

Fig. 8 Simulation results at the S7 output node of 6bit decimation filter.

VI. LAYOUT OF 6 BIT DECIMATION FILTER

In Fig 9, Layout of the decimation filter is shown. The circuit has an area of 47μm x 100μm area. It uses three metal layers M1, M2 and M3, M1 for horizontal connections, M2 for vertical connections and M3 for clock distribution.

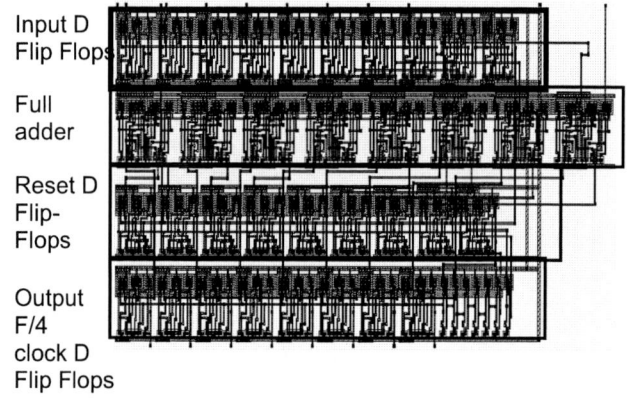

Fig. 9 Layout of the 6-bit Decimation Filter

VII. DISCUSSION AND CONCLUSION

In this paper, we have examined the operation of a 6-bit decimation filter in sub threshold region using TSMC 0.18u technology. Modeling of a static inverter in sub threshold region is discussed. Modeling of sub threshold circuit is different because of exponential dependence of current on voltage. In Sub threshold circuits, we should avoid a lot of stacks of transistors to account for V_T drop in each transistor. Thereby, static circuits have good performance. Our circuit is running at 50 kHz clock frequency and average power consumed is 2.5nW. Total area occupied is 47μm x 100μm.

1-4244-0267-0/06/$25.00 ©2006 IEEE

ACKNOWLEDGMENT

We would like to thank Kindness Israel for extensive IT support, Linda Shnell of Cadence Design Systems for access to the Cadence IC50 tool set.

REFERENCES

[1]   Benton H. Calhoun, Alice Wang, and Anantha Chandrakasan, "Device Sizing for Minimum Energy Operation in Sub threshold Circuits," *IEEE Journal of Solid-State Circuits,* vol. 40, no. 9, pp.1779-1786, September, 2005.

[2]   http://kabuki.eecs.berkeley.edu/~cjb/masters.pdf

[3]   Benton H. Calhoun and Anantha Chandrakasan, "Characterizing and Modeling Minimum Energy Operation for Sub threshold Circuits," *Massachusetts Institute of Technology, Cambridge, MA, 02139.*

[4]   [MASSACHUSETTS INSTITUTE OF TECHNOLOGY Department of Electrical Engineering and Computer Science 6.374: Analysis and Design of Digital Integrated Circuits, Problem Set #2 Solutions

[5]   [Hanno Lee and Gerald E. Sobelman, "A New Low-Voltage Full Adder Circuit", *Department of Electrical Engineering, University of Minnesota200 Union St. S.E.Minneapolis, MN 55455, U.S.A*

[6]   Jan M Rabey, "Digital Integrated Circuits, A Design Perspective", Prentice Hall

[7]   Neil H.E. Weste, David Harris, "CMOS VLSI Design, A circuits and Systems Perspective", *3rd Edition, Addison Wesley.*

[8]   Kang and Leblebici, "CMOS DIGITAL INTEGRATED CIRCUITS" *ISBN: 0072460539, San Francisco, McGraw-Hill (2002).*

1-4244-0267-0/06/$25.00 ©2006 IEEE

Frequency Multiplier Design Based on Multiple-Peak R-BJT-NDR Devices Fabricated by SiGe Technology

Dong-Shong Liang*, Kwang-Jow Gan, Chun-Da Tu,
Cher-Shiung Tsai, and Yaw-Hwang Chen

Department of Electronic Engineering, Kun Shan University
949 Da-Wan Rd., Yung-Kang City, Tainan County, 710 Taiwan, R.O.C.

E-mail: suln@mail.ksu.edu.tw

Tel: +886(6)2050194; Fax: +886(6)2050250

ABSTRACT-The negative differential resistance (NDR) device studied in this work is composed of the resistors (R) and bipolar junction transistors (BJT) devices. We regard this NDR device as R-BJT-NDR. Comparing to the resonant tunneling diode (RTD), this novel NDR device is made of resistors and transistors. Therefore, we can fabricate this NDR device by standard Si-based CMOS or SiGe-based BiCMOS process. A circuit with two NDR regions is obtained by combining two R-BJT-NDR devices in vertical integration. We can obtain two-peak I-V characteristics in the combined circuit. Circuit fabricated from the combination exhibits three stable operating points for frequency multiplier that can multiply the input signal frequency by three. The R-BJT-NDR device and frequency multiplier are implemented by the standard 0.35μm SiGe BiCMOS process.

KEYWORDS: negative differential resistance, R-BJT-NDR device, frequency multiplier, 0.35μm SiGe BiCMOS process

1. Introduction

In the past few years, several novel applications based on the negative differential resistance (NDR) have been developed [1]-[5]. Due to their folding current-voltage (I-V) characteristics, the NDR devices have high potential as functional devices. The most famous NDR device is the resonant tunneling diode (RTD). The fabrication of the RTD is often based on the technique of compound semiconductor. Comparing to silicon-based integrated circuit, the cost of compound semiconductor is more expensive. However, we proposed a NDR device that is composed of the resistors (R) and bipolar junction transistors (BTJ) devices. We call this device as R-BJT-NDR device. Because this device is totally composed of the R and BTJ devices, it is suitable for the process of Si-based CMOS or SiGe-based BiCMOS process.

If we fabricate this R-BJT-NDR device by standard Si-based CMOS process, the structure of BJT belongs to npn homojunction. However if we fabricate this R-BJT-NDR device by standard SiGe-based BiCMOS process, the structure of BJT will be heterojunction

Multiple-valued circuits based on more than two NDR devices have been extensively studied [6]-[8].

Compare to binary logic, multiple-valued logic has a great advantage to propagate more information. We will demonstrate a frequency multiplier based on the two-peak R-BJT-NDR I-V characteristics. The R-BJT-NDR device and frequency multiplier are implemented by the standard 0.35μm SiGe BiCMOS process.

2. R-BJT-NDR Device

The R-BJT-NDR device is made two heterojunction bipolar transistor (HBT) and five resistors, as shown in Fig. 1. During suitably arranging the values of the five resistors, we can obtain the I-V curve with NDR characteristics. Fig. 2 shows the simulated I-V characteristics by Hspice program. The I-V characteristics can be divided into four regions. The first segment (I) of the I-V characteristics represents a situation when Q1 and Q2 are cut off, the second segment (II) indicates the case when Q1 is cut off but Q2 is conducted, the third segment (III) refers to the state when both Q1 and Q2 are conducted, and the forth segment (IV) corresponds to the state when Q1 is saturated but Q2 is cut off.

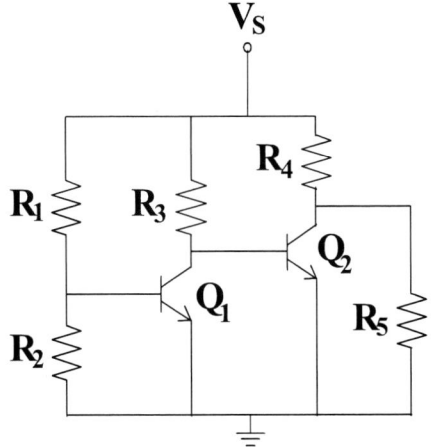

Fig. 1 The NDR device is composed of two transistors and five resistances.

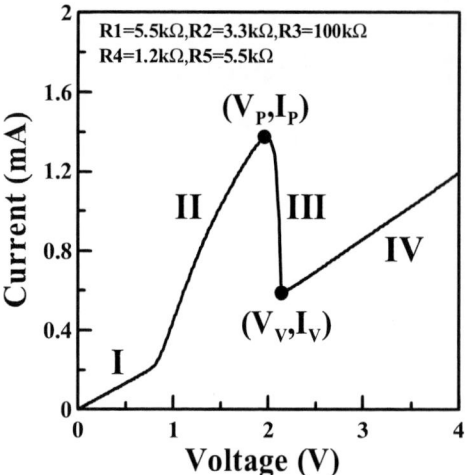

Fig. 2 The simulated I-V characteristics by Hspice program.

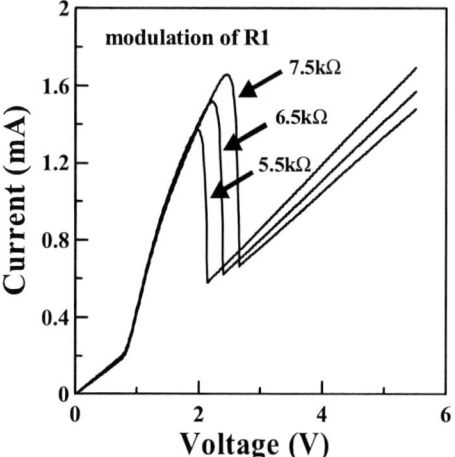

Fig. 3 The simulated I-V characteristics by modulating the values of R1 resistor.

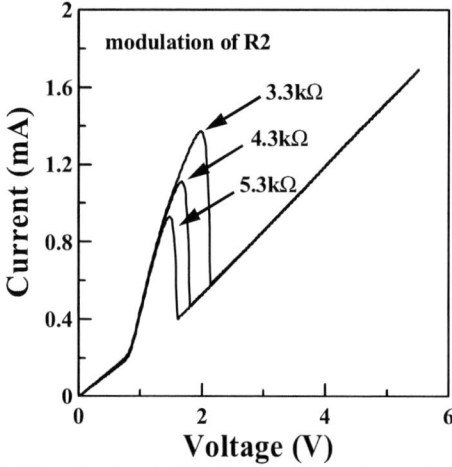

Fig. 4 The simulated I-V characteristics by modulating the values of R2 resistor.

The I-V characteristics of this R-BJT-NDR device can be modulated by the relative values of the resistors.

Figures 3 and 4 are the simulated I-V characteristics by varying the values of the resistors R1 and R2, respectively. By keeping the other parameters at some fixed values, the peak currents and voltages will be increased by increasing the values of the resistor R1. However, the peak currents and voltages will be decreased by increasing the values of the resistor R2. Based on the theory analysis, the peak voltage is approximated to the value of $(1+R1/R2)*Vr1$. The parameter $Vr1$ is the cut-in voltage of the transistor Q1. Therefore, this R-BJT-NDR device possesses the ability of adjustable I-V characteristics during suitably designing the resistance.

3. Mesured Results

The fabrication of this R-BJT-NDR device is designed based on the standard 0.35µm SiGe BiCMOS process. Fig. 5 shows the layout of the R-BJT-NDR device.

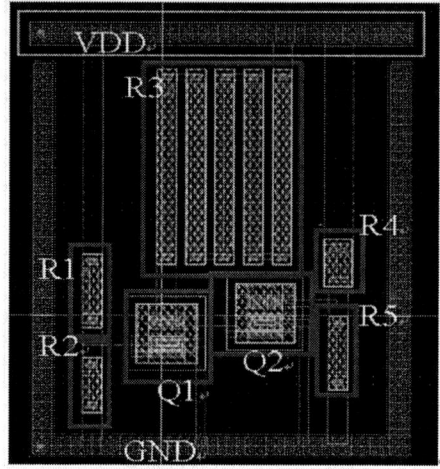

Fig. 5 The layout of the R-BJT-NDR device.

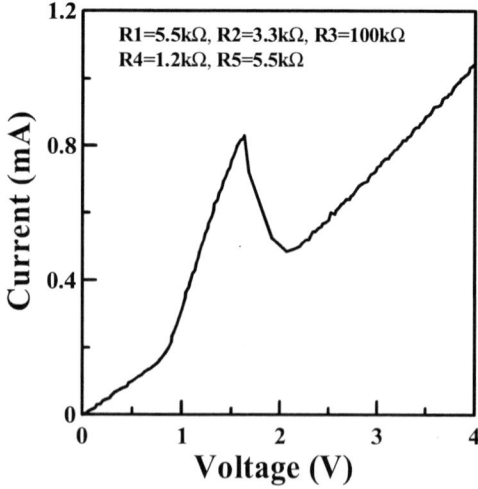

Fig. 6 The measured I-V characteristics of the R-BJT-NDR device.

Fig. 6 shows the measured I-V characteristics using the HP-4155C Programmable curve tracer. The parameters are designed as R1 = 5.5kΩ, R2=3.3kΩ, R3=100kΩ, R4=1.2kΩ, and R5=5.5kΩ. The peak current and voltage are about 0.82mA and 1.6V, respectively. The valley current and voltage are about 0.43mA and 2.1V, respectively. There is difference between the simulated and measured results. It is because of the error of the resistance in the IC design and fabrication.

Figure 7 and 8 show the measured I-V characteristics by varying the values of R1 and R2, respectively. Based on the measured results, this R-BJT-NDR device could exhibit the NDR characteristics by suitably controlling the parameters. The tendency of the measured I-V characteristics is satisfactory by comparing to the simulated results.

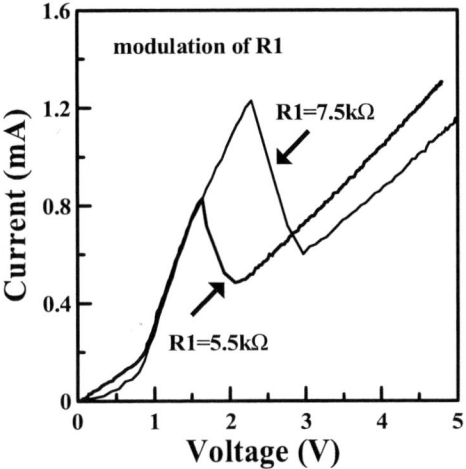

Fig. 7 The measured I-V characteristics of the R-BJT-NDR device by varying the R1 values.

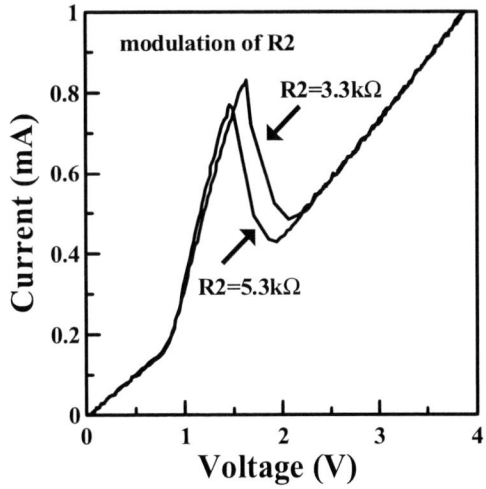

Fig. 8 The measured I-V characteristics of the R-BJT-NDR device by varying the R2 values.

4. Frequency Multiplier Design

There are many applications in signal processing for a device that exhibits more than one NDR region. To obtain the multiple-peak I-V NDR characteristics, we can connect two or more NDR devices in series. When two identical NDR devices are connected in series, there are two peaks and valleys in the combined I-V characteristics. The measured I-V curve for two vertically integrated R-BJT-NDR devices is shown in Fig. 9.

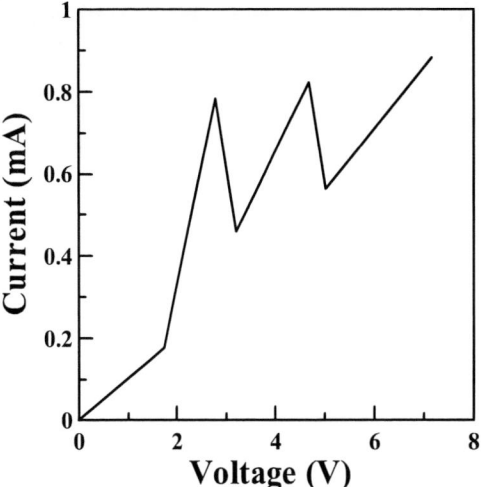

Fig. 9 The combined I-V characteristics for two identical R-BJT-NDR devices in vertical integration.

The two-peak I-V characteristics show that the individual R-BJT-NDR devices switched sequentially. This implies that the two devices have slightly different I-V characteristics. These two-peak characteristics could form three stable operating points. It can be used for signal processing applications such as frequency multiplier. An example of frequency multiplication is illustrated in Fig. 10.

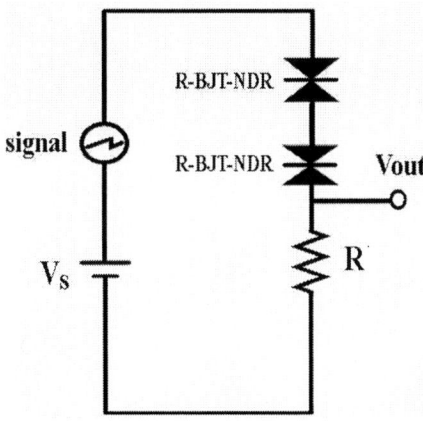

Fig. 10 Circuit configuration of the frequency multiplier.

When a sawtooth waveform was applied to the circuit, the output was also a sawtooth with a frequency three times higher than the input frequency, as shown in Fig. 11.

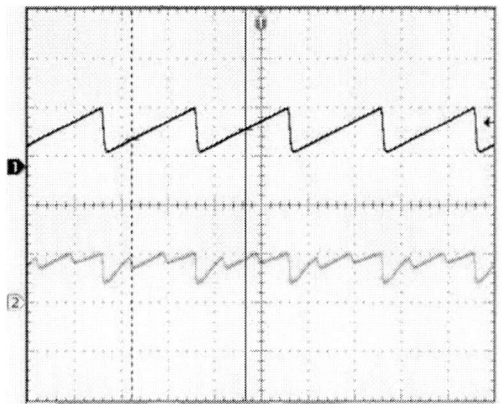

Fig. 11 The measured result of the frequency multiplier.

5. Conclusions

We have demonstrated the design and fabrication of the R-BJT-NDR device and a novel frequency multiplier based on the standard 0.35μm SiGe BiCMOS process. A frequency multiplier using two R-BJT-NDR devices and circuits can multiply the input signal frequency by a factor of three.

The modulation of the I-V characteristics of this R-BJT-NDR device is easier than that of the RTD device. We can obtain different NDR I-V characteristics by simply adjusting the values of resistors.

This R-BJT-NDR device is more convenient to integrate with other active MOS and BJT devices and passive R, L and C devices. These phenomena might provide the practical application in some NDR-based circuits. It means that the R-BJT-NDR device might have the potential to achieve the design of system-on-a-chip (SOC).

Acknowledgments

The authors would like to thank the Chip Implementation Center (CIC) for their great effort and assistance in arranging the fabrication of this chip. This work was supported by the National Science Council of Republic of China under the contract no. NSC94-2215-E-168-0012.

References

[1] S. Sen, F. Capasso, A. Y. Cho, and D. Sivco," Resonant tunneling device with multiple negative differential resistance: digital and signal processing applications with reduced circuit complexity,"IEEE Trans. Electron Devices, vol. 34, pp. 2185-2191, 1987.

[2] K. J. Chen, K. Maezawa, and M. Yamamoto, "InP-based high-performance monostable-bistable transition logic elements (MOBILE's) using integrated multiple-input resonant-tunneling devices", IEEE Electron Device Lett., vol. 17, pp. 127-129, 1996.

[3] K. Maezawa, H. Matsuzaki, M. Yamamoto, and T. Otsuji, "High-speed and low-power operation of a resonant tunneling logic gate MOBILE," IEEE Electron Device Lett., vol. 19, pp. 80–82, Mar. 1998.

[4] T. H. Kuo, H. C. Lin, R. C. Potter, and D. Shupe," A novel A/D converter using resonant tunneling diodes,"IEEE J. Solid-State Circuits, vol. 26, pp. 145-149, 1991.

[5] A. F. Gonzalez and P. Mazumder, "Multiple-valued signed-digit adder using negative differential-resistance devices", IEEE Trans. Comput, vol. 47, pp. 947-959, 1998.

[6] Z. X. Yan and M. J. Deen, "A new resonant-tunnel diode-based multivalued memory circuit using a MESFET depletion load," IEEE J. Solid-State Circuits, vol. 27, pp. 1198-1202, 1992.

[7] A. C. Seabaugh, Y. C. Kao, and H. T. Yuan, "Nine-state resonant tunneling diode memory," IEEE Electron Device Lett., vol. 13, pp. 479-481, 1992.

[8] J. P. A. van der Wagt, H. Tang, T. P. E. Broekaert, A. C. Seabaugh, and Y. C. Kao, "Multibit resonant tunneling diode SRAM cell based on slew-rate addressing," IEEE Trans. Electron Devices, vol. 46, pp. 55-62, 1999.

Studying the Etch Rates and Selectivity of SiO₂ and Al in BHF Solutions

Meow Yen Sim and Stacy Gleixner
Chemical and Materials Engineering
San Jose State University
One Washington Square, San Jose, CA 95192-0082

Abstract— One limitation to fabricating MEMS devices in some academic labs is the lack of polysilicon deposition technology. This limits the devices that can be fabricated because polysilicon is a common structural material in MEMS devices. In order to enhance the MEMS fabrication capabilities, expanding the use of Al as a structural layer has been researched. The main difficulty is the lack of a selective wet etch between the Al structural layer and the SiO₂ sacrificial layer.

In this study, seven etching solutions were studied on SiO₂ and Al for their etch rates and selectivity. Two were buffered hydrofluoric acid (BHF) solutions with different concentrations of NH₄F. Four were BHF solutions with propylene glycol or glycerin at different concentrations. The seventh solution was a commercial solution, Pad Etch 4.

Among the seven etching solutions, 5:1 BHF gave the best selectivity between SiO₂ and Al. Increasing the NH₄F concentration from 5 to 7 parts did not increase the selectivity, but selectivity increased by adding NH₄F in HF solution. Adding propylene glycol or glycerin to the 7:1 BHF solution did not increase the selectivity. When glycerin was added to 7:1 BHF solution it provided superior selectivity than was obtained from adding propylene glycol to 7:1 BHF. This paper will present the etch rates and selectivity of the 7 solutions along with comparisons with other published results.

I. INTRODUCTION

Microelectromechanical systems (MEMS) are integrated devices combining electrical, mechanical, fluidic, and optical components. MEMS range in size from micrometers to millimeters and are fabricated using integrated circuit batch-processing techniques. These devices show great promise as fabrication technologies increasingly enable the definition of smaller geometries, greater reliability, higher yields and lower cost per device. Current devices being manufactured include accelerometers, inkjet print heads, actuators, fluid pumps and pressure sensors. MEMS also show great potential for biomedical products. Medical diagnostics and drug delivery systems are being developed from submicron to nano scale, for a number of uses [1].

In MEMS fabrication, silicon, silicon dioxide (SiO₂), silicon nitride, polysilicon, and aluminum are commonly used [1]. Silicon was the earliest and is still the principal material used in MEMS fabrication. The main advantages of silicon are strength and lightness, with a yield strength of 7×10^9 N/m²

and density of 2.3×10^3 kg/m³. Silicon dioxide is frequently used as the sacrificial layer on the wafer in the etching step to fabricate MEMS microstructures. This material is most often chosen because deposition and etch processes for silicon dioxide have been extensively researched by the microelectronics community. Polysilicon has excellent mechanical properties and can be doped for various electrical applications, it's deposition and etch processes have also been extensively developed. Thus, it is the most common material used for the structural component in surface micromachined devices. Al, due to its high conductivity, is also sometimes used as an electrical component in MEMS devices. Al, which offers a cheaper and simpler deposition procedure, can also be used as a structural component when the high mechanical strength of polysilicon is not necessary.

San Jose State University's Microelectronics Processing Laboratory, as with many small academic labs, does not have the capabilities to deposit or etch polysilicon. To expand the MEMS capabilities of this lab, a procedure to make surface micromachined devices using Al as the structural component needs to be developed. Typical wet etching of the SiO₂ sacrificial layer cannot be done because the buffered HF solution etches the aluminum as well as the oxide. Literature results have shown that mixtures of HF and NH₄F have been effective in selectively etching SiO₂ and not Al [2-4]. The literature has not extensively quantified the etch rate and selectivity of these chemistries. The effects of chemistry and concentration of HF and NH₄F baths on the etch rate and etch selectivity will be quantified.

Williams *et al.* studied fifty-three materials, which could be used in the fabrication of microelectromechanical systems (MEMS) and integrated circuits [5]. These materials were etched with thirty-five different etchants to determine their etch rates. Among these etchants, Pad Etch 4 produced from Ashland and 5:1 BHF had the greatest selectivity of SiO₂ over Al. The Pad Etch 4 solution consisted of 11%-15% ammonium fluoride (NH₄F), 30%-34% acetic acid (CH₃COOH), 47%-51% water (H₂O) and 4%-8% propylene glycol (C₂H₈O₂) and surfactant. The etching process was conducted at room temperature. The etch rate of the SiO₂ layer deposited by wet oxidation was 310 Å/min and Al layer deposited by evaporation was 19 Å/min. The etch rate of a sputtered Al + 2% Silicon (Si) layer was less than 50 Å/min. The 5:1 BHF consisted of 5 parts of 40% NH₄F and 1 part 49% HF. The etching process was done at room temperature. The etch rate of SiO₂ deposited by wet oxidation was 1000 Å/min. The etching rates of evaporated Al and sputtered Al+2% Si layer were 110 and 1400 Å/min, respectively.

1-4244-0267-0/06/$25.00 ©2006 IEEE

Also, in comparing Williams *et al.* data for an HF etch versus a BHF (which has NH_4F added), the addition of NH_4F decreased the Al+2% Si etch rate and increased the oxide etch rate, thus increasing the selectivity.

Goosen *et al.* reviewed work on the selectivity between SiO_2 and Al in which the selectivity of SiO_2 was increased by adding glycerol or glycerin to a BHF solution [2]. According to Tilmans *et al.*, the standard HF solution would completely etch the Al layer [3]. By substituting some of the water with glycerol, the etch selectivity against Al was improved. The BHF solution with glycerol added consisted of 40g of NH_4F, 20ml HF, 40ml glycerol and 60ml of water. The etch rates of Al and the thermal oxide layer were 5.5 Å/min and 950 Å/min, respectively. The other chemical method discussed consisted of the addition of glycerin in a BHF solution. Gajda reported that by replacing water with glycerin in the BHF solution, many different types of glass could be removed without etching the metal layer [4]. This solution was composed of 4 parts 40% NH_4F, 1 part 48% HF, and 2 parts 87% glycerin. The etch rates reported were 60 Å/min for Al and 2000 Å/min for thermal oxide layer.

Bühler *et al.* investigated the etch rates of SiO_2 and Al with etching solutions of 7:1 BHF and Pad-etch solution [6]. The Pad etch solution consisted of 13.5 wt% NH_4F, 31.8 wt% acetic acid (CH_3COOH), 4.2 wt% ethylene glycol ($C_2H_6O_2$) and 50.5 wt% water. Three methods were used to deposit SiO_2 layers, wet oxidation, PECVD, and chemical vapor deposition (CVD). For the wet thermal oxidation method, the operating furnace temperature was at 1100°C. The etch rates of SiO_2 etched with the two solutions are listed in Table I.

TABLE I
THE ETCH RATES OF SiO_2 DEPOSITED IN THE DIFFERENT DEPOSITION METHODS AND ETCHED IN PAD-ETCH AND 7:1 BHF [6]

Materials	Pad-etch (Å/min)	7:1 BHF (Å/min)
SiO_2, thermal	220	620
SiO_2 BPSG	450	600
SiO_2, PECVD	700-1600	1200-1500

The Pad etch solution had a high selectivity for the SiO_2 layer and would not attack the Al layer [6]. The Al thickness removed from the Pad-etch after 5 minutes was less than 100 Å and after 30 minutes of removal, the thickness of Al removed was less than 200 Å. This gives an approximate etch rate of less than 10 Å/min. They did not report the actual BHF etch rates but reported that the etch rate was three times faster than the Pad etch solution.

II. PROCEDURES

The purpose of these experiments were to verify results found in multiple experiments in the literature in a single set of experiments using the same materials and conditions. The experiments were performed at San Jose State University's Microelectronics Process Engineering Laboratory using ten (111) silicon wafers. The wafers were initially cleaned using a standard RCA method. SiO_2 was grown on five of the wafers using a wet thermal process at 1100°C for twelve hours. This resulted in an oxide thickness of approximately two microns. 0.5 µm of Al was evaporated on to the other five silicon wafers. Part of the wafers were masked prior to evaporation to create a step in the Al layer (to measure with a profilometer).

Each wafer with the SiO_2 and Al films was cleaved into four pieces to reduce the amount of samples required for preparation; this resulted in twenty SiO_2 and twenty Al samples. Each quarter piece was dipped in seven different etchants, with two replicates per condition. The seven etchants used in this study are listed Table II.

TABLE II
RATIOS (IN PARTS) OF ETCH SOLUTIONS

Solution Number	Ammonium fluoride (NH_4F) (40%)	Hydrofluoric acid (HF) (49%)	Propylene glycol ($C_3H_8O_2$)	Glycerin (87%)
1	5	1	0	0
2	7	1	0	0
3	7	1	1	0
4	7	1	3	0
5	7	1	0	1
6	7	1	0	3
7	Pad Etch 4,manufactured by Ashland*			

Note: The content of Pad Etch 4 consists of 11%-15% NH_4F, 30-34%CH_3COOH, 47%-51% H_2O, 4%-8% $C_3H_8O_2$, and surfactant.

250 milliliters of each etching solution was used in the etch rate experiment. Both SiO_2 and Al samples were placed in the Teflon holders and etched simultaneously at various time intervals from two to five minutes. After each time interval, both samples were dipped in a Teflon beaker with distilled water for a quick rinse. Then, both of the samples were rinsed again with distilled water and sprayed dry with an air gun. The etching experiments were done at the room temperature. The replicates of the samples were etched with fresh baths and at the different time intervals.

A Nanometrics NanoSPEC 210 was used to measure the oxide thickness before and after etching. Three thickness measurements were taken from each quarter piece of the wafers. A Profilometer (Tencor P-1 long scan profiler) with five micrometers stylus was used to measure the step height and thus the thickness of the Al before and after etching with each solution.

III. RESULTS AND DISCUSSION

Table III summarized the SiO_2 and Al etch rates and selectivity obtained from Solutions 1 through 7. The SiO_2 and Al etch rates in Table III are the average etch rates of the six locations from the two runs at 4 minutes. The selectivity was calculated using the average SiO_2 etch rates of the six locations from the two runs divided by the average Al etch rates of the six locations from the two runs.

TABLE III
THE ETCH RATES OF SiO_2 AND AL AND SELECTIVITY BETWEEN SiO_2 AND AL IN SOLUTIONS 1 THROUGH 7 AT 4 MINUTES.

Solution	Material	Etch Rate (Å/min)	Selectivity
1 5:1 BHF	SiO_2	1416.3	11:1
	Al *	133.46	
2 7:1 BHF	SiO_2	937.17	8:1
	Al	115.33	
3 7:1 BHF + 1 part propylene glycol	SiO_2	750.71	4:1
	Al	201.58	
4 7:1 BHF + 3 parts propylene glycol	SiO_2 **	550.74	3:1
	Al	202.19	
5 7:1 BHF + 1 part glycerin	SiO_2	961.21	5:1
	Al	175.96	
6 7:1 BHF + 3 parts glycerin	SiO_2	653.42	6:1
	Al	111.2	
7 Pad Etch	SiO_2	383.48	4:1
	Al	95.08	

* Average etch rates of the three locations from Run 1 were taken at 5 minutes.

** Average etch rates of the three locations from Run 1 were taken at 3 minutes.

When comparing the 5:1 BHF with the 7:1 BHF, the SiO_2 etch rate decreased from 1416.3 Å/min to 937.17 Å/min, and the Al etch rate decreased from 133.46 Å/min to 115.33 Å/min. Increasing NH_4F concentrations in BHF solution decreased the selectivity between the SiO_2 and Al from 11:1 to 8:1. These results compare well with those of Williams *et al.* who found a 5:1 BHF solution etched wet SiO_2 at 1000 Å/min and evaporated Al at 110 Å/min [5]. The selectivity reported in Williams *et al.* was 9:1 for 5:1 BHF [5]. The SiO_2 and Al etch rates in Williams *et al.* indicated that adding NH_4F in HF solution increased the SiO_2 etch rate and decreased the Al etch rate. In Williams *et al.*, the etch rates of wet SiO_2 in 5:1 BHF was 1000 Å/min and 10:1 BHF was 500 Å/min which showed adding more NH_4F in BHF decreased the SiO_2 etch rate [5]. These experiments verify this; the SiO_2 etch rates in 5:1 BHF and 7:1 BHF decreased with the additional NH_4F.

Adding 1 part of propylene glycol to a 7:1 BHF solution decreased the SiO_2 etch rates from 937.17 Å/min to 750.71 Å/min but increased the Al etch rate from 115.33 Å/min to 201.58 Å/min. The increased Al etch rate and decreased SiO_2 etch rate caused the selectivity to decrease. The increased propylene glycol concentration in BHF solution from 1 to 3 parts decreased the SiO_2 etch rate from 750.71 Å/min to 550.74 Å/min and slightly increased the Al etch rates. The selectivity between the SiO_2 and Al decreased as the propylene glycol concentration was increased in 7:1 BHF solution. It was hypothesized that adding propylene glycol

would increase the selectivity. This is based on propylene glycol being an ingredient in some Pad Etch 4 solutions shown in the literature to increase the selectivity. However, this experiment showed that adding propylene glycol and increasing the concentration of propylene glycol in BHF solution did not increase the selectivity between SiO_2 and Al.

Glycerin is another component which was added to the 7:1 BHF solution in this study. By adding 1 part glycerin to the 7:1 BHF solution, the SiO_2 etch rate increased from 937.17 Å/min to 961.21 Å/min and also the Al etch rate increased from 115.33 Å/min to 175.96 Å/min. The increase of both etch rates decreased the selectivity from 8:1 to 5:1. By increasing glycerin concentration from 1 to 3 parts, the etch rate of SiO_2 decreased from 961.21 Å/min to 653.42 Å/min, Al etch rates also decreased from 175.96 Å/min to 111.2 Å/min. The decrease of both etch rates increased the selectivity between Al and SiO_2, which was 6:1. According to Goosen *et al.*, the selectivity between SiO_2 and Al was increased when two parts glycerin were added to 4:1 BHF [2]. This experimental results show the selectivity between SiO_2 and Al was not improved by just adding glycerin in 7:1 BHF solution, but the selectivity was improved with the increased glycerin concentration in 7:1 BHF solution. The mixtures in Solutions 5 and 6 have greater NH_4F concentration and glycerin concentration was only one part different from the etching solution reported in Goosen *et al*. The results in this experiment showed that the SiO_2 etch rate in 7:1 BHF with one part of glycerin was a little higher than the literature value and the SiO_2 etch rate in 7:1 BHF with 3 parts of glycerin was lower than the literature value. The experimental Al etch rates for Solutions 5 and 6 were much higher than the literature value.

Pad Etch 4 is another etching solution showing promising results in etching SiO_2 and not Al [2, 5, 6]. The experimental etch rates of SiO_2 and Al were 383.48 Å/min and 95.08 Å/min, respectively. The etch rate of SiO_2 from the experiment was close to Williams *et al.* which was 310 Å/min. For Al etch rate, the experimental value was very different from Williams *et al.* which was 19 Å/min [5]. The high experimental Al etch rate resulted in a low experimental selectivity compared to 16:1 for Williams *et al.* (using a wet oxide) and 22:1 for Buhler *et al.* (using a thermal oxide) [5, 6]. Repeat experimentation is needed to investigate this large discrepancy in the results.

IV. CONCLUSIONS

The selectivity and etch rates of SiO_2 and Al were studied with seven etching solutions. The experimental selectivity between SiO_2 and Al was 11:1 in the 5:1 BHF solution and the selectivity between SiO_2 and Al was 8:1 in 7:1 BHF solution. Increasing the NH_4F concentration from five to seven parts in a BHF solution decreased the etch rate of SiO_2 and Al, therefore, the selectivity decreased. The results were consistent with Williams *et al.* findings [5].

For 7:1 BHF solution with propylene glycol, experimental selectivity was 4:1 in the BHF solution with one part propylene glycol and was 3:1 in the BHF solution with three

1-4244-0267-0/06/$25.00 ©2006 IEEE

parts propylene glycol. The experimental selectivity for 7:1 BHF solution without any components added was 8:1. The results show that the selectivity did not improve by adding propylene glycol to the solution. The increase of propylene glycol concentration in the 7:1 BHF solution decreased etch rate of SiO_2 and slightly decreased the Al etch rate. Therefore, the selectivity between SiO_2 and Al was decreased.

Glycerin was another component that was added to BHF solutions to determine its effectiveness in etching. The selectivity was 5:1 in 7:1 BHF solution with one part glycerin and the selectivity was 6:1 in 7:1 BHF with three parts of glycerin. The selectivity of SiO_2 and Al obtained from 7:1 BHF with glycerin showed that adding glycerin to the BHF solution did not improve the selectivity. The experimental selectivity did not agree with those reported by Goosen *et al.*[2]. However, the selectivity depends on not just the glycerin concentration in BHF solution but also concentration of NH_4F in the HF solution. When comparing the different concentrations of glycerin in the BHF solution, the experimental selectivity improved by just increasing the glycerin concentration from one part to three parts in the 7:1 BHF solution.

For the Pad Etch 4 solutions, the experimental selectivity was 4:1 and Williams *et al.* reported 16:1 [5]. The experimental selectivity for the Pad Etch 4 solution was not the best among the seven solutions, but the selectivity obtained from Pad Etch 4 was better than the selectivity obtained from the BHF solution with three parts propylene glycol.

Of the SiO_2 and Al etch rates and selectivity obtained from the seven solutions, 5:1 BHF produced the best selectivity. Comparing the 5:1 BHF and 7:1 BHF solutions, the 5:1 BHF solution gave the best selectivity. Of the four 7:1 BHF solutions with one additional component, two solutions with propylene glycol and two solutions with glycerin, the 7:1 BHF solution with three parts of glycerin had the best selectivity between SiO_2 and Al.

REFERENCES

[1] R. R. Tummala, *Fundamentals of Microsystems Packaging*, San Francisco, CA: McGraw-Hill, 2001, pp. 544-547.

[2] J. F. L. Goosen, B. P. van Drieenhuizen, P.J. French, and R.F. Wolffenbuttel, "Problems of sacrificial etching in the presence of aluminum interconnect," *Sensors and Actuators A*, vol. 62, pp. 692-697, 1997.

[3] H. A. C. Tilmas, K. Baert, A. Verbist, and R. Puers, "CMOS foundry-based micromaching," *Journal of Micromechanical and Microengineering*, vol. 6, pp. 122-127, 1996.

[4] J. Gadja, "Techniques in failure analysis of MOS devices," in *Proc. 6th Micromechanics Europe Workshop (MME'95)*, pp. 183-187, 1995.

[5] K.R. WilliamsK. Gupta, and M. Wasilik, "Etch rates for micromachining processing- Part II," *Journal of Micromechanical Systems*, vol. 12, pp. 761-778, 2003.

[6] J. Buhler, F-P. Steiner, and H. Baltes, "Silicon dioxide sacrificial layer etching in surface micromachining," *Journal of Micromechanical and Microengineering*, vol. 7, pp. R1-R13, 1997.

Implementation of On-line Error Detecting, Constant Delay, Carry Free Adder

Yugandhar Asmath

Electrical Engineering, San Jose State University

yugandhar.asmath@gmail.com

Abstract:

This work is derived from the implementation of design proposed by Whitney J. Townsend, Mitchell A. Thornton and Parag K. Lala in 'On-line Error Detection in a Carry-Free Adder'. The circuit was implemented using AMI06 process to function at 200MHz clock. The observations made from the implementation were studied and a more generalized, simple and cost effective error detecting technique is proposed.

I. Introduction:

As the circuits are shrinking in accordance with Moore's law. The dimensions of VLSI circuits are shrinking in both size and voltage levels [3].

With the onset of nano-meter technologies [13], the vulnerability of circuits to faults is increasing manifold.

The faults could be caused by cross talk, power supply noise, alpha particles [4]. These faults will lead to errors in data flow, and ultimately corrupt the output.

The errors occurring during the data transmission or data storage could detected and correcting by introducing the parity information [14].

But the errors being induced due to the faults in combinational circuit are difficult to rectify because the data is dynamically changing as it goes through the circuit, and it is not possible to identify the errors [6]. Such kind of data can be checked for errors using redundancy [12].

Most commonly, we employ either time redundancy [4] or information redundancy. But either of them is not economical. The redundancy has to be reduced.

The proposed method [1] [2] uses redundant complimentary bits at each cell in a combinational circuit for concurrent error detection, and would require a re-transmission of data when an error is detected.

The concept was developed from the observations made by implementing an online error detecting carry free adder proposed in [1] & [2].

Adders are essential parts of every computer. Therefore the design of self-checking adders is of special interest.

The organization of the paper is as: Part II gives a basic background in adder arithmetic and error codes, Part-III explain the logic of the circuit that was implemented. Part IV gives an insight into the results and Part V gives the observations obtained from the implementation. Part VI gives the conclusion and proposes a more generalized method for error detection.

II. Background:

The Adder implemented follows a constant delay, carry free arithmetic using signed bit representation [11] with numerical encoding and online error detection [1] [2]. Carry free additions let the addition to be made in parallel and hence eliminate the issue of the carry propagation delay [1] [2][15]16].

The numerical encoding has always been of special interest in error detection as well as error correction. One such code used for error detection is m-out-of-n code, described in [10]. Single bit errors are detectable by the presence of non-code word in the data. Self checking circuits for m-out-of-n codes are described in [5][8][10][14]. A subset of such a code, where m is 1 and n is 3 is used for the design, as proposed in [1] & [2].

III. Methodology

An overview of the design, in terms of block diagram is given in figure.1 [1]. Two four bit signed bit binary representations were encoded into 1-out-of-3 code, as {-1(MSB), 0, 1} are encoded as {[100], [010], [001]}.

This would suggest that all the n-1 bits are encoded as either 010 or 001, and MSB bits as 010 or 100 for 0 and 1 respectively. [1], [2].

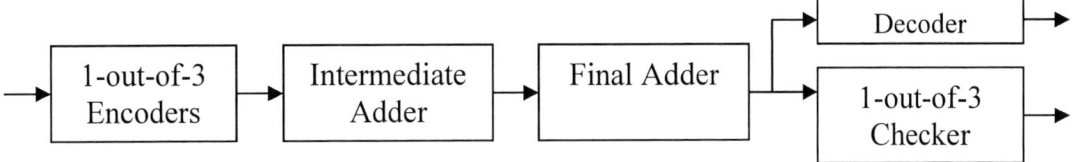

Figure.1 – Block diagram of On-line error detecting adder

IV. Implementation and Results

The combinational circuits that implemented the block diagram as suggested in [1] and [2] were developed for the encoder and a the two stage adder blocks.

The decoder was a straight extraction of the middle bit of the 1-out-of-3 code, hence did not require additional logic circuit. The checker circuit was however indigenously designed combinational circuit, and was not self checking in nature.

The circuit was implemented using AMI 06 process and was designed to function at 200MHz. Figure-2 shows the layout for the implemented circuit.

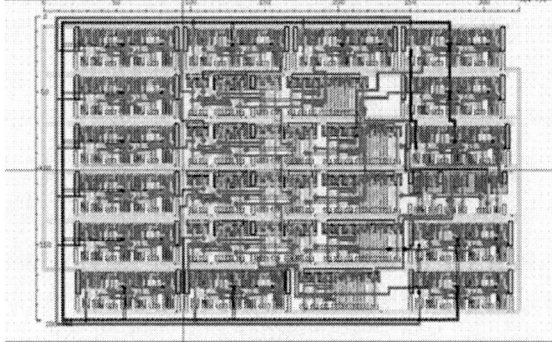

Figure. 2

The circuit after being tested for logic using NC-Verilog was laid out and simulated. The layout occupied 200x324 mm^2 and met the timing specifications with a very decent margin, taking just 3.4ns for a stable transition, including the delay due to the Flip-Flops.

V. Observations

i. Implementation observations

The junk of the circuit the increased the propagation delay in the circuit was due to the bulky checker circuit. A totally self checking circuit as proposed in [17] could possibly occupy lesser space and would perhaps be much faster than the combinational circuit employed in the implementation.

If the implementation would have changed accordingly much higher speeds, up to 200-250% the obtained speeds could be attained, using a circuit occupying a much lesser space.

The layout can easily be expanded to add words of any size with out increasing the complexity of the layout and design, while least affecting the timing of the circuit.

ii. Design observations

The logical circuit blocks actually used subsets of the code for the design i.e. the n-1 bits always used either of {010,001} for encoding and the MSB used {010,100}. The union of these two was a 1-out-of-3 code containing the super set of three valid code words, {100,010,001}.

VI. Conclusions

From the above design observation, we observe that, when representing an un-signed binary numbers, the 1-out-of-3 code cuts down to a 1-out-of-2 code, which actually is a two rail complimentary code. Totally self checking circuits can be designed for two-rail complementary codes. A circuit that detects the errors from such a code is described in [1-8, 13].

As seen from the implemented adder circuit, we can infer that it is possible to design on-line error detecting circuits using the two rail complementary code[4] for unsigned binary number system that occupy little space and are fast.

Further work necessitates implementing other complex arithmetic circuits using the two rail complimentary codes with totally self checking circuit embedded into the lay out. Those implementations would give more interesting results in tackling errors occurring in circuits due to stuck-at faults.

References

[1] On-Line Error Detecting Constant Delay Adder - Whitney J. Townsend, Jacob A. Abraham & Parag K. Lala, *Proceedings of the 10th IEEE International On-Line Testing Symposium (IOLTS'04).*

[2] On-Line Error Detection in a Carry-Free Adder- Whitney J. Townsend and Mitchell A. Thornton. Parag K. Lala, *Proceedings of Thirteenth International Workshop on Logic and Synthesis. June 2-4, 2004.*

[3] 'Cramming More Components onto Integrated Circuits' Gordon E. Moore, *Electronics, April 19, 1965.*

[4] C. Metra, M. Favalli, and B. Ricco, "On-line Detection of Logic Errors due to Crosstalk, Delay and Transient Faults," in *Proceedings of ITC*, pp. 524–533, 1998.

[5] Self-Checking Code-Disjoint Carry-Select Adder with Low Area Overhead byUse of Add1-Circuits - V. Ocheretnij D. Marienfeld E. S. Sogomonyan M. Gössel, *Proceedings of the 10th IEEE International On-Line Testing Symposium (IOLTS'04)*

[6] K. Mohanram and N. A. Touba, "Cost-Effective Approach for Reducing Soft Error Failure Rate in Logic Circuits," in *Proceedings of ITC*, vol.1,pp. 893–901, 2003.

[7] Evolution of fault-tolerant and noise-robust digital designs - M. Hartmann and P.C. Haddow.

[8] On the Design of Combinational Totally Self-checking 1-out-of-3 Code Checkers JIEN-CHUNG LO AND SUCHAI THANAWASTIEN. *IEEE TRANSACTIONS ON COMPUTERS, VOL. 39, NO. 3, MARCH 1990*

[9] A MOS Implementation of Totally Self-checking Checker for the 1-out-of-3 Code - D. L. TAO, PARAG K. LALA AND CARLOS R. P. HARTMAN, *IEEE JOURNAL OF SOLID-STATE CIRCUITS, VOL. 23, NO. 3, JUNE 1988.*

[10] D. A. Anderson and G. Metze, "Design of totally self-checking check circuits for *m*-out-of-*n* codes", *IEEE Trans. on Computers*, vol. 22, pp. 263-269, March 1973.

[11] A. Avizienis, "Signed-digit number representations for fast parallel arithmetic", *IRE Trans. on Electronic Computers*, vol. 10, pp. 389-400, September 1961.

[12] H. Edamatsu, T. Taniguchi, T. Nishiyama, and S. Kuninobu, "A

1-4244-0267-0/06/$25.00 ©2006 IEEE

33 mflops floating point processor using redundant binary representation", In *IEEE International Solid-State Circuits Conf.*, pp. 152-153 and 342-343, February 1988.

[13] 65 nm Technology for High Performance and Low Power, Mark Bohr, *in Intel Developer Forum, August 25, 2005.*

[14] P. K. Lala. *Self-Checking and Fault-Tolerant Digital Design.*Morgan Kaufmann Publishers, 2001.

[15] P. K. Lala and A.Walker. On-line error detectable carry-free adder design. In *IEEE International Symposium on Defect and Fault Tolerance in VLSI Systems*, pages 66–71, October 24-26, 2001.

[16] W. J. Townsend, M. A. Thornton, and P. K. Lala. On-line error detection in a carry-free adder. In *11th IEEE/ACM International Workshop on Logic & Synthesis*, pages 251-54, June 4 -7, 2002.

[17] On the Design of Combinational Totally Self-checking 1-out-of-3 Code Checkers, JIEN-CHUNG LO AND SUCHAI THANAWASTIEN, *IEEE TRANSACTIONS ON COMPUTERS, VOL. 39, NO. 3, MARCH 1990*

[18] J. C. Lo, "Novel area-time efficient static cmos totally self-checking comparator", *IEEE Journal of Solid-State Circuits*, vol. 28, pp. 165-168, February 1993.

[19] S. J. Piestrak, "Design method of a class of embedded combinational self-testing checkers for two-rail codes", *IEEE Trans. on Computers*, vol. 51, pp. 229-234, February 2002.

1-4244-0267-0/06/$25.00 ©2006 IEEE

Pixel Level Analog to Digital Converter

Nguyen Phong, Chung Joseph, Mariavanessa Pascua, and Scott Tarkul, Eric Vasham, and
David Parent
EE167 Microelectronics Manufacturing Methods
San Jose State University
One Washington Square, San Jose, CA 95192

Abstract— A semi custom design flow was used to implement and fabricate an analog circuit. The pixel level detector circuit was designed on a sea-of-gates called an Analog-Leaf-Cell. Cadence Tools was used to design the schematic, layout, and simulate the analog circuit. Once the layout and schematic has been verified (LVS) on Cadence Tools, a post extraction simulation is observed. When all specifications have been reached, the circuit design is ready to be fabricated and is sent out to MOSIS. Eight weeks later, the integrated circuit is fabricated and packaged into an IC chip and returned to students to be tested.

I. INTRODUCTION

A semi custom design flow is a design process which uses prefabricated transistors, called an Analog-Leaf-Cell (ALC). The ALC design approach is used to reduce design and fabrication time. Since a full custom design flow is time consuming, an ALC Pad Frame for layout is created on Cadence Tools. The pre-existing layout of transistors is used for the designer to quickly design analog circuits by routing one metal layer. The ALC Pad Frame consists of 8 PMOS and 8 NMOS banks (top and bottom of Pad Frame) which shares ten metal bus lines in between the banks of the transistors. Each bank consists of ten transistors with a width and length of 6.4 um.

II. PIXEL LEVEL DETECTOR DESIGN FLOW

The first step in implementing the pixel level detector circuit is to draw out the schematic, symbol, and test bench on Cadence Tools; which is called Schematic Capture. The schematic of the circuit is seen in Figure 1. In the Schematic Capture, the circuit is simulated to make sure it functions properly and meets specifications. The test bench of the circuit can be seen in Figure 2. Some widths and lengths of the PMOS and NMOS transistors are changed to meets specification.

Once the Schematic Capture is simulated and checked for specifications, a step called Pre-Layout is used to plan how to wire the circuit together in layout on an ALC Pad Frame. This step uses the bubble schematic of the leaf cell to determine which horizontal bus lines are used for Vdd,

Ground, inputs, and outputs; also which transistor gates are connected vertically on a particular bus line. The bubble schematic of the circuit is drawn on paper which acts as a reference guide when connecting metal on the ALC Pad Frame. See Figure 3 for the final layout of the pixel level detector circuit.

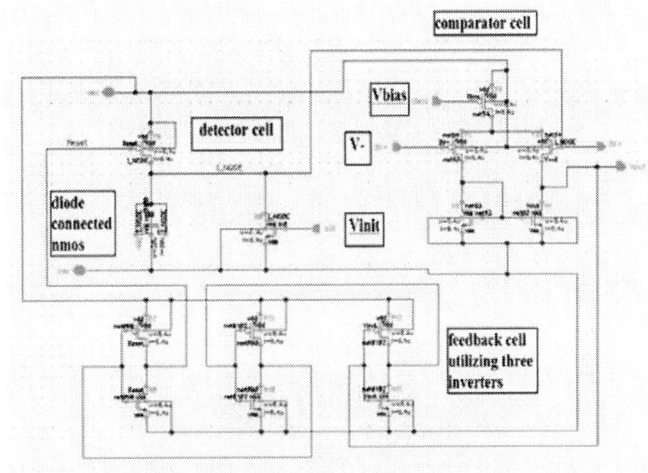

Figure 1: schematic of Pixel Level Detector circuit

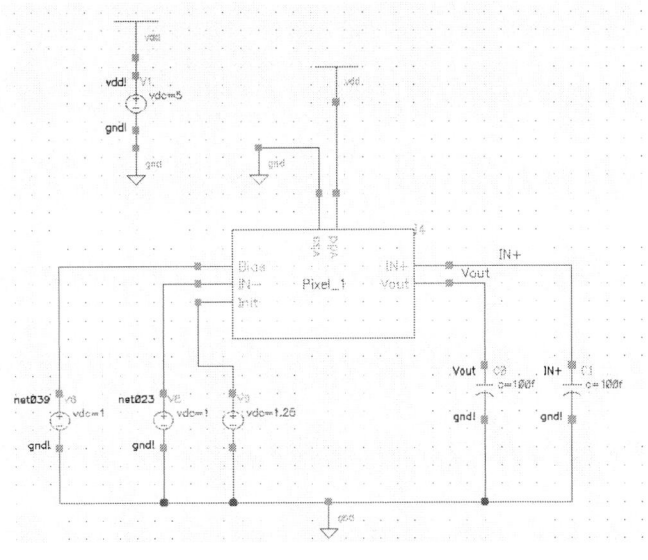

1-4244-0267-0/06/$25.00 ©2006 IEEE

After the circuit has been laid out, a design rule check (DRC) will be ran to make sure the circuit does not have any errors once it's been fabricated. With no DRC errors, a Circuit Extraction step is used to check for the electrical equivalence of the circuit. After having the extracted view of the circuit, a layout verses schematic check (LVS) is ran to verify if the electrical properties of the layout matches with the schematic of the extracted view. Once the LVS report shows that the extracted view of the circuit matches with the layout, the circuit is ready to be sent out for fabrication.

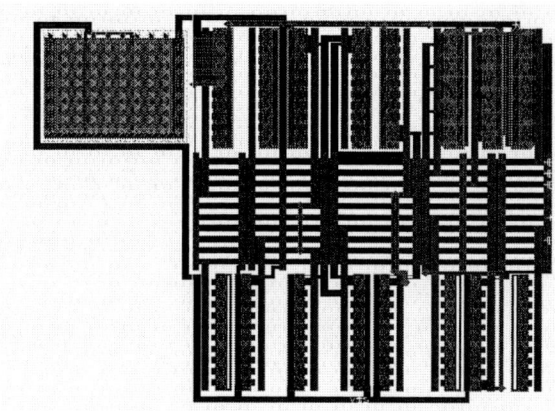

Figure 3: Final layout of pixel level detector circuit

III. FABRICATED CIRCUIT

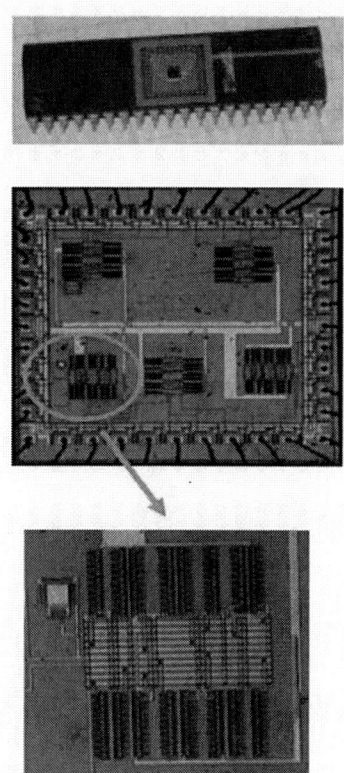

Figure 2: test bent of Pixel Level Detector circuit

IV. BACKGROUND

A pixel level detector is basically transforming light waves into electrons. This basic circuitry is a building block in today's digital camcorders and cameras. Instead of film, a camera uses a piece of silicon to convert light to electrical charges, which is used to catch a particular image. The image sensor measures the light intensity which gives off the different variations of contrast; a brighter image, a higher electrical charge, and a darker image, a lower electrical charge.

There are two kinds of image sensors being used today in digital camcorders and cameras, charge-coupled device (CCD) and CMOS (complementary metal-oxide semiconductor). When shopping for a new digital camcorder or camera, pros and cons can be observed. The CMOS image sensor tends to be popular since it's less expensive to manufacture and has lower power consumption.

V. TESTING

In order to test for frequency dependency upon variation of light intensity, the following pins are needed in the set up: Vbias, V-, and Vinit. In addition, Vdd (+5V) and gnd are needed. Figure 1 shows the pins relative to the transistors.

Vbias is used to adjust the tail current for the top PMOS transistor of the comparator. The operational start time of the comparator changes according to the corresponding Vbias. The Vbias adjusts gate voltage in the mentioned transistor that has a 2V threshold voltage; therefore, the Vbias is set to a reasonable voltage of 1V allowing enough tail current for the operation of the comparator.

The V- is the reference voltage which adjusts the operating range for the comparator. As the V- voltage is adjusted lower, further below the threshold voltage, the output swing becomes wider. The threshold voltage for the particular transistor is 2.8V; therefore, the V- is set to 1 V, creating a reasonable output range, while ensuring the PMOS is on.

The Vinit is used to calibrate the frequency for the circuit. When Vinit is increased further above the NMOS threshold voltage, the transistor's drain current is increased. Because the photodiode and the NMOS transistor (for which the Vinit controls) are in parallel, increasing the transistor's drain current will quicken the overall current drain time, which increases the integration time. The threshold voltage for the particular NMOS transistor is 1V; therefore the Vinit is set to 1.25V ensuring a minimum amount of drain current through the transistor allowing effects for increase of light intensity for the photodiode to be seen.

Using the DC values mentioned, the simulated SPICE test is shown in the following figure 4.

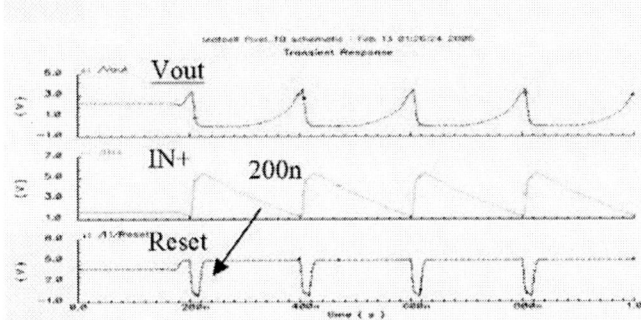

Figure 4: SPICE simulated results – Output, IN+, and Reset waveforms

In order to test the pixel level ADC's circuit dependency on light intensity, a variable light source is used. Because light intensity is dependent on the distance from the light source, the distance is fixed. The equipment that is used is a Leica model 13410311, which has four variations of intensity, setting 0 being off, then, 1, 2, and 3. Figure 5 and figure 6 shows the result of the two extremes, in setting 0 (off) and in setting 3 (maximum intensity).

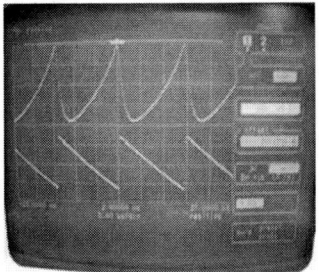

Figure 5: Setting 0, "off"   Figure 6: Setting 3, "Maximum"

Testing all the available settings resulted in the following results as shown in figure 2.

Light Intensity (knob setting) vs. Frequency

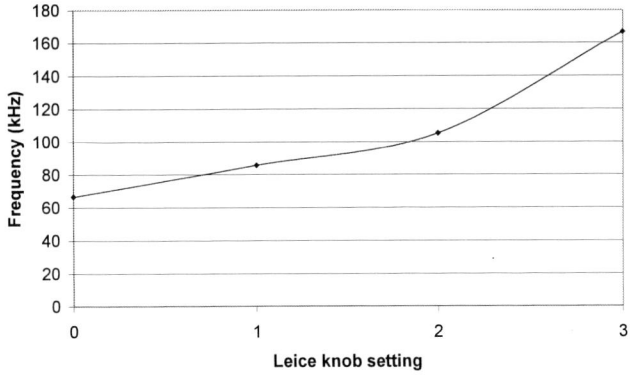

Figure 6: Light intensity vs. Frequency

V. ANALYSIS

The three waveforms in figure 4 are oscillating at the set DC voltages. The best way to describe what is going on is to start time t = 200n, when the circuit (figure 1) begins to oscillates. At this time, reset is low, which pulls the detector cell's output to Vdd. This output is fed directly to IN+ (comparator cell). The high voltage level turns the pmos transistor off, which

pulls down Vout (comparator cell) to Vss. The low voltage Vout signal is fed through the feedback cell, which inverts the signal. The reset pin at about 225n sees this high voltage signal. The high voltage level turns the pmos of the detector off. Once the pmos is off, the nmos can do its job in pulling the output to Vss. Since the size of the nmos is the minimum and the gate voltage is small, the output is pull down slowly about a 200n period. Halfway through, however, one can see Vout is slowly rising. This is evident in the fact that Vout's voltage level is control by IN+, which comes from the detector's output. Once IN+ reaches 1V, Vout reaches its highest level at about 400n. Finally, Vout is inverted again and sent to reset at about 425n, and the cycle repeats.

As shown in figure 6, the frequency increase as light intensity increases. The plot of the light intensity by stepping the knob setting relative to the frequency did not come out linear. The light intensity directly affects the diode, which is a non-linear component. The number of points that were attained through experimentation was done with equipment with limited settings, which is why a continuously variable light source is needed for future experiments. But overall, the results showed that the circuit did work in that the frequency of the circuit clearly shows a dependency on the intensity of the light source.

VI. CONCLUSION

A pixel-level detector was developed under a semi-custom IC design environment. Utilizing an existing analog leaf cell, the circuit was layout, fabricated, and tested for proper functionality. The results of the test were successful. Under "no-light" conditions, the circuit oscillated at a set frequency. With light, the frequency increased as expected. Furthermore, the outputs were a pulse and corresponding saw-tooth, which was desired. Therefore, the final results of the project were a success.

ACKNOWLEDGMENT

SJSU IC DESIGN LAB
DR. DAVID PARENT
ERIC BASHAM
CADENCE TOOLS LAB

REFERENCES

Kang, S. and Leblebici, Y. (2005). CMOS Digital Intergrated Circuits, (3rd ed., chap 1,2,3). International: McGraw Hill.
Streetman, B., and Banerjee, S. (2000). Solid State Electronic Devices. (5th ed., chap 5). Upper Saddle River, NJ: Prentice Hall.
Dr Parent Website. lecture notes ee166
Retrieved Spring 2006 from
http://www.engr.sjsu.edu/~daprent/ee166
Eric Basham Website. project notes: pixel-level
Retrieved Spring 2006 from
http://engr.sjsu.edu/dparent/dans_project

1-4244-0267-0/06/$25.00 ©2006 IEEE

1-4244-0267-0/06/$25.00 ©2006 IEEE

Improvement of a 4-mask Process Recipe

Kuang-Wai Tseng, Mariavanessa Pascua, Taslima Rahman,
Siu Kuen Leung, and Scott Echols
EE 198: Senior Design Project
San Jose State University
One Washington Square, San Jose, CA 95192

Abstract— **A fabrication process used by SJSU was altered in order to lower the threshold voltage, VT. This was done to allow the fabrication process to be used for analog integrated circuit design for future EE classes. Wafers fabricated from previous semesters using the same process recipe were measured to have about a 4-5 V threshold voltage on average. The Athena process simulator was used to observe VT for different times and temperatures of the substrate diffusion step of boron. After simulations, the times for substrate diffusion used to carry out the fabrication process were chosen to be 180, 240 (unchanged), and 300 minutes while the temperature remained unchanged. For the 180, 240, and 300 minute diffusion times, Athena measured VT to be 1.21, 1.16, and 1.12 V respectively. The fabrication process was then carried out for the three different diffusion times in SJSU's IC fabrication lab. Once fabrication was completed, the wafers were measured for their alignment, gate oxide thickness, transconductance, channel length modulation, body effect, effective length, and threshold voltage. From the fabricated wafers the average VT measured for each diffusion time was 3.54, 3.08, and 2.15 V for 180, 240, and 300 minute diffusion times, respectively. When extracting Vt, the oxide interface charge, Qi, may have been assumed to be too high for the simulations (5×10^{11} cm^{-3}). These results show that the EE/MatE 129 recipe may need to use a dopant with a smaller boron concentration or the boron diffusion step before the growth of the field oxide may need to be removed in order to reduce the substrate concentration.**

KEYWORDS: VT, diffusion, substrate, dopant, analog leaf cell, Athena, transoconductcance, channel length modulation, body effect, effective length, γ, λ.

I. INTRODUCTION

San Jose State University offers several analog IC design classes, which involve using EDA tools such as Cadence. Students can simulate and verify that their circuit functions properly. Once this is done, the design is sent out to MOSIS for fabrication. It takes about 12 weeks to do this. However, if SJSU were to do the fabrication, the turn-around time would be reduced to 1 week. Additionally, there more document-ation available to the student that wouldn't otherwise be available through MOSIS.

The 4-mask NMOS fabrication process is an in-house process that students use to manufacture simple devices such as resistors, MOS capacitors, diodes, and transistors. But it also contains ring oscillators, an inverters, and analog leaf cells, ALC (Figure 1). The ALC is an array of NMOS transistors that could be used for analog ICs. So, the layout of the 4-mask process is set up such that it ready to have analog circuits added. However, in the past several semesters, after students complete the fabrication, measurement of the threshold voltage of the transistors have been consistently high, between 4-5 V. If this process were ever to be used for building analog circuits, VT would need to be reduced. Considering that the voltage breakdown is about 10 V, a high VT gives the NMOS a small range of the gate voltage to operate in. Therefore, the process recipe would need to be changed in order for analog circuits to operate in a more reasonable range of gate voltages.

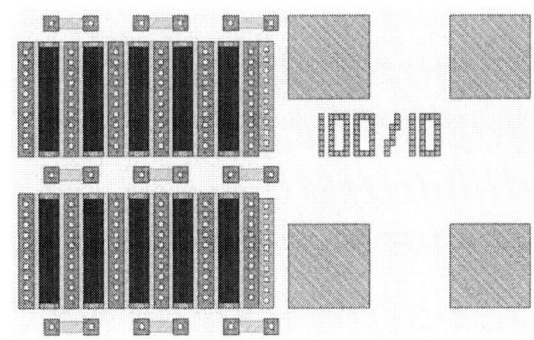

Figure 1: Layout of an Analog Leaf Cell from the 4-mask fabrication process

There are two possible ways to reduce the threshold voltage: Decrease the concentration of the p-type dopant in the substrate or reduce oxide thickness over the gate. Since VT needs to be reduced by about 3-4 V, reducing the oxide thickness at the gate would tend to be less feasible. The fabrication process grows the gate oxide to 400 nm. Reducing the oxide thickness any further would cause the transistor less stable and make it more sensitive to voltage swings. Lowering the substrate doping would be a better option. This can be done by increasing the amount of time for the deposited SOG (spin-on-glass) dopant to diffuse in the furnace.

SPICE parameters would also be needed for analog IC design

1-4244-0267-0/06/$25.00 ©2006 IEEE

documentation from wafers fabricated from previous semesters, other than the threshold voltage. Parameters that would be measured are the channel length modulation, the transconductance, body effect, effective length, gate oxide thickness, and VT.

II. DESIGN FLOW

Using the Athena process simulator, the 4-mask process is simulated at different substrate diffusion times and temperatures. The normal time and temperature for this step is 240 minutes at 1100 °C. After several simulations, three different times were chosen to carry out the fabrication process: 180, 240, 300 minutes. The temperature was chosen to remain the same since the furnace SJSU's IC lab can only go as high as 1200 °C and reducing the temperature would slow down the diffusion and cause the substrate to have large differences of dopant depending on the depth of the substrate. Three batches of wafers will be processed with four wafers per batch. Once the fabrication process is carried out and completed, the wafers can be measured for their SPICE parameters. For all tests except the effective length, 10 MOSFETs with the same width and length will be tested on each wafer. For the effective length, four MOSFETS in five different locations on a wafer will be tested. Once testing is completed, the measured data can then be compared with the data from the Athena simulations. Further modifications to the 4-mask recipe will be made based on the difference in the two sets of data.

III. TESTING AND ANALYSIS

With the exception of the gate oxide thickness, the SPICE parameters were measured using the HP4145A semiconductor parameter analyzer along with Metrics ICS software. The oxide thickness was measured using a film thickness spectrometer.

Table 1 shows the SPICE data obtained from the three batches.

Table 1:SPICE Data of Fabricated Wafers

Parameter/Batch #	ALC1 (180 min)	ALC2 (240 min)	ALC3 (300 min)
GAMMA (V)	5.00	4.96	4.84
KN (lin) (μA/V^2)	58.9	56.0	52.8
LAMBDA (mV^{-1})	8.4	13.0	17.2
LEFF* (μm)	4.27	4.34	4.81
TOX (μm)	365	392	402
VT (V)	3.54	3.08	2.15

*LEFF is the effective length measured for a transistor with a gate length of 10 μm.

The threshold voltage was obtained measuring the drain current and transconductance, GM, while sweeping a voltage over the gate and keeping the voltage at the drain constant. To

value of GM is. VT is the x-intercept of that tangent line. This can be seen in Figure 2.

Looking at Table 1, it shows that the threshold voltage is much higher than what was simulated in Athena. This may be due to an assumed constant value, the oxide interface charge, Qi, used in simulations was too high. Another reason may be that the dopant used in the fabrication process is too high. But the data clearly shows that more substrate diffusion time lowers VT.

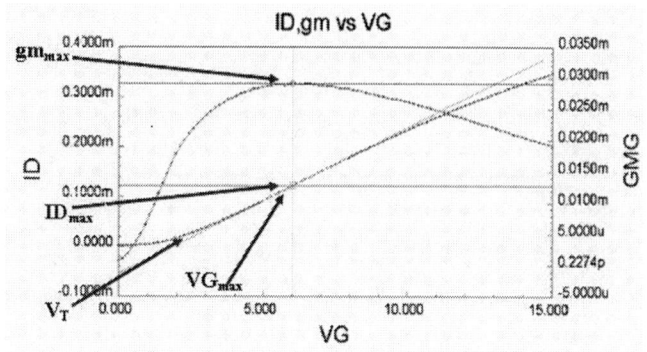

Figure 2: Measurement of Drain Current and Transconductance measurement. VG is swept while VD is constant.

The channel length modulation, λ, was measured by sweeping the drain voltage while keeping the voltage constant at the gate. Measurements had shown that λ varied as the gate voltage changed. Ideally, it should be the same for any gate voltage that the MOSFET is turned on. So instead, λ was measured only at a gate voltage that was 0.5 V above VT. This was done with the thought that this would be the gate voltage that would be most frequently used in analog circuits.

Table 1 shows that λ increases as the substrate diffusion time increases. A greater channel length modulation is an un-wanted characteristic in a MOSFET. So, even though VT is lowered by increasing the diffusion time, more current leaks through the MOSFET while in saturation.

The effective length, L_{eff}, data was obtained by measuring the maximum transconductance, GM-max, of five transistors with the same gate length, but different widths. The inverse of the GM-max was plotted with their respective lengths (Figure 3), which showed to have a linear relation. A trendline was extracted from the data, which was used to find the difference between the gate length and the channel length, ΔL. Knowing the length of the MOSFET, L_{eff} can be found by subtracting ΔL from the gate length, L, or $L_{eff} = L - \Delta L$.

L_{eff} is about ⅓ smaller than what was measured in the Athena simulations, about 6.5 μm. This may be due to the fact that wet etching is used in the fabrication process, which may make the source and drain wider than what Athena would produce. The calculated ΔL, which averaged to be about 5

work. The source and n-wells would be touching, essentially shorting the transistor. This parameter could be improved if plasma etching were used instead of wet etching.

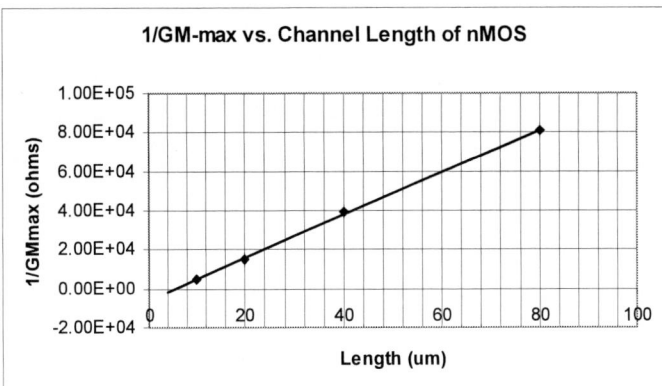

Figure 3: Plot of the inverse of GM-max versus the Channel Length of an NMOS. This is used to find L_{eff}.

The body effect, γ, was obtained using the same test that was used to find VT. ID and GM were measured while VG is swept and VD is held constant. The difference is that several different biases are applied to the substrate. γ can be obtained by using the following equation:

$$\gamma = \frac{V_T\,(V_{SB}) - V_{T0}}{\sqrt{\varphi_S\,(inv.) + V_{SB}} - \sqrt{\varphi_S\,(inv.)}},$$

where V_{SB} is the applied bias to the substrate, $\varphi_S(inv.)$ is the surface potential, V_{T0} is the threshold voltage without an applied substrate bias, and $V_T\,(V_{SB})$ is the threshold voltage with an applied V_{SB}.

Table 1 shows that γ only decreases slightly as the substrate diffusion time increases. 5 V is too high for a transistor if it is to be used for analog design. Like VT, it should also be about 1 V.

The transconductance parameter in the linear region, k_n (linear), was calculated by taking the slope of the tangent line of ID obtained from the VT measurements and dividing it by the constant drain voltage, VD. Table 1 shows that k_n (lin) deacreases when the diffusion time decreases. This is due to the fact that more of the p-type dopant diffuses through the substrate, making it less conductive.

The measurement of the oxide thickness was close to what the simulated value of 414 nm. The first batch, ALC1, was the furthest off by about 10% with an average of 365 nm. Overall the gate oxide thicknesses were consistent for all the wafers. This is important for analog circuits. Too much variation of the gate oxide would cause transistors to have large differences in their VTs. If transistors like this were to be used for current mirrors, there would be differences in the input and output current. Also, when the gate voltage is near

and the other would be off. For this process, this would only become a problem if the differences in thicknesses on the same wafer were in the hundreds of nanometers.

The difference between the VT data of the fabricated wafers and the Athena simulations may be due to the assumed value of the oxide interface charge, Qi, that was used in the simulations. The original value used was 5×10^{11} cm^{-3}. After, further simulations and using the VT measured from the wafers, it was found that the Qi that better fit the measured data was found to be 3×10^{10} cm^{-3}. However, this only fits the data of the last batch, ALC3 (see Table 2). Further reducing Qi doesn't change VT significantly.

There may be other reasons for the higher VT. One may be that the concentration used in the boron dopant is too high. This could be fixed by adding the dopant after growing the screen oxide in order to reduce the concentration. Or, a dopant with a lower concentration could be used.

* The concentration of the old Qi is 5×10^{11} cm^{-3}.

Table 2: Threshold Voltage with Different Oxide Interface Charges, Qi

Diffusion time (min)	Na (cm^{-3})	Simulated Vt with old Qi * (V)	Simulated Vt with new Qi ** (V)
180	2.11×10^{17}	1.21	2.11
240	2.00×10^{17}	1.16	2.05
300	1.90×10^{17}	1.12	2.01

** The concentration of the new Qi is 3×10^{10} cm^{-3}

VI. CONCLUSION

The measured data shows that the threshold voltage and body effect are still high. While increasing the substrate diffusion time reduces VT, it changes the body effect very little and also increases the channel length modulation. The oxide interface charge may have been assumed to be too high. But even after adjusting Qi, the VT data only agreed with the last batch, ALC3, which had the longest diffusion time (Table 2). So, there may be other factors involved in the high VT. The initial concentration of the dopant may be too high. It may need to be changed to a boron dopant with lower concentration. What could be done to compensate for the high concentration of the dopant is to only deposit it with a screen oxide over the substrate. This would allow the screen oxide to absorb some of the dopant and reduce its concentration in the substrate.

ACKNOWLEDGMENTS

PROFESSOR DAVID PARENT
MPE FACULTY
DR. ALLEN, GLEIXNER, YOUNG, AND LEE
NEIL PETERS, LAB ENGINEER
KASEM TANTANASIRIWONG, EE GRAD STUDENT
ERIC BASHAM, PHD. STUDENT
CADENCE LAB & FABRICATION LAB

REFERENCES

[1] Parent, David W.. "Mask Documentation of the EE/MatE 129 Process Wafers." Documentation on Masks and Testing. Gregory L. Young. 2005. San Jose State University. Accessed 21 April 2006. <http://www.engr.sjsu.edu /MatE129/Process%20Handouts/Mask_Testing /mask_testing.pdf>.

[2] Pascua, M.. Resistors, Diodes, and Transistors. EE/MatE129 Intro IC Fabrication and Processing (2005).

[3] Sedra, Adel S., Smith, Kenneth C.. Microelectronic Circuits. New York: Oxford University Press, 2004.

[4] Streetman, Ben G., Banerjee, Sanjay. Solid State Electronic Devices. Upper Sadler River, New Jersey: Prentice Hall, 2000.

[5] Sze, S.M.. Physics of Semiconductor Devices. New York: John Wiley & Sons, 1981.

[6] Van Zeghbroeck, Bart J.. Principles of Semiconductor Devices. 1997. University of Colorado at Boulder. Accessed 17 May 2006. <http://astha.ee.ui.ac.id/courses/ts/teksem /title.htm>.

[7] Wolf, S.. Microchip Manufacturing. Lattice Press, Sunset, CA, 2004).

Design of an Ultra-low Power Receiver for 2.4GHz Applications in 90nm CMOS

Maryam Tabesh and Koorosh Aflatooni
Department of Electrical Engineering, San Jose State University
maryam_tabesh@hotmail.com

Abstract- An ultra low power receiver operating in 2.4GHz ISM band is designed using 90nm CMOS PTM with a 1V supply voltage. This direct conversion receiver has an integrated noise figure of 4.15 dB in the 1kHz to 10MHz band. The input referred IP_3 for offset of 100MHz is -22.4dBm. The LO leakage to the RF port is less than -148dB. A base-band gain stage and buffer are used to provide a reasonable output impedance and power gain. With this, input and output are matched to 50 Ω and 500 Ω with both S_{11} and S_{22} being better than -25dB. The LNA, mixers and the base-band stages with the biasing network consume less than 1.1 mW .

I. INTRODUCTION

The surge in low cost mobile application created a large interest in RF subsystems fabricated on bulk CMOS technology. The fully integrated CMOS subsystem will offer many advantages over existing technology where bipolar based RF side is integrated to CMOS base-band modules. Some of these advantages include price, reduction in minimum feature sizes (compactness) that in turn offers others advantages including speed improvements and the ability to add logic at minimum cost. Therefore, the integration of the whole system on a single chip is becoming more and more attractive [10], [11]. Beyond this, it is predicted that in the near future "digital calibration" techniques that make use of the low cost digital side to tune the more complex and demanding RF side will become more and more widespread. As a result integration will eventually become a "must" especially in applications where CMOS has the required speed capabilities.

In this work, the design of a low cost CMOS based RF front end in 90nm PTM model [1] is presented. The frequency of operation was chosen to be 2.4GHz in the ISM band where many of the wireless systems operate. The main application would be in WLAN or WPAN systems where the cost is an important factor and the requirements are possible to meet using a bulk-CMOS process. The main goal at system level design has been to reduce the power and therefore the complexity of the system and also not to use off-chip components.

The low-power goal of this work has an impact on block level decision making as well as that of circuit and topological choices. Another important goal of this design has been integration of an RF radio on-chip .Although this goal enables significant cost reduction, but adds complexity to the design. For example, on chip inductors (and in general all passives)

have higher loss and lower quality factor which make design more challenging. With these two initial goals in mind we can see that the direct-conversion receiver is an attractable solution that provides the highest level of integration with the elimination of the need for bulky off-chip band pass filters. These band pass filters usually are used to mitigate the image problem (which is of less concern in a Direct Conversion Receiver or DCR) [2]. Direct conversion scheme also requires a single local oscillator (LO) tuning. Power consumption is also low and acceptable in DCR's. However DCR is susceptible to noise, and requires tight restrictions at circuit level for noise reduction. This is because of the fact that the received signal is down converted to around DC and here the flicker noise plays a dominant role.

To overcome the problem of folding [2], as illustrated in Fig.1 the DCR will consist of two Quadrature paths that will provide the "phase" information to the base-band circuitry for data recovery. The difference between this Quadrature system and that of a dual conversion image reject type (e.g. Hilbert Architecture [2]), is the relaxed constraints on the matching of the two paths.

Apart from the flicker noise issue that DCR has, LO self-mixing could also cause a problem. This means that when LO and RF are close in frequency (Direct Conversion is the extreme case) then LO leakage to the RF port causes a "self-mix" where the LO signal of the RF port is also mixed down and causes some DC offset like problem. Hence such receivers require a tight spec on LO leakage to the RF port.
Linearity is closely related to power consumption. For example, to provide a 1dB compression point of -30dBm or equivalently 10mV at the input with the power gain of 30dB to the mixer output, one requires the ability to sink/source enough current from the supply at the output of the mixer to provide 0dBm of power. This in turn means from a class-A style circuit we cannot expect DC power consumption below 1mW. This example shows, in low-power design, meeting the linearity requirements can be challenging.

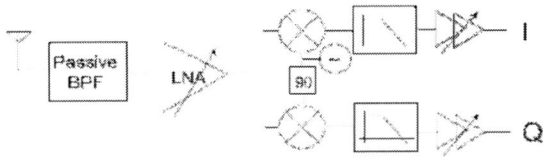

Fig.1. The general block diagram of a direct conversion receiver

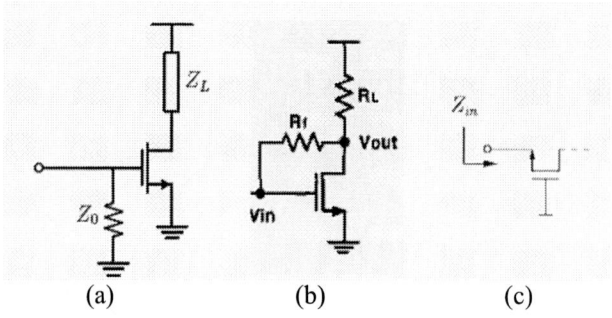

Fig.2. LNA topologies: Common source stage with shunt termination (a), the shunt feedback topology (b) and also the common-gate stage (c)

To overcome this limitation several techniques have been used in this design including a low-pass filter at the mixer output to suppress unwanted harmonics, a passive mixer to provide linearity and a passive mixer with band-pass characteristics. In the section, we present the design trade offs of the noise amplifier, follows by sections on buffer and mixer design. Then simulations results are presented, followed by conclusion.

II. THE LOW NOISE AMPLIFIER

The first stage in the signal path is the low noise amplifier (LNA). The main factors in the design of an LNA are noise figure, linearity, gain, input matching and power consumption.

Fig.2 shows some of the possible LNA topologies. The first candidate is the input terminated common-source LNA shown in Fig.2a. This topology provides wideband matching as the termination could be a resistance but suffers from a poor NF. To show this briefly lets assume that the matching is done through Z_O and therefore it is equal in value to R_S the source resistance. Then by definition the NF is at least 3dB because now Z_O produces as much noise as R_S. This 3dB noise figure (note this is without the contribution of the transistor) makes this topology unsuitable for low noise applications.

The second topology shown in Fig.2 is a shunt feedback topology [3]. This topology provides the input matching through shunt feedback. This might be quite attractive for wideband applications but the power consumption is pretty high because no method of tuning the input has been used. Intuitively if there happens to be no tuned circuit at the input, then the Gm gain is sustained throughout the band and no additional boost through the LC elements is provided. This means that for a given gain the required DC current of this topology is higher.

The other possibility would be the use of common-gate stage in the input [4]. As shown in Fig.2c, no tuning element is used; moreover, the noise factor is higher by $1+\gamma/\alpha$ where for short channel devices this factor is at least 3. This means a minimum achievable noise figure of 4.7dB (10log3) that is unacceptable.

The topology employed in this work is a common source stage with inductive degeneration as shown in Fig.3b. This topology provides the closest NF to NF_{min} of the device and the degeneration provides real input impedance that is highly

desirable for matching. Inductive degeneration has been widely used in Bipolar, GaAs and CMOS LNAs [4]. Little additional noise is injected for the case of low loss inductors.

Let's assume a voltage of v_S to be driving the input (gate) with a source resistance of R_S. Then we would have:

$$
\begin{aligned}
i_{DRAIN} &= g_m.v_{GS} \\
&= g_m.\frac{v_S}{R_S + R_G + \dfrac{1}{j\omega C_{GS}}}.\left(\dfrac{1}{j\omega C_{GS}}\right) \\
&\approx v_S.\frac{g_m}{j\omega C_{GS}.(R_S + R_G)}
\end{aligned}
\tag{7}
$$

In which R_G is the total gate series resistance. This equation assumes the frequency to be high enough so that the approximation can be used. The above equation could be used to express G_m or the equivalent transconductance of the stage as i_{DRAIN}/v_S. Also it should be noted that $\dfrac{g_m}{C_{GS}}$ is equal to the ω_t of the device (assume negligible C_{GD} as we will use a cascode in the end). For noise figure calculations we could write:

$$
\begin{aligned}
F &= \frac{Total\_Output\_Noise}{Total\_Output\_Noise\_from\_R_S} \\
&= 1 + \frac{v_G^2}{v_S^2} + \frac{i_N^2}{G_m^2.v_S^2} \\
&= 1 + \frac{R_G}{R_S} + \left(\frac{\gamma}{\alpha}\right).g_m.R_S.\left(\frac{\omega}{\omega_t}\right)^2
\end{aligned}
\tag{8}
$$

The significant point in this expression is that the noise factor relies heavily on ω_t of the device. The optimum source impedance to achieve lowest noise can be found from (8) as,

$$
R_{S\_OPT} = 1 + 2(\frac{\omega}{\omega_t})\sqrt{\left(\frac{\gamma}{\alpha}\right).g_m.R_G}
\tag{9}
$$

Note that without taking the correlation effects into consideration the above value has no imaginary part. Shaeffer et al. [4] have proposed a parallel correlated noise current source to model the gate induced noise (distributed effect of the gate). For this work, the total gate series resistance is approximated to be a simple resistor with value of R_G, similar to what Pospieszalski has proposed [9]. This resistor value also models the contribution from the loss of the inductor and the distributed gate induced effect as well. In other words, R_G will include a $1/(5g_m)$ term that captures the gate induced effect [5]. We actually will model the gate side with a series pair and use a voltage noise model to produce effect of the extra noise seen in a real FET. This way, instead of a shunt-shunt model for the FET device a hybrid-pi model will be used. It can be shown [9] that some of the correlation that results from a shunt current model is actually artificial and therefore the model used here is quite accurate. The value of

R_{S_OPT} is calculated to be about 20 Ohms, if f_t of 50GHz (matching our prediction for achievable ft), g_m of 10mS and R_G of 20 Ohms are used. This shows the trade off between noise and power. Clearly, this value does not match to the 50 Ohms required for power matching. Generally, the optimum impedance for noise performance found from (9) will be different from that required to match the input. To see this let's assume that with a good layout the series gate resistance is minimized and R_G is dominated by the gate induced effect (i.e. $R_G=1/(5g_m)$). This is generally a good assumption LNA design as special attention is paid to the layout of the main device to minimize it's noise and maximize the maximum oscillation frequency of the device (f_{MAX}) [5]. R_{S_OPT} can be found from (9) by replacing the R_G. Thus R_{S_OPT} depends only on ω_t. Therefore no extra tuning parameters are available for making R_{S_OPT} equal to the source impedance; therefore simultaneous noise and power matching cannot be achieved.

If an inductive degeneration is used then the input sees a real impedance of $\omega_t x L_S$ [4] where L_s is the series degeneration inductance (Fig. 3a). The input impedance seen from the gate with this degeneration is given by:

$$Z_{IN} = s(L_S) + \frac{1}{s(C_{GS})} + \omega_t.L_S .$$

In longer channel devices this real input impedance helps reduce the discrepancy between R_{S_OPT} and R_{Source}. In practice, device is sized so that it's R_{S_OPT} (note here that L_s will slightly shift R_{S_OPT}) and R_{Source} are made to be equal. L_s is then picked to match the imaginary parts [4], [6]. However, this approach is not directly applicable to this work due to these considerations:

1) The power dissipation of LNA should be limited and this affects the choice of device sizes.
2) With the small channel length device used here the value of C_{GS} is very small and therefore matching to such a high Q series pair will be difficult.
3) The linearity requirement is difficult to meet with power dissipation requirements and therefore the overdrive voltage has the constraint of satisfying the linearity as well.

With all these factors in mind, we decided to introduce additional free parameters by using a Pi input-matching network. A Pi-match network has been used to provide a wideband characteristic [8] but here the purpose is different and two folds. First it will convert the impedances with an extra degree of freedom (see Fig.3a). It also reduces the resistance from the loss of the inductor L seen from both sides. This has a positive impact on noise performance.

To ease the matching requirements of LNA, it is desirable to reduce the S_{12} of the LNA. Therefore, a cascode LNA has been used (Fig.3a) instead of a common source to have a closer unilateral structure. It can be shown that with high impedance at the source of device (M1), this cascode structure adds no extra noise [5]. In the layout, a shared junction can be used to reduce the effective area and perimeter of source/drain at this node. This will reduce the capacitance at this node and therefore prevent the increase of noise figure at high frequencies.

Another technique used in the LNA is the addition of an extra capacitance (C_P) in parallel to C_{GS} to reduce the Q of

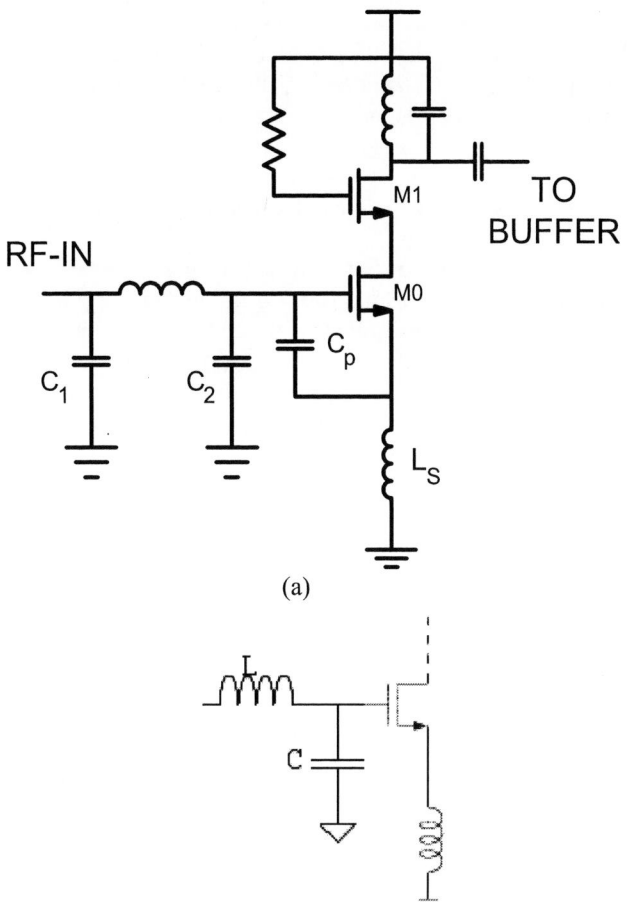

(a)

(b)

Fig 3: (a) The proposed Structure, (b) L-Matched CS stage

this series resonant circuit. This will be clarified using a numerical example. Let's assume gate capacitance of 30fF, inductor value of 1nH and also 10 Ohms of total series resistance. With these values and without adding a parallel capacitance to the gate the quality factor for this series tank would be about 220. This would transform the series resistance of the FET to an equivalent parallel resistance with a value of 500 KOhms ($R_P \approx R_S.Q^2$). This large value of R_P will sophisticate matching. There are two solutions to this, one is to increase the W of the device which would lead to more power consumption with a similar overdrive voltage and the other is to add the capacitance in parallel.

A current source is used to bias the gate of the main transistor for the required DC current. The ratios of the absolute sizes of mirror transistors are picked to be 0.08 so that the current of M0 is about 500µA (with reference of 25µA) regardless of the absolute size. There is a resistor between the mirrors to reduce the effect of the noise of the current source. This noise would be attenuated as it sees a voltage divider before getting to the gate of M0. Also, a capacitor is added to the gate of the reference transistor to further knock down the high frequency noise.

The output tank is chosen to resonate at 2.4GHz taking the capacitance at the drain of M1 into account. The value of the load inductor is picked to be (12nH) so that the gain is maximized.

1-4244-0267-0/06/$25.00 ©2006 IEEE

The real part of the input impedance seen from the gate of the cascode device (M1) could potentially become negative and cause oscillatory behavior. To solve this issue, a small resistor is added to this point to de-Q this parasitic tank which is formed by C_{GS} of M1 in series with the total impedance from the drain of M0 to ground. This resistor has minimal effect on LNA performance and the thermal noise associated with it will be rejected similar to the channel noise from M1 [5].

Linearity is a good starting point in the actual design. The required P_{-1dB} being -30dBm a 10mV of swing at the input has to be sustained. For this to be true in the first approximation the inductive degeneration has to support 10mV voltage across it with the nominal DC current. In this case, the DC current of the LNA is limited to about 500µA. This gives a value of 1.4nH for the degeneration. Now, the resonant input circuit will boost V_{GS} by nearly Q which degrades linearity further. A 2nH degeneration was seen to be appropriate. Note that the nonlinearity due to C_{GS} and other elements and also the memory effects introduced by the inductor have been neglected for the first cut approximations here.

In square law regime increasing the overdrive will decrease g_m nonlinearities. With the fixed current for the LNA, the overdrive is inversely proportional to W of the device. So it may seem a very small W will provide larger overdrive which will help linearity and also increase ω_t of the device as seen below:

$$\omega_t \approx \frac{g_m}{C_{GS}+C_P} = \frac{g_m}{\frac{2}{3}.WLC_{OX}+C_P}$$
$$= \frac{\sqrt{2I_D.\frac{W}{L}.C_{OX}}}{\frac{2}{3}.WLC_{OX}+C_P} \qquad (10)$$

Here C_P is the added parallel capacitance and the equation shows that for a fixed current decreasing W will increase ω_t. Also, as seen previously higher ω_t mean less noise figure.

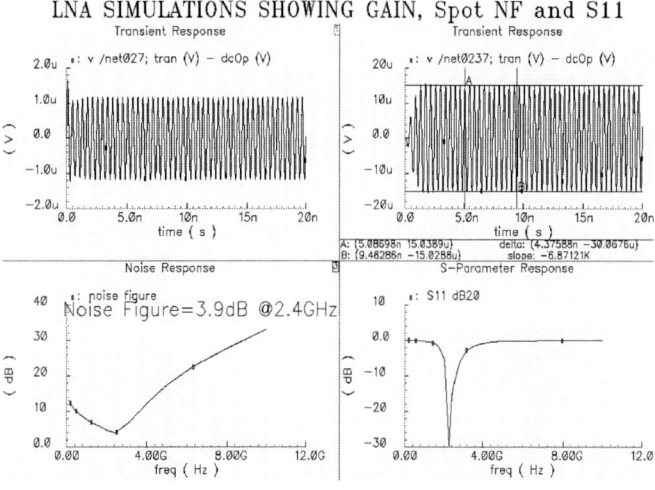

Fig.4. Simulation results for the LNA showing voltage gain at 2.4GHz, spot NF and S_{11}

Table 1: Simulation results for the LNA

Simulation Results for LNA	
VOLTAGE GAIN	14.5
P-1dB (IIP3)	-14.4dBm (-4.8dBm)
NOISE FIGURE	3.9dB
INPUT MATCH (S11)	-25.5 dB
POWER CONSUMPTION	504.9µW

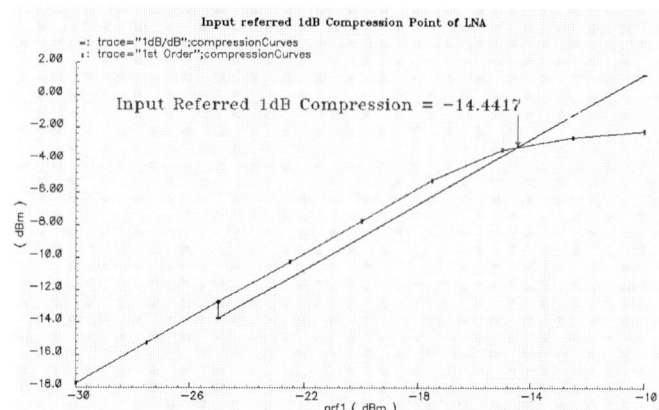

Fig.5. Input referred 1dB compression point for LNA

The main problem in decreasing W is the decrease in g_m and therefore the decrease in the voltage gain of the stage. Also, C_{GS} would decrease with the decrease of W and Q of the series input tank would go up requiring a higher C_P to compensate. Higher C_P will eventually result in the lowering of ω_t and therefore the decrease of W will not increase ω_t further.

If we assume R_G is dominated by the gate induced effect then by first order R_{S_OPT} is just a function of ω_t. A constraint on C_P (e.g. $0.1C_{GS}<C_P<2C_{GS}$) was chosen and W was optimized for a fixed gain (of 15-20) and maximum ω_t. The values were re-iterated using simulations for simultaneous matching and minimum noise figure. Also, the inductor value was set to be the minimum possible so that the loss would not degrade noise figure further. This way the optimum value for W came to be 47µ. Table1 summarizes the simulation results for the LNA and Fig.4 illustrates the simulated gain, noise figure and S_{11} of the LNA. Fig.5 shows the 1dB input referred compression point for the LNA.

III. THE BUFFER

After the LNA there is conventionally a mixer stage that performs the frequency translation. The choice of the mixer determines the requirement of a buffer. An active mixer uses a Gilbert multiplier topology and in general requires less drive for the mixing transistors. This type of mixer will further knock down the effect of the noise of the following stages by providing gain. However, it has more noise as the flicker is present in an active mixer. Also, the power consumption of the mixer may dominate as there are two mixers in the system.

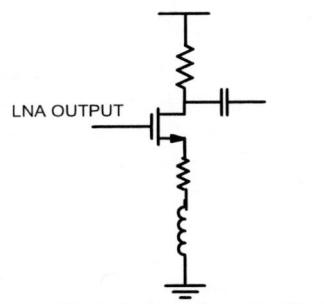

LNA OUTPUT

Fig.6. (a) The schematic of the Buffer stage

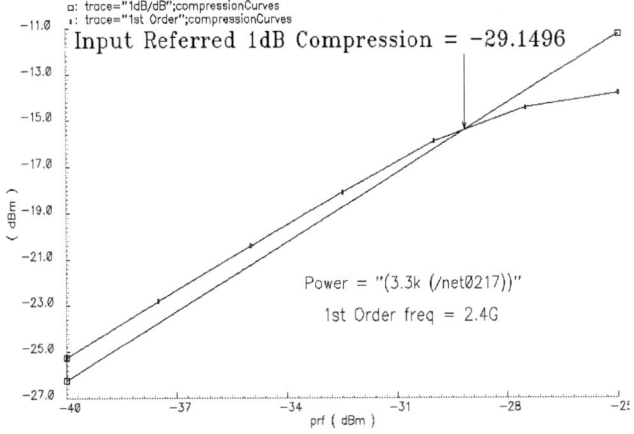

P-1dB Simulation for the LNA and Buffer
Periodic Steady State Response

□: trace="1dB/dB";compressionCurves
+: trace="1st Order";compressionCurves

Input Referred 1dB Compression = -29.1496

Power = "(3.3k (/net0217))"
1st Order freq = 2.4G

Fig.7. Linearity simulation of the LNA and Buffer stages

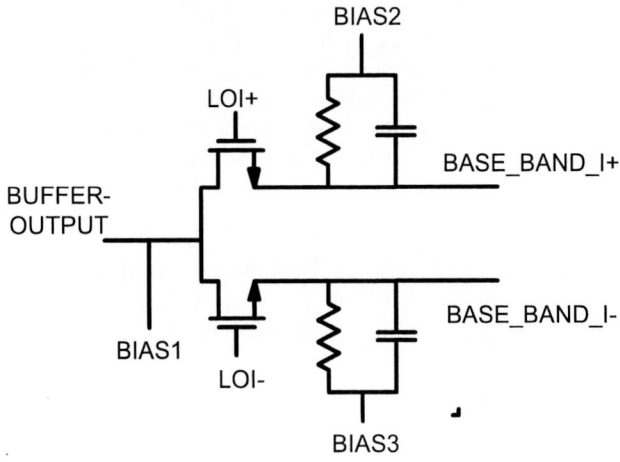

Fig.8. The schematic of the I-mixer (the Q-mixer is similar)

load resistance chosen. To pick an appropriate R_L we should keep in mind that a large R_L will result in a smaller V_{DS} and the transistor may be pushed into triode on large swings. In that case the linearity suffers. To see the appropriate value of R_L the signal swing that has to be sustained is calculated (knowing the gain and the input swing being 10mV) and the margin is chosen to be large enough to support this swing. The stand-alone buffer provides a voltage gain of 4.4. Also, simulations show a gain of 14 for the "loaded" LNA and overall gain of 24 to the buffer output.

IV. THE MIXER STAGE

As previously stated the mixer is chosen to be a passive switching mixer. The differential gain of the mixer is $4/\pi$ [5]. In [8], Shahani *et al* have discussed the details of effect of having a sinusoidal excitation with various DC levels. Here we will not discuss the details.

The device sizes are chosen so that the R_{ON} is sufficiently small and the mixer acts as a good switch. If the devices are made too big the LO power required to drive them will increase. Assuming a pulse excitation (which is not the case here) the power requires to "switch" the mixers is $C_{gate}.f.V_{swing}^2$ where f is the frequency. Similar tradeoffs can be seen in the choice of V_{swing}. In general the required swing for a passive mixer is large. The subtle point here is that if the swing is chosen to be too large with a zero DC value, then the actual conversion gain of the mixer reduces but the linearity improves [8]. This is a key observation in trading voltage gain with linearity and from here the chosen swing is 0.5V (peak) whilst the devices are made 10μ wide with minimum channel length.

Another issues in picking the device sizes is the LO feed-through. The large LO signal can feed to the output of the buffer through the overlap capacitances. Sizing up the devices results in larger leakage and this may exceed the required specifications.

The DC biasing of the passive mixer is quite important in this design. First of all to prevent any DC current from flowing into the device the bias points of the drain and source are chosen to be the same. In this design, they are taken from the same reference and as a result track each other in case of temperature or supply variations. Apart from this we have introduced a low pass filter at the output of the mixer (Fig.8) to eliminate the spurious harmonics and therefore relax the linearity requirement of the base-band stage. The low pass filter requires a low impedance point at its bias point. This bias voltage could provide

Getting gain from an active mixer will always have some linearity penalty and it is common to heavily degenerate the Gm stage of a Gilbert cell mixer. Also, to overcome the problem of flicker noise device sizes could be increased which in turn will increase the power required to drive the larger capacitances.

A passive mixer requires more drive on the gates but introduces no flicker noise as the DC current through it is zero. Also, it is very linear and is often used for extreme linearity requirement applications. One of the problems with a passive mixer is that it loads the LNA.

This above line of reasoning led us into using a passive mixer with an inter-stage buffer that provides a low gain and prevents the mixer from loading the LNA. The buffer will also reduce the LO leakage to the RF port acting as a "shield" stage. The problem is with the nonlinearity that the buffer may introduce. Unlike noise, linearity is dominated by later stages and therefore small nonlinearity in the buffer may lead to a large overall nonlinearity as the signal is gained up when it sees the buffer. To overcome this problem, in this circuit we have used a heavy degeneration of 12nH plus 100 Ohms of resistance. Also, for linearity purposes a relatively high overdrive voltage is used and at the same time to keep the current low the device is sized down. A key advantage in using a buffer stage is the possible non-linearity cancellation with the LNA buffer pair. Because of the different bias mechanisms used, the nature of the nonlinearities of the LNA and the mixer are different. So potentially the compressing and expansive nonlinearities could partially cancel out.

Fig.6 shows the schematic of the buffer stage. The linearity of the LNA buffer pair is shown in Fig.7 for the final value of the required bias point for the base-band amplifiers as well. The other side of the mixer however should not be biased with a low-impedance biasing scheme as this will load the buffer and result in loss of gain. Simulations show that the total single-ended gain to base-band is 30.5 in terms of voltage. The simulated LO leakage to the RF port is -148dB in terms of attenuation.

V. RESULTS

To provide sufficient power gain to base-band and also output matching to 500 Ohms a simple pseudo-differential gain-stage and a subthreshold operating buffer were employed.

As previously pointed out, to satisfy linearity specifications, the band-pass characteristics of the passive mixer at RF have been combined with an explicit low-pass filter at the output of the stage to knock down the third harmonic components that can produce in-band interfering signals. Thus, the linearity will be mainly determined by the linearity of the LNA and buffer stage and also the passive mixer.

To see the total noise figure, the IEEE form of noise factor was integrated from 1KHz to 10MHz and then averaged. This way the total NF of 4.15dB was obtained. Fig.9 illustrates IEEE noise figure and noise factor vs frequency in the MHz range. Table2 summarizes the simulation results for the whole front-end.

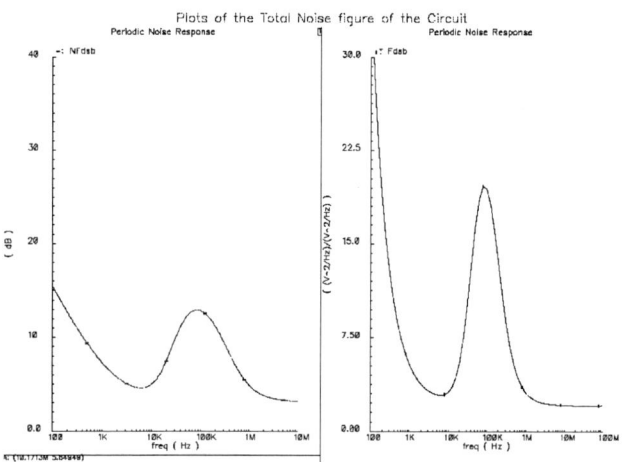

Fig9. NF$_{IEEE}$ and F$_{IEEE}$ of the whole circuit

Table2: Simulation results for the direct-conversion receiver

Parameter	Simulation
Power gain to base-band	52.9dB
Max-in-band-ripple	0.5dB
Integrated Noise figure	4.15dB
S11(input)	-25.5dB
S22(output)	-27dB
Power Consumption(LNA, Mixers, Base-band and biasing)	1.081mw
P$_{-1dB}$	-32dBm
LO leakage (Attenuation to RF Port)	-148dB

VI. CONCLUSIONS

In this project an entire direct conversion front-end was designed that is totally integrable and requires no off chip components. The 90nm predictive transistor models (PTM) were used in this design. At the architectural level various trade-offs inherent in the choice of topology of the LNA and the mixer stages were described.

In the LNA, with the loss of the inductor dominating the noise, the use of pi-matching network provided better noise figure. Also, a novel approach in using a buffer stage and partially canceling the nonlinearities of the LNA buffer pair was described. The buffer prevented the passive mixer from loading the LNA and provided a small gain itself.

One of the key observations in this project was the potential merits of a passive mixer in providing a low-noise, high-linearity solution in low-power applications. A novel approach in taking advantage of the band-pass characteristics of the passive mixer for mitigating the spurious harmonics was illustrated.

With the down-scaling of technology and also the demand for low-power devices in CMOS, we believe passive mixers will be used more and more and the traditional arguments in favoring active mixers will eventually loose strength.

REFERENCES

[1] Predictive Transistor Models available at: http://www.eas.asu.edu/~ptm

[2] B. Razavi, *RF Microelectronis*, Prentice Hall 1998

[3] Wang, S.B.T.; Niknejad, A.M.; Brodersen, R.W., "A sub-mW 960-MHz ultra-wideband CMOS LNA," *RFIC Symposium, 2005. Digest of Papers* pp. 35- 38, 12-14 June 2005

[4] Shaeffer, D.K.; Lee, T.H., "A 1.5-V, 1.5-GHz CMOS low noise amplifier," *Solid-State Circuits, IEEE Journal of* , vol.32, no.5pp.745-759, May 1997

[5] T. H. Lee, *The Design of CMOS Radio-Frequency Integrated Circuits*. Cambridge, U.K.: Cambridge Univ. Press, 1998

[6] Trung-Kien Nguyen; Chung-Hwan Kim; Gook-Ju Ihm; Moon-Su Yang; Sang-Gug Lee, "CMOS low-noise amplifier design optimization techniques," *Microwave Theory and Techniques, IEEE Transactions on* , vol.52, no.5pp. 1433- 1442, May 2004

[7] Bevilacqua, A.; Niknejad, A.M., "An ultrawideband CMOS low-noise amplifier for 3.1-10.6-GHz wireless receivers," *Solid-State Circuits, IEEE Journal of*, vol.39, no.12pp. 2259- 2268, Dec. 2004

[8] Shahani, A.R.; Shaeffer, D.K.; Lee, T.H., "A 12-mW wide dynamic range CMOS front-end for a portable GPS receiver," *Solid-State Circuits, IEEE Journal of* , vol.32, no.12pp.2061-2070, Dec 1997

[9] Pospieszalski, M.W., "Modeling of noise parameters of MESFETs and MODFETs and their frequency and temperature dependence," *Microwave Theory and Techniques, IEEE Transactions on* , vol.37, no.9pp.1340-1350, Sep 1989

[10] Magoon, R.; Molnar, A.; Zachan, J.; Hatcher, G.; Rhee, W., "A single-chip quad-band (850/900/1800/1900 MHz) direct conversion GSM/GPRS RF transceiver with integrated VCOs and fractional-n synthesizer," *Solid-State Circuits, IEEE Journal of* , vol.37, no.12pp. 1710- 1720, Dec 2002

[11] Dow, S.; Ballweber, B.; Ling-Miao Chou; Eickbusch, D.; Irwin, J.; Kurtzman, G.; Manapragada, P.; Moeller, D.; Paramesh, J.; Black, G.; Wolischeid, R.; Johnson, K., "A dual-band direct-conversion/VLIF transceiver for 50GSM/GSM/DCS/PCS," *Solid-State Circuits Conference, 2002. Digest of Technical Papers. ISSCC. 2002 IEEE International* , vol.2, no.pp.182-462, 2002

1-4244-0267-0/06/$25.00 ©2006 IEEE

AUTHOR INDEX

A

Abebe, H.	113, 119
Aflatooni, Koorosh	191, 241
Aghahassan, Houshang Amir	195
Aquilino, Michael	7
Arquette, Maureen	23
Asmath, Yugandhar	229
Avci, I.	41

B

Balasingam, P.	41
Barber, John P.	181
Basham, Eric J.	125
Belisle, Chuck	207
Berger, Paul R.	109
Biswas, Abhijit	145
Blondell, Scott P.	23
Bokor, Jeffrey	175
Bowman, Lawrence	1
Braga, N.	51
Bunzow, David A.	1

C

Campbell, Steve	101
Chan, J.	137
Chen, Yaw-Hwang	189, 221
Chi, Jane	195
Cho, YoungKeun	71
Choi, Munkang	63
Cibuzar, Greg	101
Couillard, J. Greg	161
Crain, Mark	89
Cumberbatch, E.	113, 119

D

Dawson-Elli, David	161
Dickinson, John	1

E

Echols, Scott	237
El Sayed, K.	41
Ewbank, Dale E.	23

F

Farshchi, Shahin	105
Fenger, Germain	161
Fernández, Bautista	131
Fichtner, W.	41, 51
Flickinger, Michael	101
Flounders, A. William	19
Fuller, Dr. Lynn F.	7, 13, 23

G

Gan, Kwang-Jow	189, 221
Gharib, J.	41
Gleixner, Stacy	225
Gonzales, R. Bryan	77
Greenwood, Bruce	207
Gruener, Charles	23
Gurtovoy, Frank	77
Guttal, Pratibha	215

H

Halverson, Karen	101
Hartin, O.L.	33
Haugstad, Greg	101
Hawkins, Aaron R.	175, 181
Hirschman, Karl D.	23, 109, 161
Hunt, David J.	81
Hunt, John	1

I

Itamura, Lisa	89

J

Jackson, Michael A.	23
Jain, Ritu	215
Jeon, SangCheol	71
Johnson, M.D.	41
Joseph, Chung	233
Judy, Jack W.	105

K

Karulkar, Pramod C.	1, 145
Kells, K.	41
Kemelgor, Bruce	89
Keynton, Robert	89
Kim, Donghyun	47
Kim, Jeoung Woo	97
Kim, JinSoo	71
Kim, KiNam	71
Kim, KwangHee	71
Kiralyfalvi, G.	41
Koltyzhenkov, V.	41
Kown, Sunghoon	175
Krishnamohan, Tejas	47
Kubby, Joel A.	131
Kucherov, A.	41
Kurinec, Santosh K.	23, 109

L

Lane, Rickard L.	23
Law, Mark E.	213
Lee, HeeChurl	71
Lee, HeeMok	71
Lee, S.J.	137
Leung, Siu Kuen	237
Liang, Dong-Shong	189, 221
Liddle, Alexander	175
Lin, Xi Wie	63
Lin, Xiao	65
Liu, Tsu-Jae King	155
Lowther, Mark A.	175
Lyumkis, E.	41

M

Manley, Robert G.	161
Maung, K.J.	137
Measor, Philip	181
Meyyappan, Meyya	183
Mickevicius, R.	51
Moroz, Victor	63
Morris, H.	113, 119

N

Neff, George R.	143
Neff, Jimmie K.	143
Newberry, Deb	101

O

Oh, JaeSub	71
Oh, Saeroonter	47
Opp, Micheal	101

P

Papalias, Tamara A.	77, 201
Parent, D.W.	57, 215
Parent, David W.	125, 233
Park, Jong Wan	97
Pascua, Mariavanessa	233, 237
Pawlik, David J.	109
Pearson, Robert E.	13, 23
Penzin, O.	41
Pethe, Abhijit	47
Phong, Nguyen	233
Polsky, B.	41
Pramanik, Dipu	63, 65
Puchades, Ivan	13

Q

Qureshi, Naser	175

R

Rahman, Taslima	237
Rao, V.	41, 51
Reed, Jack C.	195
Rink, Karl K.	143
Rollins, Greg	65
Rommel, Sean L.	23, 109

S

Saha, Samar	169
Saraswat, Krishna	47
Saxer, Robert L.	161
Schmidt, Holger	175, 181
Seballos, Leo	181
Severs, Philip	1
Shakouri, Ali	183
Sim, Meow Yen	225
Simeonov, S.D.	41
Singaraju, Pavan	169
Smith, Bruce W.	23
Smith, Scott	89
Strecker, N.	41
Sudirgo, Stephen	109
Sun, Xuhui	183

T

Tabesh, Maryam	241
Tan, Z.	41
Tarkul, Scott	233
Thirupapuliyur, Sunderraj	159
Thomas, David	1
Thomason, Mike	207
Thompson, Phillip E.	109
Tierney, Joan	23
Tsai, Cher-Shiung	189, 221
Tseng, Kuang-Wai	237
Tu, Chun-Da	189, 221
Tyree, V.	113, 119

V

Varadarajan, Vidya	155
Vasham, Eric	233
Venkat, Rama	169
Verma, Vivek	201
Villablanca, L.	41
Voros, Katalin	19

W

Walsh, Kevin	89
Wang, Suqin	175
Wang, Xi	183
Warrick, Jami	1
Wen, Chun-Min	189
Widlund, Sara	23
Wiegand, Maria	23
Williams, Brett	207
Wong, H.S. Philip	47
Woodard, Eric M.	161

X

Xu, Xiaopeng	65

Y

Yen, Albert	195
Yin, Dongliang	181
Yu, Bin	183
Yu, Evan	33
Yu, Jonathan	191

Z

Zhang, Jin	181
Zhou, Zhiping	151